AF862859

Skeletal Biomineralization: Patterns, Processes and Evolutionary Trends

Volume II. Atlas and Index

Skeletal Biomineralization: Patterns, Processes and Evolutionary Trends

Volume II. Atlas and Index

Joseph G. Carter, *Editor*

VNR SPRINGER SCIENCE+BUSINESS MEDIA, LLC

© Springer Science+Business Media New York 1990
Originally published by van Nostrand Reinhold in 1990
Softcover reprint of the hardcover 1st edition 1990

Library of Congress Catalog Card Number 89-14866

ISBN 978-1-4899-5393-3 ISBN 978-1-4899-5391-9 (eBook)
DOI 10.1007/978-1-4899-5391-9

All rights reserved. No part of this work covered by the copyright hereon may be reproduced or used in any form or by any means—graphic, electronic, or mechanical, including photocopying, recording, taping, or information storage and retrieval systems—without written permission of the publisher.

Manufactured in the United States of America

16 15 14 13 12 11 10 9 8 7 6 5 4 3 2 1

Library of Congress Cataloging-in-Publication Data

Skeletal biomineralization.

Vol. 2 published by: New York, N.Y. : Van Nostrand Reinhold.
Includes bibliographical references and indexes.
Contents: v. 1. [without special title] -- v. 2. Atlas.
1. Biomineralization. I. Carter, Joseph Gaylord. II. Title.
QP533.S54 1989 591.4'71 89-14866

Cover: Upper left: Depositional surface of calcitic fibrous prismatic shell layer of the bivalve *Mytilus edulis* Linné, 1758, cleaned in 5% sodium hypochlorite, UNC 13620; photograph is 12.7 microns wide. Upper right: Oblique fracture through the nacreous middle layer of the bivalve *Pinctada imbricata* Röding, 1798, Recent, Bahamas, UNC 6889; photograph is 4.7 microns wide. Lower left: Holothurian ossicles freed by digestion of the soft tissue in 5% sodium hypochlorite; unidentified species nestling or boring into coral in Onslow Bay, North Carolina, UNC 13349; photograph is 271 microns wide. Lower right: Polished and then HCl-etched tangential section through the enamel of a third molar of modern *Homo sapiens*, UNC 14886; photograph is 15.2 microns wide.

Skeletal Biomineralization:

Patterns, Processes and Evolutionary Trends

Volume II. Atlas

CONTENTS

Skeletal Biomineralization: Patterns, Processes and Evolutionary Trends
Volume II

Part 1

Introduction

Joseph G. Carter

Department of Geology, University of North Carolina
Chapel Hill, NC 27514

This volume provides supplementary illustrations of skeletal microstructures for many of the invertebrate and vertebrate taxa discussed in Volume I. It also illustrates certain molluscan groups not included in Volume I, *i.e.*, the classes Polyplacophora, Scaphopoda, Archaeogastropoda and Paragastropoda, and the bivalvian subclasses Lucinata, Heteroconchia and Anomalodesmata. Other contributions in this volume, notably those by Gaspard for terebratulids and by Wendt for corals and coralline sponges, include examples of diagenetic modifications of skeletal microstructures.

Descriptive terminology. The descriptive terminology used in this Atlas is defined in the *Glossary* in Volume I. The sections on molluscs (Parts 2 and 3) utilize the standard reference terminology in that glossary.

Scale bars. The scale bars on all plates are labelled in microns unless indicated otherwise. Where only two scale bars are provided for a plate with several figures (as in Plate 77), the left and right scale bars refer to the left and right columns of figures, respectively.

Sample preparation. Sample preparation methods are indicated by number, as follows, in the plate explanations:

1. SEM of a gold- or gold/palladium-coated fracture surfaces.
2. SEM of a gold/palladium-coated surface which has been embedded in epoxy, sectioned, polished, etched with weak acid, and then peeled with an acetate slab. These illustrations show the peeled surface, not the acetate peel.
3. Light microscopy of an acetate peel of an epoxy-embedded, sectioned, polished, and then acid-etched surface.
4. Light microscopy of a natural surface of deposition viewed with reflected light.
5. SEM of a gold/palladium-coated surface which has been embedded in Araldite, sectioned, polished, and then etched for a few seconds in 12.5% RDO or RDC.
6. SEM of a gold- or gold/palladium-coated natural skeletal surface prepared cryotechnically.
7. SEM of a gold/palladium-coated section of specimen immersed in picroformol Bouin for several hours, then dehydrated in graduated alcohol.
8. SEM of a gold- or gold-palladium-coated, internal skeletal surface partly freed of organic matrix by immersion in 10% lauryl sodium sulfate for 30 minutes, then dehydrated in graduated alcohol.
9. SEM of a gold-coated fragment of natural skeletal surface immersed in 10% proteinase, activated by a detergent for several hours, then dehydrated in graduated alcohol.

10. SEM of a gold/palladium-coated, natural skeletal surface prepared by the critical point method.
11. SEM of a gold/palladium-coated, otherwise unprepared specimen.
12. SEM of a gold/paladium-coated, natural skeletal surface freed of organic coats by immersion in sodium hypochlorite for several hours and (or, if in fossil state) freed of adherent particles by brief sonication in a weak detergent followed by acetone.
13. SEM of a gold/palladium-coated fracture surface freed of adherent particles by brief sonication in a weak detergent followed by acetone.
14. SEM of a gold/palladium-coated section embedded in Araldite, cut in a preferred plane, polished with tin oxide or alumina, immersed in 2% EDTA for 20 to 30 minutes, depending on geologic age and texture of the shell, and then air-dried.
15. SEM of a gold-coated fracture surface etched for 3 hours in EDTA (disodium salt) and washed in deionized water.
16. SEM of a gold-coated section; polished, etched for three hours in EDTA (disodium salt), washed in deionized water, and then air-dried.
17. TEM of material fixed for one hour in 2.5% glutaraldehyde in 0.1 M sodium cacodylate mixed 1:1 with sea-water; decalcified for 68 hours at 4 degrees C in 0.2 M EDTA (disodium salt); post-fixed in 1% osmium tetroxide; stained in 2% uranyl acetate in 70% ethanol, and in 0.3% lead citrate in 0.1 N sodium hydroxide.
18. SEM of a gold/palladium-coated fracture surface etched in 5% HCl for 5 seconds (Recent teeth) or 10 seconds (fossil teeth).
19. SEM of a gold/palladium-coated surface polished with 3-micron levigated alumina grit and etched with 5% HCl for 5 seconds (Recent teeth) or ten seconds (fossil teeth).
20. SEM of a gold/palladium-coated skeleton with adherent soft tissue structures partially digested in dilute sodium hypochlorite and then critical-point dried.

Abbreviations. The following abbreviations are used in this volume:

ac: acicular crystallite.
AMNH: American Museum of Natural History.
AOL: apatitic outer layer (Arthropoda).
ap: aragonitic prism.
ca: canopy of punctum (Brachiopoda).
CCF: complex crossed foliated (Mollusca).
CCL: complex crossed lamellar (Mollusca).
CCSF: complex crossed semi-foliated (Mollusca).
CF: crossed foliated (Mollusca).
cg: calcitic granules.
ch: channel in primary layer (Brachiopoda).
CL: crossed lamellar (Mollusca).
CLZ: central laminate zone (Arthropoda).
CP: composite prismatic (Mollusca).
CSF: crossed semi-foliated (Mollusca).
CZ: calcified zone (Arthropoda).
D: depositional surface.
DCP: denticular composite prismatic (Mollusca).
E: exterior.
Ep: epicuticle (Arthropoda).
Ex: exocuticle (Arthropoda).
F: fascicular stereom (Echinodermata).
F: foliated (Mollusca).
fc: terminal face of fiber.
f.c.: calcitic fibers.
FCCL: fine complex crossed lamellar (Mollusca).
fe: fiber.
FP: fibrous prismatic (Mollusca).
gc: granular calcite.
gw: growth band or growth line.
H: horizontal section or fracture.
HOM: homogeneous (Mollusca).
I: imperforate stereom (Echinodermata).
I: interior (Mollusca).
ICCL: irregular complex crossed lamellar (Mollusca).
ICM: isolated crystal morphotypes.
I/COMPCL: irregular to compressed CL (Mollusca).
ILZ: inner laminate zone (Arthropoda).
ISP: irregular simple prismatic (Mollusca).
IVPP: Inst. Vert. Paleont. and Paleoanth., Beijing.
kb: keeled blade (Brachiopoda).
Lb: labyrinthic stereom (Echinodermata).
le: lens.
Lm: laminar stereom (Echinodermata).
M: matrix (Arthropoda).
M: middle (Mollusca).
mi: dividing fiber (Brachiopoda).
ml: mural pore (Brachiopoda).
ML: membranous layer (Arthropoda).
mp: micropunctum (Brachiopoda).
N: nacre (Mollusca).
O: oblique section or fracture.
OLZ: outer laminate zone (Arthropoda).
op: organic partition.
os: organic sheet.
OL: outer layer.
P: periostracum (Mollusca, Brachiopoda).
pd: pad (of calcite in shell repair).
pe: periostracum (Brachiopoda, Mollusca).
PL: perforate stereom layer (Echinodermata).
pl: primary layer (Brachiopoda, Bryozoa).
PL: principal layer (Arthropoda).
PL: prismatic layer (Mollusca).
pn: pseudopunctum (Brachiopoda).
pu: punctum (Brachiopoda).
R: radial section or fracture.
RESP: radially elongate simple prismatic (Mollusca).
RF: regularly foliated (Mollusca).
RL: retiform stereom layer (Echinodermata).
RSP: regular simple prismatic (Mollusca).
S: periostracal spike or granule (Mollusca).
SEM: scanning electron microscopy.

SF: semi-foliated (Mollusca).
sl: secondary layer (Brachiopoda, Bryozoa).
sn: diagenetic siliceous nodule.
SphP: spherulitic prismatic (Mollusca).
su: substrate.
T: transverse section or fracture.
ta: taleola (Brachiopoda).
tb: tablet.
te: tubercle.
TEM: transmission electron microscopy.
tl: tertiary layer (Brachiopoda, Bryozoa).
tr: transgression in growth banding.
UCD: University of California, Davis.
UCMP: Univ. California Museum of Paleontology.
UMMP: Univ. Michigan Museum of Paleontology.
UNC: Univ. North Carolina, Chapel Hill.
USNM: U.S. National Museum, Washington, DC.
ve: vesicle (brachiopod periostracum).
YPM: Yale Peabody Museum, New Haven, Conn.
Z: zone of structural intergradation.
zm: zooecial chamber (Bryozoa).
ZPAL: Inst. Paleobiol., Polish Acad. Science.

Part 2
Bivalvia (Mollusca)
Plates 1-121

Joseph G. Carter

Department of Geology, University of North Carolina,
Chapel Hill, NC, 27514

Richard A. Lutz

New Jersey Agricultural Experiment Station, Cook College,
Rutgers University, New Brunswick, NJ 08903

Introduction

The accompanying illustrations document taxonomic variations in shell microstructure for a variety of modern and fossil bivalves. Unlike the accompanying chapter in Volume I, which focuses on the subclasses Palaeotaxodonta, Pteriomorphia and Isofilibranchia, the present illustrations represent primarily the subclasses Heteroconchia and Anomalodesmata. Sections 1-8 are arranged by microstructural categories to facilitate comparisons of similar microstructures among various taxa. The plates within each of these sections follow the taxonomic order of the *Treatise on Invertebrate Paleontology* (Cox *et al.*, 1969), with the exception of the anomalodesmatan genus *Poromya* (species no. 85). The species numbers, explained below, are arranged in the same taxonomic order.

Section 9 illustrates certain Paleozoic pterioid, pectinoid and veneroid bivalves which appear to be important for understanding microstructural evolution within their respective orders. Ordovician *Pterinea demissa* Conrad, 1842 (Plate 101) is the earliest member of the family Pterineidae for which microstructural data are available. The Middle Devonian species of the pterinopectinid *Pseudaviculopecten* (Plates 102-103) are in some respects microstructurally transitional between *Pterinea demissa* and later Paleozoic pectinoid bivalves. Middle Devonian *Eodon tenuistriata* (Hall, 1870) is microstructurally similar to Middle Devonian *Ptychodesma knappianum* (Hall and Whitfield, 1872), suggesting a close relationship between this astartid and the arcoid family Cyrtodontidae. *Eodon tenuistriata* is the only veneroid bivalve presently known to have a nacreous or partially nacreous shell.

Section 10 documents microstructural diversity in the superfamily Mytiloidea. Mytiloideans are extraordinary for the variety of microstructures and mineralogies present in their periostracum and outer shell layer. This variety may well provide clues to their evolutionary relationships.

Shell mineralogy. Unless indicated otherwise, all of the calcium carbonate structures illustrated in the accompanying plates are, or were, originally aragonitic except for structures of the foliated category (*e.g.*, regularly foliated, crossed foliated, complex crossed foliated, crossed semi-foliated, complex crossed semi-foliated).

Arrows. The thick, short arrows in Plates 1-121 point toward the shell interior and/or the depositional surface. The thin, long arrows point in a radial direction toward the nearest shell margin. The small, short arrows point to a feature indicated in the figure caption.

Specimen preparation methods. Specimen preparation methods are indicated by number in the explanations to the plates. The preparation methods are described in the introduction to this volume.

Species numbers. The specimens illustrated in Plates 1-105 are identified on the plates themselves by species numbers, *e.g.*, "1. *Nucula*". These species numbers refer to the following list. This list follows the taxonomic order of the *Treatise on Invertebrate Paleontology* (Cox *et al.*, 1969), except for species no. 85 (*Poromya* sp.).

1. *Nucula* (*Nucula*) *proxima* Say, 1822: Recent, Long Island Sound, New York, YPM 10014. Plates 1A, 19A,B. Family Nuculidae, superfamily Nuculoidea.
2. *Solemya* (*Solemyarina*) *parkinsoni* Smith: Recent, Awarua Bay, New Zealand, YPM 5364. Plate 20A-F. Family Solemyidae, superfamily Solemyoidea.
3. *Barbatia obtusoides* (Nyst): Recent, Japan, YPM 6420. Plate 21A,B. Family Arcidae, superfamily Arcoidea.
4. *Arca ventricosa* Lamarck, 1819: Recent, Palau, YPM 6251. Plates 54A, 77A. Family Arcidae, superfamily Arcoidea.
5. *Cucullaea granulosa* Jonas: Recent, Japan?, YPM 10033. Plates 54B, 77B. Family Cucullaeidae, superfamily Arcoidea.
6. *Noetia* (*Eontia*) *ponderosa* (Say, 1822): Recent, Florida, YPM 6634. Plates 54C, 77C. This is the type species of *Eontia*. Family Noetiidae, superfamily Arcoidea.
7. *Limopsis sulcata* Verrill and Bush, 1898. Recent, Maine, YPM 9637. Plates 55A, 78A. Family Limopsidae, superfamily Limopsoidea.
8. *Glycymeris pectinata* (Gmelin, 1791). Recent, Tobago, West Indies, YPM 6039. Plates 55B, 78B. Family Glycymerididae, superfamily Arcoidea.
9. *Pinctada imbricata* Röding, 1798: Recent, Bimini, Bahamas, YPM 6889. Plates 22A,B. Family Pteriidae, superfamily Pterioidea.
10. *Vulsella vulsella* (Linné, 1758): Recent, New Caledonia, YPM 6933. Plate 23A. Family Malleidae, superfamily Pterioidea.
11. *Pinna bicolor* Gmelin, 1791: Recent, Calapan, Mindoro, Philipines, YPM 6964. Plates 4A,B; 23B; 24A,B. Family Pinnidae, superfamily Pinnoidea.
12. *Oxytoma* (*Oxytoma*) *inequivalve* (J. Sowerby, 1819): Oxfordian, Jurassic, Bucks, England, UNC 4527. Plates 8B, 25A. This is the type species of the genus. Family Oxytomidae, superfamily Monotoidea.
13. *Propeamussium dalli* (E.A. Smith, 1886): Recent, West of Martinique, Windward Islands, YPM 8387. Plates 7A,B; 25B,C; 55C; 78C. Family Propeamussiidae, superfamily Pectinoidea.
14. *Plicatula gibbosa* Lamarck, 1801: Recent, St. Thomas, Virgin Islands, YPM 9631. Plate 56A. Family Plicatulidae, superfamily Plicatuloidea.
15. *Spondylus gussoni* O.G. Costa: Recent, Gulf of Mexico, YPM 7233. Plate 56B. Family Spondylidae, superfamily Pectinoidea.
16. *Anomia simplex* d'Orbigny, 1842: Recent, New Haven, Connecticut, YPM 9715. Plates 6A,B; 25D; 56C; 79A-C. Family Anomiidae, superfamily Anomioidea.
17. *Crassostrea virginica* (Gmelin, 1791): Recent, New Haven, Connecticut, YPM 7057. Plates 5A,B; 26A,B. This is the type species of the genus. Family Ostreidae, superfamily Ostreoidea.
18. *Ctenoides scabra* (Born, 1778): Recent, Biscayne Bay, Florida, YPM 9702. Plates 27A-D, 57A. This is the type species of the genus. Family Limidae, superfamily Limoidea.
19. *Crenella glandula* (Totten, 1834): Recent, Cape Cod, Massachusetts, YPM 9194. Plate 28A-C. Family Mytilidae, superfamily Mytiloidea.
20. *Neotrigonia gemma* Iredale: Recent, Sydney, Australia, UNC 5427. Plates 3A,B; 15A; 29A,B. Family Trigoniidae, superfamily Trigonioidea.
21. *Obliquaria reflexa* Rafinesque, 1820: Recent, Meramec River, St. Louis, Missouri, YPM 9749. Plates 1B, 30A-C. This is the type species of the genus. Family Unionidae, superfamily Unionoidea.
22. *Parvilucina* (*Cavilinga*) *lampra* (Dall, 1901): Recent, La Paz, Gulf of California, YPM 8709. Plate 57B. Family Lucinidae, superfamily Lucinoidea.
23. *Myrtea* (*Myrtea*) *spinifera* (Montagu, 1803): Recent, England, YPM 9086. Plate 57C. Family Lucinidae, superfamily Lucinoidea.

24. *Codakia (Codakia) orbicularis* (Linné, 1758): Recent, Abaco Cays, Bahamas, YPM 10007. Plate 8A. Family Lucinidae, superfamily Lucinoidea.
25. *Divalucina cumingi* (Adams and Angas, 1863): Recent, New Zealand, YPM 9065. Plates 31A,B; 58A; 80A,B; 81A. This is the type species of the genus. Family Lucinidae, superfamily Lucinoidea.
26. *Thyasira flexuosa* (Montagu, 1803): Recent, Scarborough, England, YPM 1125. Plates 58B, 81B. Family Thyasiridae, superfamily Lucinoidea.
27. *Fimbria elegans* (Deshayes): Recent, Mindoro, Philippines, YPM 9700. Plates 32A; 58C; 81C. Family Fimbriidae, superfamily Lucinoidea.
28. *Felaniella (Zemysia) sericata* (Reeve, 1850): Recent, Gulf of California, YPM 9629. Plates 59A, 82A. Family Ungulinidae, superfamily Lucinoidea.
29. *Chama iostoma* Conrad, 1837: Recent, Palau, YPM 9744. Plates 59B, 82B. Family Chamidae, superfamily Chamoidea.
30. *Arcinella cornuta* (Conrad, 1866): Recent, Sanibel, Florida, YPM 9698. Plates 59C, 82C. Family Arcinellidae, superfamily Chamoidea.
31. *Aligena elevata* (Stimpson, 1851): Recent, Wellfleet, Cape Cod, Massachusetts, YPM 9730. Plate 60A. Family Kelliidae or Lasaeidae, superfamily Galeommatoidea.
32. *Montacuta ferruginosa* (Montagu): Recent, Woods Hole, Massachusetts, YPM 9647. Plates 9A,B; 32B; 60B. Family Montacutidae or Lasaeidae, superfamily Galeommatoidea.
33. *Galeomma (Galeomma) turtoni* Sowerby, 1825: Recent, England, YPM 9728. Plate 60C. This is the type species of the genus. Family Galeommatidae, superfamily Galeommatoidea.
34. *Cyamium mosthoffii* Pfeffer: Recent, Antarctic Drift Line, USNM 252868. Plate 61A. Family Cyamiidae, superfamily Cyamioidea.
35. *Anisodonta lutea* Dall: Recent, Keokea Hilo, Hawaii, USNM 337470. Plate 61B. Family Sportellidae, superfamily Cyamioidea.
36. *Beguina semiorbiculata* (Linné, 1758): Recent, Philippines, YPM 9709. Plate 61C. This is the type species of the genus. Family Carditidae, superfamily Carditoidea.
37. *Glans incrassata* (Sowerby): Recent, Carnarvon, Sharks Bay, Australia, YPM 9694. Plates 62A, 83A. Family Carditidae, superfamily Carditoidea.
38. *Astarte (Astarte) undata* Gould, 1841: Recent, Maine, YPM 9727. Plates 62B, 83B. Family Astartidae, superfamily Crassatelloidea.
39-40. *Eucrassatella (Hybolophus) speciosa* (Adams): Recent, from Fort Walton, Florida, YPM 9734. Plate 62C and Plate 83C. Family Crassatellidae, superfamily Crassatelloidea.
41. *Fragum (Lunulicardia) subretusum* (Sowerby, 1841): Recent, Calapan, Mindoro, Philippines, YPM 9676. Plates 33A-C; 84A,B. Family Cardiidae, superfamily Cardioidea.
42. *Nemocardium (Microcardium) peramabile* (Dall, 1881): Recent, Delroy Tank, Palm Beach Co., Florida, YPM 9670. Plate 34A,B. This is the type species of the subgenus *Microcardium.* Family Cardiidae, superfamily Cardioidea.
43. *Acrosterigma leucostoma* (Bruguière): Recent, from Cebu, Philippine Islands, Yale Peabody Museum no. 9745. Plate 63A and Plate 85A. Family Cardiidae, superfamily Cardioidea.
44. *Hippopus hippopus* (Linné, 1758): Recent, Philippines, YPM 9743. Plates 63B, 85B. Family Tridacnidae, superfamily Tridacnoidea.
45. *Lutraria philippinarum* Deshayes: Recent, Calapan, Mindoro, Philippines, YPM 9742. Plates 13A,B; 35A,B; 63C; 85C. Family Mactridae, superfamily Mactroidea.
46. *Anapella adelaide* (Angas): Recent, Adelaide, Australia, YPM 9179. Plate 64A. Family Mesodesmatidae, superfamily Mactroidea.
47. *Ensis siliqua* (Linné, 1758): Recent, from the North Sea, west end of Ameland, Netherlands, YPM 9716. Plates 15B; 64B; and 86A,B. Family Cultellidae, superfamily Solenoidea.
48. *Ensis directus* Conrad, 1843: Recent, Beverly, Massachusetts, YPM 541. Plate 64C. Family Cultellidae, superfamily Solenoidea.
49. *Siliqua japonica* (Dunker): Recent, Japan, YPM 9741. Plate 65A. Family Cultellidae, superfamily Solenoidea.
50. *Tellina (Peronidia) lutea* Wood, 1828: Icy Cape, Alaska, YPM 9611. Plate 65B. Family Tellinidae, superfamily Tellinoidea.
51. *Florimetis intastriata* (Say, 1826): Recent, Tobago, West Indies, YPM 9607. Plate 65C. This is the type species of the genus. Family Tellinidae, superfamily Tellinoidea.
52. *Hemidonax donaciformis* (Bruguière, 1792): Recent, from Matte Davao, Philippine Islands, Yale Peabody Museum no. 9612. Plate 66A, Plate 87A-D, and Plate 88A. This

is the type species of the genus *Hemidonax*. Family Hemidonacidae, superfamily Cardioidea.

53. *Donax variabilis* Say, 1822: Recent, Sanibel, Florida, YPM 10074. Plate 36A,B. Family Donacidae, superfamily Tellinoidea.
54. *Sanguinolaria olivacea* Say: Recent, Japan, YPM 9603. Plates 66B, 88B. Family Psammobiidae, superfamily Tellinoidea.
55. *Sanguinolaria boeddinghausi* (Lischke, 1870): Recent, Japan, YPM 9605. Plates 66C, 88C. Family Psammobiidae, superfamily Tellinoidea.
56. *Gari maculosa* (Lamarck, 1818): Recent, Calapan, Mindoro, Philippines, YPM 9604. Plate 37A-C. Family Psammobiidae, superfamily Tellinoidea.
57. *Semele decisa* (Conrad, 1837): Recent, San Onofre, California, YPM 10006. Plate 38A-C. Family Semelidae, superfamily Tellinoidea.
58. *Solecurtus divaricatus* (Lischke, 1869): Recent, Japan, YPM 9608. Plates 67A,B; 89A,B. Family Solecurtidae, superfamily Tellinoidea.
59. *Tagelus dombeii* (Lamarck, 1818): Recent, Paita, Peru, YPM 9606. Plate 67C. Family Solecurtidae, superfamily Tellinoidea.
60. *Mytilopsis leucophaeatus* (Conrad, 1831): Recent, Whitehall Creek, Choptank River, Maryland, YPM 10075. Plates 68A, 90A. This is the type species of the genus. Family Dreissenidae, superfamily Dreissenoidea.
61. *Trapezium (Trapezium) oblongum* (Linné, 1758): Recent, Kanai, Hawaii, YPM 9722. Plate 90B. This is the type species of the genus. Family Trapeziidae, superfamily Arcticoidea.
62. *Trapezium (Neotrapezium) sublaevigatum* (Lamarck, 1819): Recent, Bahia, Philippines, YPM 9717. Plates 11A-C; 12A,B; 68B,C; 90C; 91A,B. This is the type species of the subgenus *Neotrapezium*. Family Trapeziidae, superfamily Arcticoidea.
63. *Corallipohaga (Coralliophaga) coralliophaga* (Gmelin, 1791): Recent, Bermuda, YPM 9633. Plates 69A, 93A. This is the type species of the genus. Family Trapeziidae, superfamily Arcticoidea.
64. *Glossus vulgaris* (Reeve): Recent, China, YPM 9718. Plates 39A-C; 69B; 92A-C; 93B. Family Glossidae, superfamily Glossoidea.
65. *Batissa* sp.: Recent, Philippines, YPM 9699. Plates 69C; 93C; 94A,B. Family Corbiculidae, superfamily Corbiculoidea.
66. *Dosinia ponderosa* (Gray, 1838): Recent, Gulf of California, YPM 4138. Plates 70A, 96A. Family Veneridae, superfamily Veneroidea.
67. *Cyclina (Cyclina) sinensis* (Gmelin, 1791): Recent, Japan, YPM 9695. Plates 40A-C; 70B; 95A-D; 96B. This is the type species of *Cyclina*. Family Veneridae, superfamily Veneroidea.
68. *Circe stutzeri* Donovan: Recent, Matte Davao, Philippines, YPM 9738. Plates 70C, 96C. Family Veneridae, superfamily Veneroidea.
69. *Compsomyax subdiaphana* (Carpenter, 1864): Recent, San Juan Island, Washington State, YPM 9705. Plates 41A-C, 97A. Family Veneridae, superfamily Veneroidea.
70. *Paphia cuneata* (Deshayes): Recent, Fiji, YPM 2060. Plate 97B. Family Veneridae, superfamily Veneroidea.
71. *Tivela trigonella* Lamarck: Recent, Grassy Key, Florida, YPM 9630. Plate 97C. Family Veneridae, superfamily Veneroidea.
72. *Meretrix undulosa* Lamarck: Recent, Western Australia, YPM 9740. Plate 71A. Family Veneridae, superfamily Veneroidea.
73. *Tivela (Tivela) byronensis* (Gray, 1838): Recent, Guaymas, Gulf of California, YPM 9737. Plate 71B. Family Veneridae, superfamily Veneroidea.
74. *Saxidomus nuttalli* (Conrad, 1837): Recent, Newport Bay, California, YPM 9696. Plate 71C. Family Veneridae, superfamily Veneroidea.
75. *Protothaca staminea* (Conrad, 1837): Recent, Salmon Bay, Puget Sound, YPM 9678. Plates 16A, 42A,B. Family Veneridae, superfamily Veneroidea.
76. *Chione (Chione) californiensis* (Broderip, 1835): Recent, Eastern Pacific, YPM 3012. Plate 72A. Family Veneridae, superfamily Veneroidea.
77. *Cooperella (Cooperella) subdiaphana* (Carpenter, 1864): Recent, Orange County, California, YPM 9645. Plate 72B. This is the type species of the genus. Family Cooperellidae, superfamily Veneroidea.
78. *Glauconome virens* (Linné, 1767): Recent, Sumatra, YPM 10080. Plate 72C. Family Glauconomidae, superfamily Veneroidea.

79-80. *Mya (Arenomya) arenaria* Linné, 1758: Number 79 = Recent, New Haven Connecticut, YPM 10082; Plate 73A. Number 80 = Recent, Great Egg Harbor, New Jersey, YPM 4311; Plates 98A,B; 99A. Family Myidae, superfamily Myoidea.

81. *Cryptomya* (*Cryptomya*) *californica* (Conrad, 1837): Recent, Morro Bay, California, YPM 9723. Plates 73B,C; 99B. This is the type species of the genus. Family Myidae, superfamily Myoidea.
82. *Paramya subovata* (Conrad, 1845): Recent, off Cape Hatteras, North Carolina, at 17 fathoms, YPM 9599. Plate 99C. This is the type species of the genus. Family Myidae, superfamily Myoidea.
83. *Platyodon* (*Platyodon*) *cancellatus* (Conrad, 1837): Recent, Anaheim Bay, California, YPM 9704. Plates 43A-D; 74A. This is the type species of the genus. Family Myidae, superfamily Myoidea.
84. *Corbula* (*Juliacorbula*) *bicarinata* (Sowerby, 1833): Recent, South America, YPM 2326. Plate 74B. Family Corbulidae, superfamily Myoidea.
85. *Poromya* (*Poromya*) sp.: Recent, from an unspecified, Eastern Atlantic deep sea locality, YPM 9636. Plates 1C; 10A,B; and 44A-D. Family Poromyidae, superfamily Poromyoidea.
86. *Corbula nucleus* Lamarck: Recent, Britain, YPM 636. Plate 74C. Family Corbulidae, superfamily Myoidea.
87. *Rocellaria hawaiensis* Dall, Bartsch and Rehder, 1938: Recent, Mokuoloe Island, Oahu, Hawaii, USNM 337313. Plate 75A. Family Gastrochaenidae, superfamily Gastrochaenoidea.
88. *Gastrochaena* sp.: Pliocene, Tamiami Formation, near Nocatee, Florida, YPM 9799. Plate 75B. Family Gastrochaenidae, superfamily Gastrochaenoidea.
89. *Spengleria rostrata* (Spengler, 1783): Recent, Soldier Key, Florida, YPM 9473. Plates 17A-C; 18A,B; 45A-C; 75C. Family Gastrochaenidae, superfamily Gastrochaenoidea.
90. *Hiatella* (*Hiatella*) *arctica* (Linné, 1767): Recent, Eastport, Maine, YPM 9724. Plate 76A. This is the type species of the genus. Family Hiatellidae, superfamily Hiatelloidea.
91. *Cyrtodaria siliqua* (Spengler, 1793): Recent, off Martha's Vineyard, Massachusetts, YPM 4995. Plates 14B; 16B; 46A,B; 47A,B. This is the type species of *Cyrtodaria*. Family Hiatellidae, superfamily Hiatelloidea.
92. *Panopea generosa* (Gould, 1850): Recent, Puget Sound, Washington, YPM 9746. Plates 14A,C; 48A,B. Family Hiatellidae, superfamily Hiatelloidea.
93. *Barnea truncata* (Say, 1822): Recent, Scituate Harbor, Massachusetts, YPM 9952-11. Plate 76B. Family Pholadidae, superfamily Pholadoidea.
94. *Zirfaea crispata* (Linné, 1776): Recent, Scituate Beach, Massachusetts, YPM 9712. Plate 76C. This is the type species of the genus. Family Pholadidae, superfamily Pholadoidea.
95. *Pholas campechiensis* Gmelin, 1791. Recent, Western Atlantic, YPM 10083. Plate 100A. Family Pholadidae, superfamily Pholadoidea.
96. *Cyrtopleura* (*Scobinopholas*) *costata* (Linné, 1758): Recent, Florida (YPM 9513) and Texas (UNC 6749). Plate 100B. This is the type species of the subgenus *Scobinopholas*. Family Pholadidae, superfamily Pholadoidea.
97. *Zirfaea pilsbryi* Lowe, 1931: Recent, Dana Point, California, YPM 9514. Plate 100C. Family Pholadidae, superfamily Pholadoidea.
98. *Parapholas californica* (Conrad, 1837): Recent, Monterey Bay, California, YPM 10258. Plates 49A,B; 50A-D. This is the type species of the genus. Family Pholadidae, superfamily Pholadoidea.
99. *Laternula* cf. *L. elliptica* King and Broderip, 1830-1831: Recent, Kerguelen Islands or Gough Island, YPM 4956. Plate 51A-C. Family Laternulidae, superfamily Pandoroidea.
100. *Thracia* (*Thracia*) *pubescens* (Pulteney, 1799): Recent, England?, USNM 171297. Plates 2A,B; 52A,B. This is the type species of the genus. Family Thraciidae, superfamily Pandoroidea.
101. *Euciroa* (*Euciroa*) *elegantissima* (Dall, 1881): Recent, Caribbean, YPM 9653. Plate 53A-C. This is the type species of the genus. Family Verticordiidae, superfamily Poromyoidea.

EXPLANATION OF PLATES

Section 1: Laminar Structures.

PLATE 1. A. Sheet nacreous structure. Radial, vertical fracture through the inner shell layer of *Nucula proxima*. Preparation method 1. B. Sheet nacreous to columnar nacreous structure. Radial, vertical fracture through the middle shell layer of *Obliquaria reflexa*. Preparation method 1. C. Sheet nacreous (N) structure of the middle shell layer and prismatic (PL) structure of the outer shell layer of *Poromya* (*Poromya*) sp.. Radial, vertical fracture. Preparation method 1.

PLATE 2. A. Sheet nacreous (N) and complex crossed lamellar (CCL) structure separated by a thin zone (Z) of homogeneous to irregular simple prismatic pallial myostracum. Radial, vertical fracture through the middle and inner shell layers of *Thracia pubescens*. Preparation method 1. B. Closer view of the nacreous (N) and inner complex crossed lamellar (CCL) layers of *Thracia pubescens* separated by the pallial myostracum (Z). The major structural arrangement of the CCL layer is not apparent at this high magnification. Preparation method 1.

PLATE 3. A. Columnar nacreous structure. Transverse, vertical fracture through the middle shell layer of *Neotrigonia gemma*. Preparation method 1. B. Sheet nacreous structure. Transverse, vertical fracture through the inner shell layer of *Neotrigonia gemma*. Parts of this shell layer show a stair-step arrangement of the nacre tablets. Preparation method 1.

PLATE 4. A. Row stack nacreous structure. Vertical fracture oblique to the length direction of the nacre row stacks of *Pinna bicolor*. Preparation method 1. B. Same shell layer as in Figure A, only viewed along a fracture parallel to the length of the nacre row stacks. Preparation method 1.

PLATE 5. A. Calcitic regularly foliated structure. Transverse, vertical fracture through the middle shell layer of *Crassostrea virginica*. Preparation method 1. B. Same layer as in Figure A, only viewed in a horizontal fracture. Preparation method 1.

PLATE 6. A. Calcitic regularly foliated structure. Horizontal fracture through the outer shell layer of the left valve of *Anomia simplex*. Preparation method 1. B. Higher magnification of the regularly foliated structure in Figure A. Preparation method 1.

Section 2: Crossed Foliated Structure.

PLATE 7. A. Branching crossed foliated structure. Horizontal fracture through the foliated layer of the left valve of *Propeamussium dalli*. Portions of only two first-order lamellae are visible at this magnification. Preparation method 1. B. Higher magnification of the branching crossed foliated structure in Figure A. Preparation method 1.

Section 3: CCL and CCF Structures.

PLATE 8. A. Irregular complex crossed lamellar structure. Composite figure showing three perpendicular views of the inner shell layer of *Codakia orbicularis*. Each view shows numerous first-order lamellae. Preparation method 3. B. Irregular complex crossed foliated structure. Composite figure showing three perpendicular views of the inner layer of Jurassic *Oxytoma* (*Oxytoma*) *inequivalve*. Each view shows numerous first-order lamellae. Preparation method 3.

PLATE 9. A. Irregular complex crossed lamellar (CCL) structure of the inner layer of *Montacuta ferruginosa*, also showing the periostracum (P), a thin irregular simple prismatic layer (ISP), a fibrous prismatic layer (FP), and an extremely thin crossed lamellar layer (CL). Transverse, vertical section near the umbo. Preparation method 2. B. Closer view of the irregular CCL structure in Figure A. Preparation method 2.

PLATE 10. A. Irregular complex crossed lamellar (CCL) structure of the inner shell layer of *Poromya* (*Poromya*) sp. Nearly horizontal (slightly oblique) section through the inner shell layer, with the sheet nacreous middle shell layer (N) appearing in the upper right-hand corner of the photograph. Preparation method 2. B. Closer view of the irregular CCL structure in Figure A. Preparation method 2.

PLATE 11. A. Cone complex crossed lamellar structure of the inner shell layer of *Trapezium sublaevigatum*. Composite figure showing three perpendicular views. Preparation method 3. B. Cone complex crossed lamellar structure of the inner shell layer of *Trapezium sublaevigatum*. Nearly horizontal (slightly oblique) section. Preparation method 3. C. Cone complex crossed lamellar structure of the inner shell layer of *Trapezium sublaevigatum*. Radial, vertical fracture. Preparation method 1.

PLATE 12. A. Substructure of the cone complex crossed lamellar structure of the inner shell layer of *Trapezium sublaevigatum*. Radial, vertical fracture through the same shell layer illustrated in Plate 11. Preparation method 1. B. Higher magnification of the cone complex crossed lamellar structure in Figure A. Preparation method 1.

PLATE 13. A. Cone complex crossed lamellar structure immediately interior to the prismatic outer shell layer of *Lutraria philippinarum*. Radial, vertical fracture. Preparation method 1. B. Higher magnification of the cone complex crossed lamellar structure in Figure A. Preparation method 1.

Section 4: Finely Textured Structures.

PLATE 14. A. Transitional homogeneous to fine complex crossed lamellar structure of the inner shell layer of *Panopea generosa*. Radial, vertical section. Preparation method 2. B. Transitional homogeneous to crossed acicular structure immediately interior to the prismatic outer shell layer of *Cyrtodaria siliqua*. Radial, vertical fracture. Preparation method 1. C. Transitional crossed acicular to crossed lamellar structure immediately interior to the prismatic outer shell layer of *Panopea generosa*. Radial, vertical section. Preparation method 2.

PLATE 15. A. Homogeneous structure of the outer part of the homogeneous to prismatic outer shell layer of *Neotrigonia gemma*. Transverse, vertical fracture. See also Plate 29. Preparation method 1. B. Homogeneous structure of the inner shell layer of *Ensis siliqua*. Nearly horizontal (slightly oblique) fracture. Preparation method 1.

PLATE 16. A. Homogeneous structure locally developed between the prismatic outer layer and the crossed lamellar middle layer of *Protothaca staminea*. Radial, vertical section. Preparation method 2. B. Homogeneous structure immediately exterior to the pallial myostracum of *Cyrtodaria siliqua*. Radial, vertical fracture. Preparation method 1.

Section 5: Periostracal Structures.

PLATE 17. A. Aragonitic periostracal spike freed from the posterior periostracum of *Spengleria rostrata* by dissolution of the organic periostracum in a solution of 5% sodium hypochlorite (Clorox). B. Higher magnification of the aragonitic periostracal spike in Figure A. C. Posterior periostracum of *Spengleria rostrata* near the ventral shell margin, showing several aragonitic periostracal spikes (S). Transverse, vertical fracture. Desiccation has caused the periostracum (P) to peel away from the underlying prismatic shell layer (PL). Preparation method 1.

Section 6: Isolated Crystal Morphotypes.

PLATE 18. A. Aragonitic isolated crystal morphotypes in the space between the periostracum and the underlying prismatic shell layer in the posterior of *Spengleria rostrata*. Radial, vertical fracture. B. Higher magnification of the isolated crystal morphotypes in Figure A.

Section 7: Prismatic Structures.

Plates 19 through 53 illustrate prismatic structures developed in the outer shell layer near the posterior margin of the adult shell.

PLATE 19. A. Denticular composite prismatic (DCP) outer shell layer of *Nucula proxima*. Composite figure showing three perpendicular views. The radial and transverse views also show the underlying nacreous (N) middle shell layer. Preparation method 3. B. Radial, vertical fracture through the denticular composite prismatic (DCP) outer shell layer (above) and the nacreous (N) middle shell layer (below) of *Nucula proxima*. Preparation method 1.

PLATE 20. A. Radial, vertical fracture through the radially elongate simple prismatic (RESP) outer layer and the underlying irregular simple prismatic (ISP) myostracum of *Solemya parkinsoni*. Desiccation has caused the periostracum (P) to peel away from the exterior of the RESP layer. Preparation method 1. B. Homogeneous substructure of the radially elongate simple prismatic outer layer of *Solemya parkinsoni*. Radial, vertical fracture. Preparation method 1. C. Closer view of the irregular simple prismatic middle shell layer of *Solemya parkinsoni*. Radial, vertical fracture. Preparation method 1. D. Homogeneous substructure of the irregular simple prismatic middle shell layer of *Solemya parkinsoni*. Radial, vertical fracture. Preparation method 1. E. Horizontal section through the radially elongate simple prismatic outer shell layer of *Solemya parkinsoni*. Preparation method 3. F. Transverse, vertical section through the radially elongate simple prismatic (RESP) outer shell layer and the periostracum (P) of *Solemya parkinsoni*. Preparation method 3.

PLATE 21. A. Composite figure showing three perpendicular views of the reclined fibrous prismatic to simple lamellar fibrous prismatic outer shell layer (PL) of *Barbatia obtusoides*. The radial and horizontal views also show the underlying crossed lamellar (CL) middle layer, and the transverse view also shows the periostracum (P). Preparation method 3. B. Prismatic (PL) outer layer (toward bottom of photograph) and crossed lamellar (CL) middle layer (upper left-hand corner of photograph) of *Barbatia obtusoides*. Radial, vertical fracture. Preparation method 1. In this view, the prismatic layer shows simple lamellar fibrous prismatic structure intergradational with crossed lamellar structure.

PLATE 22. A. Calcitic regular simple prismatic (polycrystalline variety) outer shell layer of *Pinctada imbricata*. Composite figure showing three perpendicular views. Preparation method 3. B. Radial, vertical fracture through the calcitic regular simple prismatic outer shell layer of *Pinctada imbricata*. Preparation method 1.

PLATE 23. A. Calcitic regular simple prismatic (RSP) (polycrystalline variety) outer shell layer of *Vulsella vulsella*. Composite figure showing three perpendicular views. The radial view also shows the underlying nacreous (N) middle shell layer. Preparation method 3. B. Calcitic regular simple prismatic (monocrystalline variety) outer shell layer of *Pinna bicolor*. Composite figure showing three perpendicular views. See also Plate 24. Preparation method 3.

PLATE 24. A. Calcitic regular simple prismatic (RSP) (monocrystalline variety) outer shell layer and underlying nacreous (N) middle layer of *Pinna bicolor*. Oblique, vertical fracture through the ventral shell margin. Preparation method 1. B. Contact between the calcitic regular simple prismatic (RSP) (monocrystalline variety) and underlying nacreous (N) shell layers of *Pinna bicolor*. Same fracture surface and preparation method as in Figure A.

PLATE 25. A. Calcitic regular simple prismatic (RSP) (polycrystalline variety) outer shell layer of the right valve of Jurassic *Oxytoma inequivalve*. Composite figure showing three perpendicular views. The radial and transverse (actually slightly oblique) views also show the underlying crossed foliated (CF) middle shell layer. Preparation method 3. B. Calcitic, lathic regular simple prismatic (RSP) (polycrystalline variety) outer shell layer of the right valve of *Propeamussium dalli*. The RSP layer is underlain by a predominantly calcitic fibrous prismatic to irregular spherulitic prismatic (FP) middle layer and an aragonitic crossed lamellar (CL) inner layer. Preparation method 3. C. Radial, vertical section through the calcitic, lathic regular simple prismatic (RSP) outer shell layer and underlying calcitic fibrous prismatic (FP) to irregular spherulitic prismatic middle layer of *Propeamussium dalli*. Preparation method 2. D. Radial, vertical fracture through the calcitic, lathic regular simple prismatic (RSP) (polycrystalline variety) outer shell layer and the aragonitic crossed lamellar (CL) inner shell layer of the right valve of *Anomia simplex*. Preparation method 3.

PLATE 26. A. Calcitic regular simple prismatic (RSP) (polycrystalline variety) outer shell layer of the right valve of *Crassostrea virginica.* Composite figure showing three perpendicular views. The radial and transverse views also show the underlying regularly foliated (RF) middle shell layer. Preparation method 3. B. Radial, vertical fracture through the calcitic regular simple prismatic outer shell layer of the right valve of *Crassostrea virginica.* Preparation method 1.

PLATE 27. A,B. Rod type fibrous prismatic structure of the outer part of the calcitic outer shell layer of *Ctenoides scabra.* Radial, vertical fractures. Preparation method 1. C,D. Inner portion of the calcitic outer layer of *Ctenoides scabra*, showing a structure transitional between crossed bladed and regularly foliated. Horizontal fractures. Preparation method 1.

PLATE 28. A. Calcitic fibrous prismatic (FP) outer shell layer of *Crenella glandula.* Composite figure showing three perpendicular views. The radial and transverse views also show the overlying periostracum (P) and the underlying nacreous (N) middle shell layer. Preparation method 3. B. Calcitic fibrous prismatic (FP) outer shell layer of *Crenella glandula.* Radial, vertical section. This figure also shows the overlying periostracum (P) and the underlying sheet nacreous (N) middle layer. Preparation method 2. C. Closer view of the calcitic fibrous prismatic structure in Figure B. Preparation method 2.

PLATE 29. A,B. Nondenticular composite prismatic (CP) and simple prismatic (on the right) structure of the prismatic to homogeneous outer shell layer of *Neotrigonia gemma.* Transverse, vertical fracture. Figure A also shows the underlying nacreous (N) middle shell layer. Figure B is a closer view of Figure A. The outer shell layer locally grades from nondenticular composite prismatic to regular simple prismatic to homogeneous structure, with the latter commonly developed on the crests of the ribs. See also Plate 15. Preparation method 1.

PLATE 30. A. Nondenticular composite prismatic (CP) portion of the outer shell layer of *Obliquaria reflexa.* Composite figure showing three perpendicular views. The radial and transverse views also show the overlying periostracum (P). The outer layer of this shell locally grades from nondenticular composite prismatic to regular simple prismatic. Preparation method 3. B,C. Radial, vertical fracture through a nondenticular composite prismatic (CP) portion of the outer shell layer of *Obliquaria reflexa.* Figure C also shows the overlying periostracum (P). Preparation method 1.

PLATE 31. A. Prismatic outer shell layer of *Divalucina cumingi*, showing an outer portion with fibrous prismatic (FP) structure and an inner portion with irregular spherulitic prismatic (SphP) structure. Composite figure showing three perpendicular views. The transverse view also shows the underlying aragonitic crossed lamellar (CL) middle shell layer. Preparation method 3. B. Radial, vertical fracture through the outer shell layer of *Divalucina cumingi*, showing the outer fibrous prismatic (FP) portion and the inner spherulitic prismatic (SphP) portion. Preparation method 1.

PLATE 32. A. Prismatic outer shell layer of *Fimbria elegans*, which varies from fibrous prismatic to irregular simple prismatic, irregular spherulitic prismatic, and nondenticular composite prismatic. Composite figure showing three perpendicular views. The horizontal view shows both exterior (E) and interior (I) parts of this shell layer. Preparation method 3. B. Radial, vertical section through *Montacuta ferruginosa* showing the periostracum (P), an irregular simple prismatic outer shell layer (ISP), and an underlying fibrous prismatic layer (FP). The two prismatic layers are partially obscured by a thin film of acetate left over from the acetate peel process. Preparation method 2.

PLATE 33. A. Fibrous prismatic (FP) outer shell layer deposited on the reflected shell margin of *Fragum subretusum.* Composite figure showing three perpendicular views. The radial view also shows the overlying periostracum (P) and the underlying branching crossed lamellar (CL) middle shell layer. Preparation method 3. The outer shell layer locally shows a weak development of denticular composite prismatic structure (not shown here). B,C. Radial, vertical fracture through the inner (B) and middle (C) portions of the fibrous prismatic outer shell layer of *Fragum subretusum.* Preparation method 1.

PLATE 34. A. Prismatic outer shell layer (PL) of *Nemocardium peramabile*. Composite figure showing three perpendicular views. This layer varies from nondenticular composite prismatic and compound composite prismatic to poorly developed irregular spherulitic prismatic and fibrous prismatic. The radial and transverse views also show the middle branching crossed lamellar (BCL) shell layer, and the transverse view also shows the inner complex crossed lamellar to irregular simple prismatic shell layer (labelled CCL). Preparation method 3. B. Radial, vertical fracture through the inner part of the prismatic outer shell layer of *Nemocardium peramabile*, showing poorly developed fibrous prismatic structure near the transition zone with the underlying CL layer. Preparation method 1.

PLATE 35. A,B. Radial, vertical fracture through the irregular spherulitic prismatic to irregular fibrous prismatic outer shell layer of *Lutraria philippinarum*. Figure B is a closer view of Figure A. Preparation method 1.

PLATE 36. A. Predominantly compound composite prismatic (CP) outer shell layer of *Donax variabilis*. Composite figure showing three perpendicular views. The radial and transverse views also show the underlying crossed lamellar (CL) middle layer. Although not apparent in these figures, the most exterior portion of this prismatic layer is locally fibrous prismatic. Preparation method 3. B. Radial, vertical fracture through the middle of the predominantly compound composite prismatic outer shell layer of *Donax variabilis*. Preparation method 1.

PLATE 37. A. Denticular composite prismatic (CP) outer shell layer of *Gari maculosa*. Composite figure showing three perpendicular views. The radial and transverse views also show the underlying crossed lamellar (CL) middle layer. Preparation method 3. B,C. Radial, vertical fracture through the denticular composite prismatic outer shell layer of *Gari maculosa*, showing its fibrous prismatic substructure. Figure B is a closer view of Figure C. Preparation method 1.

PLATE 38. A. Prismatic outer shell layer of *Semele decisa*. Composite figure showing three perpendicular views. This layer is irregular spherulitic prismatic (SphP) in its outer portion and transitional denticular/nondenticular composite prismatic (CP) in its middle and inner portions. The transverse view also shows the periostracum (P) and the underlying crossed lamellar (CL) middle shell layer. The CL layer (below the dotted line in the transverse view) was not entirely replicated during the acetate peel procedure. Preparation method 3. B,C. Radial, vertical fracture through the middle (B) and inner (C) portions of the prismatic outer shell layer of *Semele decisa*, showing several first-order prisms of the transitional denticular/nondenticular composite prismatic structure. Preparation method 1.

PLATE 39. A. Composite figure showing three perpendicular views of the outer shell layer (OL) of *Glossus vulgaris*. The outer portion of this outer layer varies from homogeneous to irregular simple prismatic; the middle portion shows predominantly irregular spherulitic prismatic structure, and the inner portion is irregular simple prismatic to crossed acicular. The radial and transverse views also show the underlying crossed lamellar (CL) middle shell layer. Preparation method 3. B. Oblique, vertical fracture through the zone of structural intergradation between the irregular simple prismatic to crossed acicular inner portion of the outer layer and the underlying crossed lamellar middle layer of *Glossus vulgaris*. Preparation method 1. C. Oblique, vertical fracture through the outer layer of *Glossus vulgaris*, showing the homogeneous to irregular simple prismatic outer portion (above) and the irregular spherulitic prismatic middle portion (below). Preparation method 1.

PLATE 40. A. Irregular fibrous prismatic (FP) outer shell layer of *Cyclina sinensis* deposited on a reflected, denticulated shell margin. Composite figure showing three perpendicular views. The radial and transverse views also show the periostracum (P) and the underlying crossed lamellar (CL) middle layer. The inner portion of this prismatic outer shell layer locally approximates denticular composite prismatic structure (see horizontal view) as a result of the organizing influence of its denticulated surface of deposition. Preparation method 3. B,C. Radial, vertical fracture through the outer portion of the irregular fibrous prismatic outer shell layer of *Cyclina sinensis*. Figure C is a closer view of Figure B. Preparation method 1.

PLATE 41. A. Irregular spherulitic prismatic to irregular simple prismatic outer shell layer (OL) of *Compsomyax subdiaphana*. Composite figure showing horizontal and radial views. The radial view also shows the underlying fine complex crossed lamellar to crossed acicular middle shell layer (toward the right in the photograph). Preparation method 3. B,C. Radial, vertical section through the irregular spherulitic prismatic to irregular simple prismatic outer shell layer of *Compsomyax subdiaphana*. Figure C is a closer view of Figure B. Preparation method 2.

PLATE 42. A. Outer shell layer of *Protothaca staminea* showing predominantly nondenticular composite prismatic (CP) structure. Composite figure showing three perpendicular views. The inner portion of this layer approximates compound composite prismatic structure because of the organizing influence of the denticulated depositional surface (see horizontal view). The radial and transverse views also show the underlying homogeneous to complex crossed lamellar (labelled CCL) middle layer. Preparation method 3. B. Radial, vertical fracture through the middle portion of the predominantly nondenticular composite prismatic outer shell layer of *Protothaca staminea*. Preparation method 1.

PLATE 43. A. Outer shell layer of *Platyodon cancellatus*, showing predominantly homogeneous structure with local nondenticular composite prismatic (CP) structure (see small arrows in the radial view). Composite figure showing three perpendicular views. The horizontal view shows exterior homogeneous (HOM; upper left) and interior nondenticular composite prismatic (lower right) portions of the outer shell layer. The transverse view also shows the underlying crossed lamellar (CL) middle shell layer. Preparation method 3. B. Radial, vertical fracture through the outer homogeneous portion of the outer shell layer of *Platyodon cancellatus*. Preparation method 1. C. Radial, vertical fracture through part of the outer shell layer of *Platyodon cancellatus* showing a structure transitional between homogeneous and nondenticular composite prismatic. Preparation method 1. D. Transverse, vertical fracture through the outer homogeneous portion of the outer shell layer of *Platyodon cancellatus*. Preparation method 1.

PLATE 44. Shell microstructure of *Poromya* (*Poromya*) sp. A. Fibrous prismatic to irregular spherulitic prismatic outer shell layer of *Poromya* (*Poromya*) sp. Nearly horizontal (slightly oblique) section. The extreme lower part of this photograph also shows the underlying nacreous middle shell layer. Preparation method 2. B. Radial, vertical section through the irregular simple prismatic to fibrous prismatic to irregular spherulitic prismatic outer shell layer of the right valve of *Poromya* (*Poromya*) sp. The irregular simple prismatic portion occurs as a very thin crust at the top of the section, just below the periostracum. In addition to the prismatic layer (PL), this figure shows the overlying periostracum (P) and the underlying nacreous middle shell layer (N). Preparation method 3. C. Horizontal section through the outer portion of the prismatic outer layer of *Poromya* (*Poromya*) sp., showing several aragonitic periostracal spikes (S) partially embedded within this layer. Preparation method 3. D. Radial, vertical section through the prismatic outer layer (PL), periostracum (P) and sheet nacreous (N) middle layer of the left valve of *Poromya* (*Poromya*) sp. This figure also shows an aragonitic periostracal spike (S) cemented to the exterior of the prismatic layer. Preparation method 3.

PLATE 45. A. Nondenticular composite prismatic outer layer (PL) of *Spengleria rostrata*. Composite figure showing three perpendicular views in the shell posterior. The radial and transverse views also show the underlying crossed lamellar (CL) middle layer, and the transverse view shows an overlying layer of isolated crystal morphotypes (ICM). Preparation method 3. B. Radial, vertical section through the transitional nondenticular composite prismatic to fibrous prismatic outer shell layer of *Spengleria rostrata* as developed mediolaterally in the shell. Preparation method 2. C. Radial, vertical section through the fibrous prismatic to irregular spherulitic prismatic outer layer of *Spengleria rostrata* as developed in the anterior part of the shell. Preparation method 2.

PLATE 46. A. Nondenticular composite prismatic to simple prismatic to spherulitic prismatic outer layer of *Cyrtodaria siliqua*. Composite figure showing three perpendicular views. Preparation method 3. B. Radial, vertical fracture through the middle portion of the nondenticular composite prismatic to simple prismatic outer shell layer of *Cyrtodaria siliqua*. See also Plate 47. Preparation method 1.

PLATE 47. A,B. Radial, vertical fracture through a cavity in the porous, nondenticular composite prismatic to simple prismatic outer portion of the outer shell layer of *Cyrtodaria siliqua*. Figure B is a closer view of Figure A. See also Plate 46. Preparation method 1.

PLATE 48. A. Prismatic outer shell layer of *Panopea generosa*. Composite figure showing three perpendicular views. Predominantly irregular simple prismatic structure with individual first-order prisms showing a substructure of horizontally flattened structural subunits. The exterior part of this layer (E) is locally spherulitic prismatic. The horizontal view shows exterior (E) and middle (M) portions of this outer shell layer. Preparation method 3. B. Radial, vertical section through the middle portion of the predominantly irregular simple prismatic outer shell layer of *Panopea generosa*. Preparation method 2.

PLATE 49. A. Predominantly regular simple prismatic (RSP) outer shell layer in the posterior of *Parapholas californica*. Composite figure showing three perpendicular views. The radial and transverse views also show the overlying periostracum (P) and underlying crossed lamellar (CL) middle layer. The outer parts of these prisms are regular spherulitic prismatic to nondenticular composite prismatic. Preparation method 3. B. Radial, vertical fracture through the predominantly regular simple prismatic (RSP) outer shell layer and the overlying periostracum (P) in the posterior of *Parapholas californica*. The extreme outer portion of each prism approximates spherulitic prismatic or nondenticular composite prismatic structure (note the divergence of the structural subunits at the point of the small arrow in the upper right part of this photograph). Preparation method 1.

PLATE 50. A,B. Radial, vertical fracture through the prismatic outer shell layer in the anterior of *Parapholas californica*. This layer varies from irregular simple prismatic and nondenticular composite prismatic in its outer portion to fibrous prismatic and irregular spherulitic prismatic in its inner portion. Preparation method 1. C,D. Exterior view of the anterior part of the shell of *Parapholas californica*, showing partially abraded concentric ridges used by the bivalve in mechanical boring. Figure D is a closer view of one of the concentric ridges in Figure C. The two sublayers of the outer prismatic shell layer can be seen on the abraded ridges.

PLATE 51. A. Outer shell layer of *Laternula* sp. Composite figure showing three perpendicular views. This layer shows an outer sublayer of irregular simple prismatic structure and an inner sublayer of homogeneous to locally irregular simple prismatic and spherulitic prismatic structure. Preparation method 3. B. Radial, vertical fracture through the outer shell layer of *Laternula* sp., showing irregular simple prismatic structure. Preparation method 1. C. Radial, vertical fracture through the outer shell layer of *Laternula* sp., showing an exterior portion with irregular simple prismatic structure (above in the photograph), and an interior portion with homogeneous structure (middle and lower portion of the photograph). Preparation method 1.

PLATE 52. A. Radial, vertical fracture through a homogeneous to weakly regular simple prismatic part of the outer layer of *Thracia pubescens*. This layer shows a weak development of vertical columns, and is therefore transitional to regular simple prismatic. Other parts of this layer are more clearly regular simple prismatic and nondenticular composite prismatic. Preparation method 1. B. Closer view of the horizontally flattened structural subunits in a homogeneous to weakly regular simple prismatic part of the outer shell layer of *Thracia pubescens*. Radial, vertical fracture. Preparation method 1.

PLATE 53. A. Predominantly nondenticular composite prismatic (CP) structure in the outer shell layer of *Euciroa elegantissima*. Composite figure showing three perpendicular views. The radial view also shows the periostracum (P) and the nacreous (N) middle shell layer. Although not shown in these figures, this outer shell layer is locally spherulitic prismatic. Preparation method 3. B,C. Radial, vertical section through the prismatic outer layer of *Euciroa elegantissima* showing several nondenticular composite prisms. Note the high angle of divergence of the structural subunits within each composite prism. Preparation method 2.

Section 8: Aragonitic Crossed Lamellar Structures.

Plates 54-76 illustrate aragonitic CL structures as seen on the depositional surface (*i.e.*, "crossed lamellar signatures"). Plates 77-100 show these same structures in oriented sections and in fractures through the posterior margin of the adult shell.

PLATE 54. A. Linear crossed lamellar structure. Depositional surface of *Arca ventricosa* between the pallial line and the posterior shell margin. Preparation method 4. B. Linear crossed lamellar structure. Depositional surface of *Cucullaea granulosa* between the pallial line and the posterior shell margin. Preparation method 4. C. Linear crossed lamellar structure. Depositional surface of *Noetia ponderosa* between the pallial line and the posterior shell margin. Preparation method 4.

PLATE 55. A. Linear crossed lamellar structure. Depositional surface of *Limopsis sulcata* between the pallial line and the posterior shell margin. Preparation method 4. B. Linear crossed lamellar structure. Depositional surface of *Glycymeris pectinata* between the pallial line and the posterior shell margin. Preparation method 4. C. Linear crossed lamellar structure. Depositional surface of *Propeamussium dalli* in the posterior portion of the crossed lamellar layer. Preparation method 4.

PLATE 56. A. Linear crossed lamellar structure. Depositional surface of *Plicatula gibbosa* in the posterior portion of the crossed lamellar layer. Preparation method 4. B. Linear crossed lamellar structure. Depositional surface of *Spondylus gussoni* in the posterior portion of the crossed lamellar layer. Preparation method 4. C. Fine-width branching crossed lamellar structure. Depositional surface of *Anomia simplex* between the pallial line and the posterior shell margin. Preparation method 4.

PLATE 57. A. Linear crossed lamellar structure. Depositional surface of *Ctenoides scabra* in the posterior portion of the crossed lamellar layer. Preparation method 4. B. Medium-width branching crossed lamellar structure. Depositional surface of *Parvilucina lampra* between the pallial line and the posterior shell margin. Preparation method 4. C. Fine-width branching crossed lamellar structure. Depositional surface of *Myrtea spinifera* between the pallial line and the posterior shell margin. Preparation method 4.

PLATE 58. A. Predominantly medium-width branching crossed lamellar structure with traces of compressed crossed lamellar structure. Depositional surface of *Divalucina cumingi* in the posterior portion of the crossed lamellar layer. Preparation method 4. B. Irregular/compressed crossed lamellar structure in the posterior portion of the crossed lamellar layer of *Thyasira flexuosa*. Depositional surface. Preparation method 4. C. Branching/compressed/triangular crossed lamellar structure. Depositional surface of *Fimbria elegans* between the pallial line and the posterior shell margin. Preparation method 4.

PLATE 59. A. Very faint irregular crossed lamellar structure (upper part of photograph) and cone complex crossed lamellar structure (lower part of photograph). Depositional surface of *Felaniella sericata* near the posterior shell margin. Preparation method 4. B. Coarse-width branching crossed lamellar structure. Depositional surface of *Chama iostoma* between the pallial line and the posterior shell margin. Preparation method 4. C. Medium-width branching crossed lamellar structure. Depositional surface of *Arcinella cornuta* between the pallial line and the posterior shell margin. Preparation method 4.

PLATE 60. A. Branching to branching/irregular crossed lamellar structure in *Aligena elevata*. B. Irregular to irregular/compressed crossed lamellar structure in *Montacuta ferruginosa*. C. Branching to irregular crossed lamellar structure in *Galeomma turtoni*. All preparation method 4. Depositional surfaces between the pallial line and the posterior shell margin.

PLATE 61. A. Irregular/compressed crossed lamellar structure. Depositional surface of *Cyamium mosthoffii* in the posterior portion of the crossed lamellar layer. Preparation method 4. B. Predominantly medium-width branching crossed lamellar structure locally approximating branching/crisscross crossed lamellar structure. Depositional surface of *Anisodonta lutea* between the pallial line and the posterior shell margin. Preparation method 4. C. Medium-width branching crossed lamellar structure. Depositional surface of *Beguina semiorbiculata* between the pallial line and the posterior shell margin. Preparation method 4.

PLATE 62. A. Fine-width branching crossed lamellar structure. Depositional surface of *Glans incrassata* between the pallial line and the posterior shell margin, viewed with reflected light. Preparation method 4. B. Fine-width branching crossed lamellar structure, locally approximating branching/crisscross crossed lamellar structure. Depositional surface of *Astarte undata* between the pallial line and the posterior shell margin. Preparation method 4. C. Fine-width branching crossed lamellar structure. Depositional surface of *Eucrassatella* (*Hybolophus*) *speciosa* between the pallial line and the posterior shell margin. Preparation method 4.

PLATE 63. A. Medium-width branching crossed lamellar structure slightly transitional to branching/crisscross crossed lamellar structure. Depositional surface of *Acrosterigma leucostoma* between the pallial line and the posterior shell margin. B. Medium-width branching crossed lamellar structure. Depositional surface of *Hippopus hippopus* between the pallial line and the posterior shell margin. C. Branching/crisscross crossed lamellar structure (toward lower right). Depositional surface of *Lutraria philippinarum* between the pallial line and the posterior shell margin. All preparation method 4.

PLATE 64. A. Predominantly fine-width branching crossed lamellar structure locally approximating branching/crisscross crossed lamellar structure. Depositional surface of *Anapella adelaide* between the pallial line and the posterior shell margin. Preparation method 4. B. Crisscross crossed lamellar structure. Depositional surface of *Ensis siliqua*, family Cultellidae, between the pallial line and the posterior shell margin. Preparation method 4. C. Crisscross crossed lamellar structure. Depositional surface of *Ensis directus* between the pallial line and the posterior shell margin. Preparation method 4.

PLATE 65. A. Predominantly medium-width branching crossed lamellar structure, locally approximating branching/triangular crossed lamellar structure. Depositional surface of *Siliqua japonica* between the pallial line and the posterior shell margin. Preparation method 4. B. Intimate mixture of branching crossed lamellar, irregular crossed lamellar, and irregular/compressed/triangular crossed lamellar structures in the same general shell layer. Depositional surface of *Tellina lutea* between the pallial line and the posterior shell margin. Preparation method 4. C. Predominantly medium-width branching crossed lamellar structure locally approximating branching/crisscross crossed lamellar structure. Depositional surface of *Florimetis intastriata* between the pallial line and the posterior shell margin. Preparation method 4.

PLATE 66. A. Predominantly branching crossed lamellar structure, locally approximating branching/crisscross crossed lamellar structure. Depositional surface of *Hemidonax donaciformis* between the pallial line and the posterior shell margin. Preparation method 4. B. Branching/criss-cross crossed lamellar structure. Depositional surface of *Sanguinolaria olivacea* between the pallial line and the posterior shell margin. Preparation method 4. C. Exterior (above) and interior (below) portions of the crossed lamellar layer of *Sanguinolaria boeddinghausi* between the pallial line and the posterior shell margin. Above: triangular crossed lamellar to compressed/triangular crossed lamellar structure. Below: predominantly compressed/triangular crossed lamellar structure. Preparation method 4.

PLATE 67. A. Predominantly compressed crossed lamellar structure locally approximating compressed/triangular crossed lamellar structure. Depositional surface of the interior portion of the crossed lamellar layer of *Solecurtus divaricatus* between the pallial line and the posterior shell margin. Preparation method 4. B. Crisscross crossed lamellar structure. Depositional surface of the exterior portion of the crossed lamellar layer of *Solecurtus divaricatus* between the pallial line and the posterior shell margin. Same specimen as in Figure A. Preparation method 4. C. Predominantly irregular crossed lamellar structure. Depositional surface of *Tagelus dombeii* between the pallial line and the posterior shell margin. Preparation method 4.

PLATE 68. A. Fine-width branching crossed lamellar to slightly irregular crossed lamellar structure. Depositional surface of *Mytilopsis leucophaeatus* between the pallial line and the posterior shell margin. Preparation method 4. B. Medium-width branching crossed lamellar structure. Depositional surface of the exterior portion of the crossed lamellar layer of *Trapezium sublaevigatum* between the pallial line and the posterior shell margin. Preparation method 4. C. Fine-width branching crossed lamellar structure. Depositional surface of the interior portion of the crossed lamellar layer of *Trapezium sublaevigatum* between the pallial line and the posterior shell margin. Same specimen as in Figure B. Preparation method 4.

PLATE 69. A. Predominantly medium-width branching crossed lamellar structure, locally approximating branching/crisscross crossed lamellar structure. Depositional surface of *Coralliophaga coralliophaga* between the pallial line and the posterior shell margin. Preparation method 4. B. Medium-width branching crossed lamellar structure. Depositional surface of *Glossus vulgaris* between the pallial line and the posterior shell margin. Preparation method 4. C. Linear crossed lamellar structure. Depositional surface of *Batissa* sp. between the pallial line and the posterior shell margin. Preparation method 4.

PLATE 70. A. Coarse irregular crossed lamellar or irregular complex crossed lamellar structure. Depositional surface of *Dosinia ponderosa* near the posterior shell margin. Preparation method 4. B. Predominantly medium-width branching crossed lamellar structure locally approximating branching/crisscross crossed lamellar structure. Depositional surface of *Cyclina sinensis* between the pallial line and the posterior shell margin. Preparation method 4. C. Branching crossed lamellar structure. Depositional surface of *Circe stutzeri* near the posterior shell margin. Preparation method 4.

PLATE 71. A. Predominantly medium-width branching crossed lamellar structure. Depositional surface of *Meretrix undulosa* between the pallial line and the posterior shell margin. Preparation method 4. B. Fine-width branching crossed lamellar structure. Depositional surface of *Tivela byronensis* between the pallial line and the posterior shell margin. Preparation method 4. C. Medium-width, slightly irregular branching crossed lamellar structure. Depositional surface of *Saxidomus nuttalli* between the pallial line and the posterior shell margin. Preparation method 4.

PLATE 72. A. Rather coarse irregular crossed lamellar structure. Depositional surface of *Chione californiensis* between the pallial line and the posterior shell margin. Preparation method 4. B. Predominantly branching/irregular crossed lamellar structure, locally approximating compressed crossed lamellar structure. Depositional surface of *Cooperella subdiaphana* between the pallial line and the posterior shell margin. Preparation method 4. C. Medium-width branching/irregular crossed lamellar structure. Depositional surface of *Glauconome virens* between the pallial line and the posterior shell margin. Preparation method 4.

PLATE 73. A. Fine-width branching/irregular crossed lamellar structure. Depositional surface of *Mya arenaria* between the pallial line and the posterior shell margin. Preparation method 4. B. Branching/crisscross crossed lamellar structure. Depositional surface of the exterior portion of the crossed lamellar layer of *Cryptomya californica* between the pallial line and the shell posterior. Preparation method 4. C. Narrow-width branching to branching/irregular crossed lamellar structure. Depositional surface of the interior portion of the crossed lamellar layer of *Cryptomya californica* between the pallial line and the posterior shell margin. Same specimen as in Figure B. Preparation method 4.

PLATE 74. A. Predominantly fine-width branching/irregular crossed lamellar structure. Depositional surface of *Platyodon cancellatus* between the pallial line and the posterior shell margin. Preparation method 4. B. Predominantly medium-width branching crossed lamellar structure. Depositional surface of *Corbula bicarinata* between the pallial line and the posterior shell margin. Preparation method 4. C. Predominantly medium-width branching crossed lamellar structure, locally approximating branching/crisscross crossed lamellar structure. Depositional surface of *Corbula nucleus* between the pallial line and the posterior shell margin. The predominantly CCL inner shell layer (incorrectly labeled "N" in this figure) appears in the upper part of the photograph. Preparation method 4.

PLATE 75. A. Branching/irregular crossed lamellar structure. Depositional surface of *Rocellaria hawaiensis* between the pallial line and the posterior shell margin. Preparation method 4. B. Predominantly irregular crossed lamellar structure. Depositional surface of *Gastrochaena* aff. *G. stimpsoni* between the pallial line and the posterior shell margin. Preparation method 4. C. Fine-width branching to branching/irregular crossed lamellar structure. Depositional surface of *Spengleria rostrata* between the pallial line and the posterior shell margin. Preparation method 4.

PLATE 76. A. Predominantly irregular crossed lamellar structure, locally approximating branching crossed lamellar structure. Depositional surface of *Hiatella arctica* between the pallial line and the posterior shell margin. Preparation method 4. B. Crisscross crossed lamellar structure. Depositional surface of *Barnea truncata* between the pallial line and the posterior shell margin. Preparation method 4. C. Predominantly medium-width branching crossed lamellar structure, locally approximating branching/crisscross crossed lamellar structure. Depositional surface of *Zirfaea crispata* between the pallial line and the posterior shell margin. Preparation method 4.

PLATE 77. A. Linear crossed lamellar structure exterior to the pallial myostracum of *Arca ventricosa*. Radial, vertical section through the shell posterior viewed at lower (left) and higher (right) magnifications. Bar scales are at the bottom of the plate for the left and right photographs. Preparation method 3. B. Crossed lamellar structure exterior to the pallial myostracum of *Cucullaea granulosa*. Radial, vertical section through the shell posterior viewed at lower (left) and higher (right) magnifications. The exterior and interior portions of the crossed lamellar layer are shown in the upper and lower photographs, respectively. Bar scales are at the bottom of the plate for the left and right photographs. Preparation method 3. C. Crossed lamellar structure exterior to the pallial myostracum of *Noetia ponderosa*. Radial, vertical section through the shell posterior viewed at lower (left) and higher (right) magnifications. The exterior and interior portions of the crossed lamellar layer are shown on the right in the upper and lower photographs, respectively. Bar scales are at the bottom of the plate for the left and right photographs. Preparation method 3.

PLATE 78. A. Crossed lamellar structure exterior to the pallial myostracum of *Limopsis sulcata*. Radial, vertical section through the shell posterior viewed at lower (left) and higher (right) magnifications. Bar scales are at the bottom of the plate for the left and right photographs. Preparation method 3. B. Linear crossed lamellar structure exterior to the pallial myostracum of *Glycymeris pectinata*. Radial, vertical section through the shell posterior viewed at lower (left) and higher (right) magnifications. Bar scales are at the bottom of the plate for the left and right photographs. Preparation method 3. C. Crossed lamellar (CL) structure of the right valve of *Propeamussium dalli* viewed in a radial, vertical section through the shell posterior at lower (left) and higher (right) magnifications. The photograph on the left also shows two outer prismatic (PL) layers. Bar scales are at the bottom of the plate for the left and right photographs. Preparation method 3.

PLATE 79. A-C. Vertical fractures through the crossed lamellar structure of the right valve of *Anomia simplex*. Preparation method 1.

PLATE 80. A. Horizontal fracture through the crossed lamellar structure exterior to the pallial myostracum of *Divalucina cumingi*. Preparation method 1. B. Closer view of the fracture through the crossed lamellar structure in Figure A.

PLATE 81. A. Crossed lamellar structure exterior to the pallial myostracum of *Divalucina cumingi*. Radial, vertical section through the shell posterior viewed at lower (left) and higher (right) magnifications. Bar scales are at the bottom of the plate for the left and right photographs. Preparation method 3. B. Crossed lamellar (CL) structure exterior to the pallial myostracum of *Thyasira flexuosa*. Radial, vertical section through the shell posterior viewed at lower (left) and higher (right) magnifications. The photograph on the left also shows an irregular simple prismatic to homogeneous (labelled ISP) and a complex crossed lamellar (CCL) layer exterior to the crossed lamellar layer. Bar scales are at the bottom of the plate for the left and right photographs. Preparation method 3. C. Crossed lamellar structure exterior to the pallial myostracum of *Fimbria elegans*. Radial, vertical section through the shell posterior viewed at lower (left) and higher (right) magnifications. Bar scales are at the bottom of the plate for the left and right photographs. Preparation method 3.

PLATE 82. A. Crossed lamellar (CL) structure exterior to the pallial myostracum of *Felaniella sericata*. Radial, vertical section through the shell posterior viewed at lower (left) and higher (right) magnifications. The photograph on the left also shows the periostracum (P), a gap in the acetate slab where the outer complex crossed lamellar (black CCL letters) layer was not fully replicated, and a CCL layer interior to the crossed lamellar layer (white CCL letters). Preparation method 3. B. Coarse-width crossed lamellar layer exterior to the pallial myostracum of *Chama iostoma*. Radial, vertical section through the shell posterior viewed at lower (left) and higher (right) magnifications. Bar scales are at the bottom of the plate for the left and right photographs. Preparation method 3. C. Crossed lamellar structure exterior to the pallial myostracum of *Arcinella cornuta*. Radial, vertical section through the shell posterior viewed at lower (left) and higher (right) magnifications. Bar scales are at the bottom of the plate for the left and right photographs. Preparation method 3.

PLATE 83. A. Crossed lamellar structure exterior to the pallial myostracum of *Glans incrassata*. Radial, vertical section through the shell posterior viewed at lower (left) and higher (right) magnifications. The exterior and interior portions of the crossed lamellar layer are shown in the upper and lower photographs, respectively. Bar scales are at the bottom of the plate for the left and right photographs. Preparation method 3. B. Crossed lamellar structure exterior to the pallial myostracum of *Astarte undata*. Radial, vertical section through the shell posterior viewed at lower (left) at higher (right) magnifications. The exterior and interior portions of the crossed lamellar layer are shown in the upper and lower photographs, respectively. Bar scales are at the bottom of the plate for the left and right photographs. Preparation method 3. C. Crossed lamellar structure exterior to the pallial myostracum of *Eucrassatella speciosa*. Radial, vertical section through the shell posterior viewed at lower (left) and higher (right) magnifications. The exterior and interior portions of the crossed lamellar layer appear in the upper and lower photographs, respectively. Bar scales are at the bottom of the plate. Preparation method 3.

PLATE 84. A. Vertical fracture through the crossed lamellar structure exterior to the pallial myostracum of *Fragum subretusum*. Preparation method 1. B. Closer view of the fracture through the crossed lamellar structure in Figure A.

PLATE 85. A. Crossed lamellar structure exterior to the pallial myostracum of *Acrosterigma leucostoma*. Radial, vertical section through the ventral margin viewed at lower (left) and higher (right) magnifications. The exterior and interior portions of the crossed lamellar layer appear in the upper and lower photographs, respectively. Bar scales are located at the bottom of the plate for the left and right photographs. Preparation method 3. B. Crossed lamellar structure exterior to the pallial myostracum of *Hippopus hippopus*. Radial, vertical section through the ventral margin viewed at lower (on left) and higher (on right) magnifications. The bar scales are located at the bottom of the plate for the left and right photographs. Preparation method 3. C. Crossed lamellar structure exterior to the pallial myostracum of *Lutraria philippinarum*. Radial, vertical section through the shell posterior viewed at lower (left) and higher (right) magnifications. Bar scales are at the bottom of the plate for the left and right photographs. Preparation method 3.

PLATE 86. A. Radial, vertical fracture through the crisscross crossed lamellar structure exterior to the pallial myostracum of *Ensis siliqua*, family Cultellidae. Preparation method 1. B. Closer view of the fracture through the crossed lamellar structure in Figure A.

PLATE 87. A-D. Horizontal section through the crossed lamellar structure exterior to the pallial myostracum of *Hemidonax donaciformis*, family Hemidonacidae, superfamily Cardioidea. In Figure A, part of the crossed lamellar structure was plucked away from the etched surface during acetate peeling. Figure D is a closer view of Figure C. Preparation method 2.

PLATE 88. A. Crossed lamellar structure exterior to the pallial myostracum of *Hemidonax donaciformis*, family Hemidonacidae, superfamily Cardioidea. Radial, vertical section through the shell posterior viewed at lower (left) and higher (right) magnifications. Bar scales are at the bottom of the plate for the left and right photographs. Preparation method 3. B. Crossed lamellar (CL) structure exterior to the pallial myostracum of *Sanguinolaria olivacea*. Radial, vertical section through the shell posterior viewed at lower (left) and higher (right) magnifications. The photograph on the left also shows the outer complex crossed lamellar (CCL) layer. Bar scales are at the bottom of the plate for the left and right photographs. Preparation method 3. C. Crossed lamellar layer exterior to the pallial myostracum of *Sanguinolaria boeddinghausi*. Radial, vertical section through the shell posterior viewed at lower (left) and higher (right) magnifications. Bar scales are at the bottom of the plate for the left and right photographs. Preparation method 3.

PLATE 89. A. Transverse, vertical fracture through the crisscross crossed lamellar structure exterior to the pallial myostracum of *Solecurtus divaricatus*. Preparation method 1. B. Closer view of the fracture through the crossed lamellar structure in Figure A.

PLATE 90. A. Crossed lamellar (CL) structure exterior to the pallial myostracum of *Mytilopsis leucophaeatus*. Radial, vertical section through the shell posterior viewed at lower (left) and higher (right) magnifications. The photograph on the left also shows the overlying fibrous prismatic to irregular simple prismatic to nearly homogeneous outer layer (labelled HOM) and the underlying crossed lamellar to irregular complex crossed lamellar inner layer (labelled CCL). Bar scales are at the bottom of the plate for the left and right photographs. Preparation method 3. B. Crossed lamellar layer exterior to the pallial myostracum of *Trapezium oblongum*. Radial, vertical section through the shell posterior viewed at lower (left) and higher (right) magnifications. Bar scales are at the bottom of the plate for the left and right photographs. Preparation method 3. C. Crossed lamellar (CL) structure exterior to the pallial myostracum of *Trapezium sublaevigatum*. Radial, vertical section through the shell posterior viewed at lower (left) and higher (right) magnifications. The photograph on the left also shows the fibrous prismatic (FP) to spherulitic prismatic layer (above) and the complex crossed lamellar (CCL) layer (below), exterior and interior to the crossed lamellar layer, respectively. Bar scales are at the bottom of the plate for the left and right photographs. Preparation method 3.

PLATE 91. A. Radial, vertical fracture through the outer part of the crossed lamellar layer exterior to the pallial myostracum of *Trapezium sublaevigatum*. Preparation method 3. B. Closer view of the fracture through the crossed lamellar structure in Figure A.

PLATE 92. A. Radial, vertical fracture through the crossed lamellar layer exterior to the pallial myostracum of *Glossus vulgaris*. Preparation method 1. B. Closer view of the fracture through the crossed lamellar structure in Figure A. C. Oblique, vertical fracture through the crossed lamellar structure exterior to the pallial myostracum of *Glossus vulgaris*. Preparation method 1.

PLATE 93. A. Crossed lamellar structure exterior to the pallial myostracum of *Coralliophaga coralliophaga*. Radial, vertical section through the shell posterior viewed at lower (left) and higher (right) magnifications. Bar scales are at the bottom of the plate for the left and right photographs. Preparation method 3. B. Crossed lamellar (CL) structure exterior to the pallial myostracum of *Glossus vulgaris*. Radial, vertical section through the shell posterior viewed at lower (left) and higher (right) magnifications. The photograph on the left also shows the overlying prismatic to homogeneous layer (PL). Bar scales are at the bottom of the plate for the left and right photographs. Preparation method 3. C. Crossed lamellar structure exterior to the pallial myostracum of *Batissa* sp. Radial, vertical section through the shell posterior viewed at lower (left) and higher (right) magnifications. Bar scales are at the bottom of the plate for the left and right photographs. Preparation method 3.

PLATE 94. A. Vertical fracture through the crossed lamellar structure exterior to the pallial myostracum of *Batissa* sp.. Preparation method 1. B. Closer view of the fracture through the crossed lamellar structure in Figure A.

PLATE 95. A. Radial, vertical fracture through the nearly horizontal crossed lamellar structure layer just below the outer prismatic layer in *Cyclina sinensis*. Preparation method 1. B. Closer view of the fracture through the crossed lamellar structure in Figure A. C,D. Radial, vertical fracture through the crossed lamellar structure half-way between the shell margin and pallial myostracum of *Cyclina sinensis*. Figure D is a closer view of Figure C. Preparation method 1.

PLATE 96. A. Crossed lamellar structure deposited on the reflected margin of *Dosinia ponderosa*. Radial, vertical section through the ventral margin viewed at lower (left) and higher (right) magnifications. Bar scales are at the bottom of the plate for the left and right photographs. Preparation method 3. B. Crossed lamellar structure exterior to the pallial myostracum of *Cyclina sinensis*. Radial, vertical section through the shell posterior viewed at lower (left) and higher (right) magnifications. Bar scales are at the bottom of the plate for the left and right photographs. Preparation method 3. C. Crossed lamellar structure exterior to the pallial myostracum of *Circe stutzeri*. Radial, vertical section through the shell posterior viewed at lower (left) and higher (right) magnifications. Bar scales are at the bottom of the plate for the left and right photographs. Preparation method 3.

PLATE 97. A. Crossed lamellar (CL) structure immediately exterior to the pallial myostracum of *Compsomyax subdiaphana*. Radial, vertical section through the shell posterior viewed at lower (left) and higher (right) magnifications. The photograph on the left also shows the overlying fine complex crossed lamellar (CCL) to crossed acicular layer. Bar scales are at the bottom of the plate for the left and right photographs. Preparation method 3. B. Crossed lamellar structure exterior to the pallial myostracum of *Paphia cuneata*. Radial, vertical section through the shell posterior viewed at lower (left) and higher (right) magnifications. The exterior and interior portions of the crossed lamellar layer are shown in the upper and lower photographs, respectively. Bar scales are at the bottom of the plate for the left and right photographs. Preparation method 3. C. Crossed lamellar (CL) structure exterior to the pallial myostracum of *Tivela trigonella*. Radial, vertical section through the shell posterior viewed at lower (left) and higher (right) magnifications. The photograph on the left also shows several aragonitic periostracal spines (here labelled ICM), a very thin, predominantly fibrous prismatic (FP) layer, and a layer of crossed acicular to fine complex crossed lamellar (CCL) structure. Bar scales are at the bottom of the plate for the left and right photographs. Preparation method 3.

PLATE 98. A. Radial, vertical fracture through the crossed lamellar (CL) structure exterior to the pallial myostracum of *Mya arenaria*. The upper right corner of this figure also shows the overlying homogeneous (HOM) layer. Preparation method 1. B. Closer view of the fracture through the crossed lamellar structure in Figure A.

PLATE 99. A. Crossed lamellar (CL) structure exterior to the pallial myostracum of *Mya arenaria*. Radial, vertical section through the shell posterior viewed at lower (left) and higher (right) magnifications. The photograph on the left also shows the overlying homogeneous (HOM) layer and the underlying complex crossed lamellar (CCL) layer. Bar scales are at the bottom of the plate for the left and right photographs. Preparation method 3. B. Crossed lamellar (CL) structure exterior to the pallial myostracum of *Cryptomya californica*. Radial, vertical section through the shell posterior viewed at lower (left) and higher (right) magnifications. The photograph on the left also shows the overlying periostracum (P) and homogeneous (HOM) layers. Bar scales are at the bottom of the plate for the left and right photographs. Preparation method 3. C. Crossed lamellar (CL) structure exterior to the pallial myostracum of *Paramya subovata*. Radial, vertical section through the shell posterior viewed at lower (left) and higher (right) magnifications. The photograph on the left also shows the overlying homogeneous (HOM) and underlying complex crossed lamellar (CCL) layers. Certain other specimens of this species show a poorly developed regular simple prismatic to regular spherulitic prismatic structure in the outer shell layer, instead of homogeneous structure. Bar scales are at the bottom of the plate for the left and right photographs. Preparation method 3.

PLATE 100. A. Crossed lamellar structure exterior to the pallial myostracum of *Pholas campechiensis*. Radial, vertical section through the shell posterior viewed at lower (left) and higher (right) magnifications. Bar scales are at the bottom of the plate for the left and right photographs. Preparation method 3. B. Crossed lamellar structure exterior to the pallial myostracum of *Cyrtopleura costata*. Radial, vertical section through the shell posterior viewed at lower (left) and higher (right) magnifications. The exterior and interior portions of the crossed lamellar layer are shown on the left in the upper and lower photographs, respectively. The right photograph shows a closer view of the exterior portion of the crossed lamellar layer. Bar scales are at the bottom of the plate for the left and right photographs. Preparation method 3. C. Crossed lamellar structure (with sublayers of cone complex crossed lamellae) exterior to the pallial myostracum of

Zirfaea pilsbryi. Radial, vertical section through the shell posterior viewed at lower (left) and higher (right) magnifications. Bar scales are at the bottom of the plate for the left and right photographs. Preparation method 3.

Section 9: Paleozoic Bivalves.

PLATE 101. *Pterinea demissa* Conrad, 1842, Upper Ordovician, Maysvillian, Cincinnatti Group, Ohio, USNM 40522, family Pterineidae. A. Horizontal acetate peel through the calcitic, regular simple prismatic outer layer of the right valve. Bar scale = 50 microns. B. Transverse acetate peel through the same shell layer as in A. Bar scale = 50 microns. C. Ventral, radial acetate peel through the outwardly homogeneous "mosaic" and inwardly irregularly foliated (?) calcitic outer layer of the left valve. Bar scale = 50 microns. D. Closer view of the inner, regularly foliated (?) part of the calcitic outer layer of the left valve; same acetate peel as in Figure C. Bar scale = 100 microns. E. SEM of acid-etched, transverse section through the regular simple prismatic outer layer of the right valve, showing the epoxy matrix (above), the prismatic layer (middle) and the altered middle shell layer (below). Bar scale = 5 microns. F. SEM of acid-etched, radial, vertical section through the inner part of the outer shell layer near the ventral shell margin, showing relict laminae (dotted lines on left) of the regularly foliated (?) structure. Bar scale = 12 microns.

PLATE 102. *Pseudaviculopecten princeps* (Conrad, 1838), Middle Devonian, Lower Hamilton Group, road cut on US Highway 20, 5.8 km west of Cazenovia city limits, New York, UNC 5271, family Pterinopectinidae. A-D are acetate peels; E is an SEM of a polished and then acid-etched, horizontal section. A. Radial, vertical section through the ventral margin of the right valve, showing the calcitic, regular simple prismatic outer layer (i) and the relict structure of the underlying middle nacreous layer (ii). Bar scale = 50 microns. B. Radial, vertical section through the ventral margin of the left valve, showing the laminated, irregularly foliated (?) inner part of the calcitic, outer shell layer (iii) and the underlying relict structure of the middle nacreous shell layer (ii). Bar scale = 50 microns. C. Radial, vertical section near the ventral margin of the right valve, showing the relict nacreous structure of the middle shell layer. Bar scale = 25 microns. D. Horizontal section through the inner, laminated, irregularly foliated (?) portion of the calcitic outer layer of the left valve. Bar scale = 50 microns. E. SEM of the peeled horizontal section which produced the acetate peel in Figure D, showing the irregular orientation of the fibers (?) and blades which comprised the inner, laminated, irregularly foliated (?) portion of the calcitic outer shell layer. Relict blades are visible near the lower left center of the photograph. Bar scale = 10 microns.

PLATE 103. *Pseudaviculopecten scabridus* (Hall, 1883): Middle Devonian, Hamilton Group, probably Ludlowville Formation, Moravia, New York, USNM 14138, family Pterinopectinidae. A-D are acetate peels; E is an SEM of a polished and then acid-etched, horizontal section. A. Horizontal section through the calcitic, regular simple prismatic structure of the outer layer of the right valve. Bar scale = 10 microns. B. Radial, vertical section through the same shell layer as in A. Bar scale = 10 microns. C. Horizontal section through the middle nacreous layer of the right valve. Bar scale = 25 microns. D. Horizontal section through the calcitic, irregularly foliated (?) outer shell layer of the left valve. Bar scale = 25 microns. E. Radial, vertical section near the ventral margin of the left valve, showing the rock matrix (m), the calcitic, irregularly foliated (?) outer shell layer (i) and the relict structure of the middle nacreous shell layer (ii). Bar scale = 10 microns. F. SEM of the same acid-etched, horizontal section of the middle nacreous shell layer of the right valve which produced the acetate peel in Figure C. Bar scale = 5 microns.

PLATE 104. *Eodon tenuistriata* (Hall, 1870): Middle Devonian, Hamilton Group, near Morrisville, New York, probably Skaneateles Formation, UNC 5503, family Astartidae. All photographs show radial, vertical sections in the shell posterior of originally aragonitic layers which now show relict microstructures in diagenetic calcite. A-D are acetate peels. E-G are SEMs of acetate-peeled sections. A. Reclined, fibrous prismatic structure of the outer shell layer. Bar scale = 50 microns. B. Relict, nearly vertical fibrous prismatic structure of the outer shell layer. Bar scale = 50 microns. C. Relict, nacreous structure of the middle shell layer. Bar scale = 50 microns. D. Relict, irregular CCL structure of the inner shell layer. Bar scale = 50 microns. E. SEM of the relict, nacreous structure of the middle shell layer (same surface as the acetate peel in Figure

C). Bar scale = 10 microns. F. Relict reclined fibrous prismatic structure of the outer shell layer. Bar scale = 10 microns. G. SEM of relict, irregular CCL structure of the shell layer (same surface as the acetate peel in D). Bar scale = 10 microns. The inner shell layer also shows some nacreous structure (not shown in these figures), especially in the anterior part of the shell.

Section 10: Superfamily Mytiloidea.

PLATE 105. A. *Ischadium recurvum* (Rafinesque, 1820), Recent, Western Florida, YPM 930. Radial, vertical section through the shell posterior. The section was polished and then etched in dilute HCl. The thick, featureless periostracum (above) overlies the bimineralic outer shell layer (middle). The outer part of the middle nacreous shell layer is barely visible near the bottom of the photogaph. The thin, outer, calcitic sublayer of the outer shell layer varies from inclined, irregular simple prismatic to homogeneous. The thicker, aragonitic sublayer of the outer shell layer consists of very irregularly to radially oriented, nearly horizontal to reclined, irregular spherulitic prismatic structure (the major portion of the lower half of Figure A). Bar scale = 10 microns. B. *Mytella guyanensis* (Lamarck), Recent, Panama, YPM 731. Radial, vertical fracture through the shell posterior. The thick periostracum appears internally structureless (above), and is underlain by an aragonitic outer shell layer of slightly reclined, irregular simple prisms and very fine, irregular spherulitic prisms. The middle nacreous shell layer occupies the bottom half of the figure. Bar scale = 5 microns.

PLATE 106. *Mytilus (Crenomytilus) californianus* Conrad, 1837, Recent, California, YPM 9526. Radial, vertical section through the shell posterior. The periostracum was removed before the shell was fractured by soaking the shell for 24 hours in sodium hypochlorite. A. The outer, homogeneous sublayer of the calcitic outer shell layer, here broken away from its original position in the shell, and resting on the outer surface of this same sublayer. Bar scale = 1 micron. B. The outer homogeneous (above) and inner fibrous prismatic (below) sublayers of the calcitic outer shell layer. Bar scale = 10 microns. C. The inner sublayer of the calcitic outer shell layer, consisting of generally reclined fibrous prisms with irregular, laterally interpenetrating cross-sections. These prisms are here arranged into bundles of mutually parallel crystallites which collectively comprise a locally developed irregular CCL structure. Bar scale = 10 microns.

PLATE 107. *Mytilus (Crenomytilus) californianus* Conrad, 1837, Recent, California, YPM 9526. Radial, vertical section through the inner, calcitic fibrous prismatic shell layer, showing locally developed irregular spherulitic prisms (Figure A) and the shapes of the individual fibrous prisms (Figures B,C). Bar scales in A, B and C = 100 microns, 1 micron, and 10 microns, respectively.

PLATE 108. *Perumytilus purpuratus* (Lamarck, 1819), Recent, Peru, YPM 9120. Radial, vertical fracture through the shell posterior. A. The periostracum (above) and the outer, homogeneous sublayer of the aragonitic outer shell layer (below). Bar scale = 5 microns. B. Irregularly prismatic structure in the inner sublayer of the aragonitic, outer shell layer (above) and the underlying middle nacreous layer (below). Bar scale = 1 micron. C. Gradation from the outer, homogeneous sublayer (above) to the inner, irregularly prismatic sublayer (middle) of the aragonitic, outer shell layer, underlain by the nacreous middle shell layer (below). Bar scale = 1 micron.

PLATE 109. A. *Semimytilus algosus* (Gould, 1850), Recent, Callao-La Punta, Peru, YPM 9624. Radial, vertical fracture through the shell posterior, showing the periostracum (above), the aragonitic, slightly reclined, fibrous to irregular simple prisms of the outer shell layer (middle), and the underlying nacreous layer (below). Bar scale = 10 microns. B,C. *Septifer bilocularis*, Recent, Palau, YPM 9994. Radial, vertical section through the shell posterior which has been polished, etched in dilute HCl, and then plucked with an acetate peel. B. The periostracum (above), the aragonitic homogeneous to ISP outer shell layer (middle), and the nacreous middle shell layer (below). Bar scale = 10 microns. C. Closer view of the upper part of Figure B, showing the contact between the periostracum (above) and the outer shell layer (below). Bar scale = 5 microns.

PLATE 110. *Trichomya hirsuta* (Lamarck), Recent, Japan, YPM 9667 (Figure A) and YPM 10123 (Figures B,C). Slightly oblique (A) and radial (B,C), vertical fractures through the shell posterior. The periostracum has been removed by soaking the shell in sodium hypochlorite for 24 hours prior to fracturing. A. The two sublayers of the bimineralic outer shell layer (above and middle) and the middle nacreous shell layer (below). The calcitic, outer sublayer appears as a thin, predominantly homogeneous layer at the extreme top right portion of the photograph, and as a thicker, underlying layer of irregular fibrous prisms. The underlying aragonitic sublayer consists of very poorly defined irregular simple prisms just exterior to the middle nacreous shell layer in the lower left portion of the figure. Bar scale = 10 microns. B. A radial, vertical fracture in which the calcitic portion of the outer shell layer has an outer homogeneous and an inner fibrous prismatic component (above in the photograph); this calcitic sublayer is directly underlain by the aragonitic nacreous middle shell layer. The inner, aragonitic sublayer of the outer shell layer is not developed here. Bar scale = 10 microns. C. A different portion of the radial, vertical fracture in B, showing local development of irregular CCL structure in the calcitic outer shell layer (above), underlain by the middle nacreous shell layer (below). Bar scale = 10 microns.

PLATE 111. *Trichomya hirsuta* (Lamarck), Recent, Japan, YPM 9667. The periostracum has been removed by soaking the shell for 24 hours in sodium hypochlorite. A. Concentric fracture near the posterior shell margin, showing a thin, outermost, calcitic prismatic sublayer (upper right) underlain by the thicker, aragonitic, indistinctly prismatic sublayer of the outer shell layer. The middle nacreous shell layer appears on the lower left. Bar scale = 10 microns. B,D. Exterior surface view of the sub-periostracal shell on the crest of a radial ridge, which may be aragonitic based on Feigl staining of other specimens. Bar scale = 10 microns. C. Exterior surface view of the sub-periostracal shell showing two radial ridges and an intervening groove. The posterior shell margin is toward the lower right. Feigl staining of other specimens suggests that the ridge crests are aragonitic, whereas the grooves are calcitic. Bar scale = 50 microns. E. Exterior surface view of the sub-periostracal shell in a radial groove. Feigl staining of other specimens suggests that this surface is calcitic. Bar scale = 10 microns.

PLATE 112. A,B. *Amygdalum dendriticum* Megerle von Mühlfeld, 1811, Recent, West Indies, YPM 9997. Radial, vertical fractures through the posterior shell margin, showing nearly vertical, only slightly reclined, poorly defined, irregular simple prisms which grade laterally and vertically into homogeneous structure. Bar scale = 1 micron in both photographs.

PLATE 113. *Botula fusca* (Gmelin, 1791), Recent, Bermuda, YPM 9998 (A,B) and Recent, Florida Keys (YPM 4916) (C). Radial, vertical fractures through the shell posterior. The periostracum has been removed by soaking the shell for 24 hours in sodium hypochlorite prior to fracturing. A. Calcitic, outer sublayer (above) and aragonitic inner sublayer (middle) of the outer shell layer, and the underlying middle nacreous shell layer (below). In this view the calcitic outer shell layer is reclined, irregular simple prismatic, and the aragonitic inner sublayer is homogeneous. Bar scale = 10 microns. B. A different portion of the radial, vertical fracture in A, in which the calcitic outer sublayer appears only on the top left, and the aragonitic homogeneous inner sublayer underlies the calcitic sublayer on the middle left, and is exposed on the shell exterior on the top right. The middle nacreous shell layer appears at the bottom of the figure. Bar scale = 10 microns. C. Calcitic outer sublayer (above) and aragonitic inner sublayer (middle) of the outer shell layer, plus the middle nacreous layer (below). The calcitic outer sublayer here varies from reclined prismatic to irregular CCL. Bar scale = 5 microns.

PLATE 114. *Geukensia demissa* (Dillwyn, 1817), Recent, Western Atlantic, UNC 14893. A. Radial, vertical fracture through the shell posterior, showing the vacuolated periostracum (top), two sublayers within the outer shell layer (the thin, outer sublayer is calcitic homogeneous; the thicker, inner sublayer is aragonitic, slightly reclined to nearly vertical irregular simple prismatic to irregular spherulitic prismatic to homogeneous); and the middle nacreous shell layer (bottom). Bar scale = 5 microns. B. Another area of the radial, vertical fracture in A. Bar scale = 5 microns.

PLATE 115. *Geukensia demissa* (Dillwyn, 1817), Recent, Western Atlantic, UNC 14894. A. Radial, vertical fracture through the shell posterior, showing the periostracum (top), two sublayers within the outer shell layer (the thin, outer sublayer is calcitic homogeneous; the thicker, inner sublayer is aragonitic, slightly reclined to nearly vertical irregular simple prismatic to irregular spherulitic prismatic to homogeneous); and the middle nacreous shell layer (bottom). Bar scale = 10 microns. B. Radial, vertical fracture showing the contact between the inner, aragonitic part of the outer shell layer and the underlying nacreous layer. Bar scale = 5 microns.

PLATE 116. A. *Modiolus capax* (Conrad, 1837), Recent, Pearl Islands, Panama, YPM 2329. Acid-etched, radial, vertical section through the shell posterior, showing the inner part of the periostracum (top), the calcitic outer shell layer (middle), and the underlying nacreous shell layer (bottom). The calcitic outer shell layer has two sublayers. The outer sublayer consists of homogeneous to nearly vertical, slightly reclined ISP structure. The inner sublayer is about 15 micron thick and consists of nearly vertical, slightly reclined calcitic ISP to fibrous prismatic structure with irregular and laterally interpenetrating prism cross-sections. Bar scale = 10 microns. B,C. *Modiolus* cf. *M. agripetus* (Iredale), Recent, Japan, UNC 9990. B. Acid-etched, radial, vertical section through the shell posterior, showing the inner part of the periostracum, the outer shell layer (with two sublayers), and the underlying nacreous layer. The outer sublayer of the outer shell layer has steeply reclined, calcitic, irregular spherulitic prismatic to irregular fibrous prismatic to ISP structure; the inner sublayer, possibly also calcitic (but the mineralogy is uncertain at present) is predominantly homogeneous with local traces of radially oriented ISP and fibrous prisms. Bar scale = 10 microns. C. Closer view of the calcitic outer sublayer of the outer shell layer showing, at the extreme top center and lower left portions of the photograph, a minutely granular, homogeneous substructure of the reclined prisms. Bar scale = 5 microns.

PLATE 117. *Bathymodiolus thermophilus* Kenk and Wilson, 1985, Recent, Galapagos Deep Sea Rift, Eastern Pacific, YPM 14892. A. Radial, vertical fracture through the shell posterior, showing hirsute byssus structures and inorganic precipitates (mainly sulfide deposits) on the shell exterior (top); the periostracum (middle); and the calcitic outer shell layer with two sublayers: a very thin, outer homogeneous sublayer, and a thicker, underlying calcitic fibrous prismatic to irregular CCL sublayer. Bar scale = 10 microns. B,C. Closer views of the fibrous prismatic to irregular CCL portion of the calcitic outer shell layer. Bar scale = 10 microns in B. Bar scale = 1 micron in C.

PLATE 118. *Gregariella coralliophaga* (Gmelin, 1791), Recent, Bermuda, YPM 10132. Oblique (A) and radial (B,C), vertical fractures through the shell posterior. The periostracum was removed by soaking the valves for 24 hours in sodium hypochlorite. The valves were then rinsed in tap water for several hours, air-dried, and immersed in Feigl solution. The Feigl solution left acicular crystallites on the sub-periostracal shell exterior in Figure A, indicating the local presence of a thin, aragonitic crust exterior to the calcitic outer shell layer. Such a crust was apparently not present in Figures B and C. The calcitic outer shell layer visible in Figures A, B and C varies from radially oriented, horizontal to reclined irregular simple prisms to irregular fibrous prisms, to nearly homogeneous structure. The underlying middle nacreous shell layer appears in all three figures. Bar scales in Figures A, B and C = 10, 1, and 10 microns, respectively.

PLATE 119. A-C. *Lioberus castaneus* (Say, 1822), Recent, Sanibel, Florida, YPM 9996. A. Acid-etched, radial, vertical section through the shell posterior, showing the base of the periostracum (top), the calcitic, irregular simple prismatic outer shell layer (middle), and the underlying middle nacreous layer (bottom, here largely obscured by organic matrix). Bar scale = 5 microns. B. Another portion of the radial, vertical section in A, showing the same three shell layers. Bar scale = 10 microns. C. The minutely granular, homogeneous substructure of the calcitic, irregular simple prismatic outer shell layer shown in B. Bar scale = 1 micron. D. *Musculus (Musculus) discors* (Linné, 1767), Recent, Eastport, Maine, YPM 38. Radial, vertical fracture through the shell posterior, showing the periostracum (top), the aragonitic, largely homogeneous outer shell layer (with a very slight development of poorly defined, nearly vertical irregular simple prisms), and the underlying nacreous shell layer. Bar scale = 5 microns.

PLATE 120. *Adula falcata* (Gould, 1851), Recent, San Diego, California, YPM 10,000. Radial, vertical, polished and then acid-etched section through the shell posterior. A. The periostracum (above); the aragonitic (?) periostracal "prisms" projecting above into the periostracum and below into the calcitic, fibrous prismatic, outer shell layer (middle); and the fibrous prismatic outer shell layer (below). Bar scale = 5 microns. B. Aragonitic (?) periostracal "prisms" (above) embedded within the calcitic, fibrous prismatic outer shell layer (below). Bar scale = 10 microns. C. Inner portion of the calcitic, fibrous prismatic outer shell layer, with a few nacreous tablets from the middle shell layer appearing near the bottom of the photograph. Bar scale = 1 micron.

PLATE 121. *Lithophaga* (*Lithophaga*) *nigra* (d'Orbigny, 1842), Recent, Florida, UNC 9989. A. Radial, vertical fracture through the shell posterior, showing the middle (above) and inner (below) sublayers of the calcitic outer shell layer. The middle sublayer is probably calcitic and consists of vertical to slightly inclined to slightly reclined, blocky to irregular simple prismatic structure. The inner sublayer is calcitic and consists predominantly of slightly inclined to horizontal to reclined fibrous prisms. Bar scale = 10 microns. B. Closer view of the inner sublayer of the calcitic outer shell layer in A. Bar scale = 1 micron.

Part 3

Polyplacophora, Scaphopoda, Archaeogastropoda and Paragastropoda (Mollusca)

Plates 122-134

Joseph G. Carter and Richard M. Hall

Department of Geology, University of North Carolina
Chapel Hill, NC 27514

This chapter illustrates a few of the more common shell microstructures in chitons, scaphopods, archaeogastropods and paragastropods, and it reviews the literature of shell microstructure for selected taxa in these groups. Previously published descriptions of shell microstructure are summarized using the standard reference terminology of the *Glossary* in Volume I. New observations of shell microstructure are provided for one scaphopod, one chiton, one paragastropod, and several gastropods.

Abbreviations and symbols. The abbreviations in this section are explained in the introduction to this volume. On the accompanying plates, the shorter, thicker arrows point to the interior of the shell; the longer, thinner arrows point to the nearest shell margin or, in the case of scaphopods, to the anterior shell margin.

CLASS POLYPLACOPHORA (Text-Fig. 1; Plates 122 - 125)

Origin of shell plates. Pojeta (1980) favored the theory of independent evolution of chitons from spiculose, shell-less proto-molluscan ancestors, rather than from monoplacophorans. The shell plates of Late Cambrian *Matthevia* and *Praeacanthochiton* are externally finely tuberculate. Pojeta (1980) interpreted these tubercles as relict cuticular spicules similar to the aragonitic periostracal spikes and granules cemented to the exterior of certain bivalves. Both Pojeta (1980) and Salvini-Plawen (1985), like many previous authors, suggested that chiton shell plates originated by fusion of ancestral cuticular spicules (see Carter and Aller, 1975, for a historical review of this hypothesis). Salvini-Plawen (1985, p. 113) cited as evidence the ontogenetic development of chiton plates through coalescence of precursor fragments, and the lateral fusion of juxtaposed scales to form dorsal shields in some aplacophorans (Haas, 1981; Salvini-Plawen, 1985, Fig. 36D). However, it is also possible that chiton and other molluscan shell plates evolved independently of cuticular spicules through modification of the mechanism of spicular calcification, rather than fusion of the spicules themselves, to form subcuticular shell plates. This evolutionary pathway is suggested by the fact that cuticular calcification in certain bivalves is structurally and mineralogically distinct from the underlying shell (Carter and Aller, 1975; Carter and Lutz, this Atlas).

Salvini-Plawen (1980, 1981, 1985, p. 114) suggested that the dorsoventral muscles, the alimentary tract and the mantle structures of chitons represent a grade of evolution intermediate between aplacophorans and monoplacophorans. However, there is presently no paleontological evidence suggesting that the immediate ancestors of the Monoplacophora had multiple shell plates.

Shell microstructure. Shell microstructural data for chitons are limited to early Paleozoic *Cobcrephora* and *Matthevia* and to modern representatives of the families Lepidopleuridae, Acanthochitonidae, Chitonidae, Ischnochitonidae, Callochitonidae, Molpaliidae, and Chaetopleuridae (Text-fig. 1). Nothing is presently known about the shell microstructure of Devonian through early Cenozoic chitons.

The order Phosphatoloricata, erected by Bischoff (1981) for the Ordovician - Silurian genus *Cobcrephora* (Text-fig. 1.1), is known from phosphatic shell plates which are internally laminar but otherwise homogeneous. Phosphate has also been described in molluscan shells in the periostracum of certain bivalves (Waller, 1984; Carter and Clark, 1985), in the ammonite siphuncle (Andalib, 1972), in the pen of Jurassic and Cretaceous teuthids (Hewitt, Lazell and Moorhouse, 1983), and, presumably, in the shell of the Early Cambrian monoplacophoran or gastropod *Thorslundella* (Nyers, 1987). However, caution must be exercised in interpreting the mineralogy of *Cobcrephora* as original; many other early Paleozoic molluscs retain relict aragonitic microstructures in shells which have been replaced by calcium phosphate (Mutvei, 1983a,b).

The orders Phosphatoloricata and Paleoloricata are characterized by the presence of a tegmentum, or outer shell layer, without the projecting articulations which comprise the articulamentum of modern chitons (order Neoloricata). Among paleoloricates, Late Cambrian *Matthevia variabilis* Walcott, 1885, has vesicular shell plates but its microstructure is otherwise unknown (Runnegar *et al.*, 1979; present Text-fig. 1.2b).

Among neoloricates, the shell plates are aragonitic and predominantly prismatic, homogeneous, CL and CCL, with aragonitic spicules in the girdle cuticle (Couvreur, 1929; Bergenhayn, 1930; Haas, 1972a, 1976; Haas and Kriesten, 1974, 1975, 1977).

Bøggild (1930) described for *Chiton* sp. an outer dark, channeled layer and underlying shell layers which vary from optically homogeneous to indistinctly prismatic, composite prismatic and CL.

Haas (1972a) referred to the shell layers in the Neoloricata as the tegmentum and the "p. [polyplacophoran] hypostracum". He showed that the tegmentum varies from composite prismatic to irregular spherulitic prismatic, and the p. hypostracum contains irregular spherulitic prismatic, CL and irregular simple prismatic structures.

The shell microstructure of Recent *Acanthopleura granulata* (Gmelin, 1791) is illustrated in Plates 122-125. The outer layer of the tegmentum varies from coarsely granular with an irregular simple prismatic substructure (Plate 122A,B) to transitional homogeneous/fine CCL (Plate 122C). In the terminal shell plate the tegmentum passes directly into the CL to crossed acicular (CA) structure of the articulamentum (Plate 122A,B). In the medial plates, the tegmentum grades inward into a thin sublayer of concentric CA to transitional CA/CL structure (Plate 123A). The latter has a substructure similar to dissected crossed prismatic structure (Plate 123A). Below this CL to CA sublayer, there is locally developed a thick layer of radial fibrous and composite prisms (Plate 123B).

The underlying articulamentum has a thick, outer sublayer of predominantly rod-type CL to crossed acicular structure (Plates 123B; 124A,B; 125A). As in other modern chitons but unlike many scaphopods, bivalves and gastropods, the third-order lamellae are rod-like and do not comprise well-defined, laminar second-order lamellae. The myostracum varies from irregular simple prismatic to homogeneous (Plate 125B).

Text-fig. 1 (facing page). Shell microstructural data for modern and fossil Polyplacophora. Legend: predominantly CL (cross-hatching in groups 9, 14-18, 21); vesicular (vesicles in group 2, genus 2b); and laminar with a homogeneous substructure (stippling and horizontal dashes in group 1). The diagrammatic cross-sections of the shell plates at 1, 2a and 2b indicate the presence only of tegmentum (black). The diagrammatic cross-sections of shell plates at 10 and 19 show the presence of both tegmentum (black) and articulamentum (stippled). **1 = Phosphatoloricata:** *Cobcrephora silurica*. **2-8 = Paleoloricata:** 2a: Chelodidae (*Chelodes bergmani*). 2b: *Matthevia variabilis*. 3: Scanochitonidae (*Scanochiton jugatus*). 4: Gotlandochitonidae (*Gotlandochiton interplicatus*; here reconstructed with 8 plates). 5: *Praeacanthochiton cooperi* Bergenhayn. 6: *Yangtzechiton elongatus* Yu. 7: Llandeilochitonidae (*Llandeilochiton ashbyi*). 8: Septemchitonidae (*Septemchiton vermiformis* Bergenhayn, here reconstructed with eight plates). **9-23 = Neoloricata**: 9: Lepidopleuridae [9a = *Helminthochiton priseus* Münster, 9b = *Pterochiton eburonicus* (Ryckholt, 1845)]. 10: Acutichitonidae (*Acutichiton pyramidalis*, from

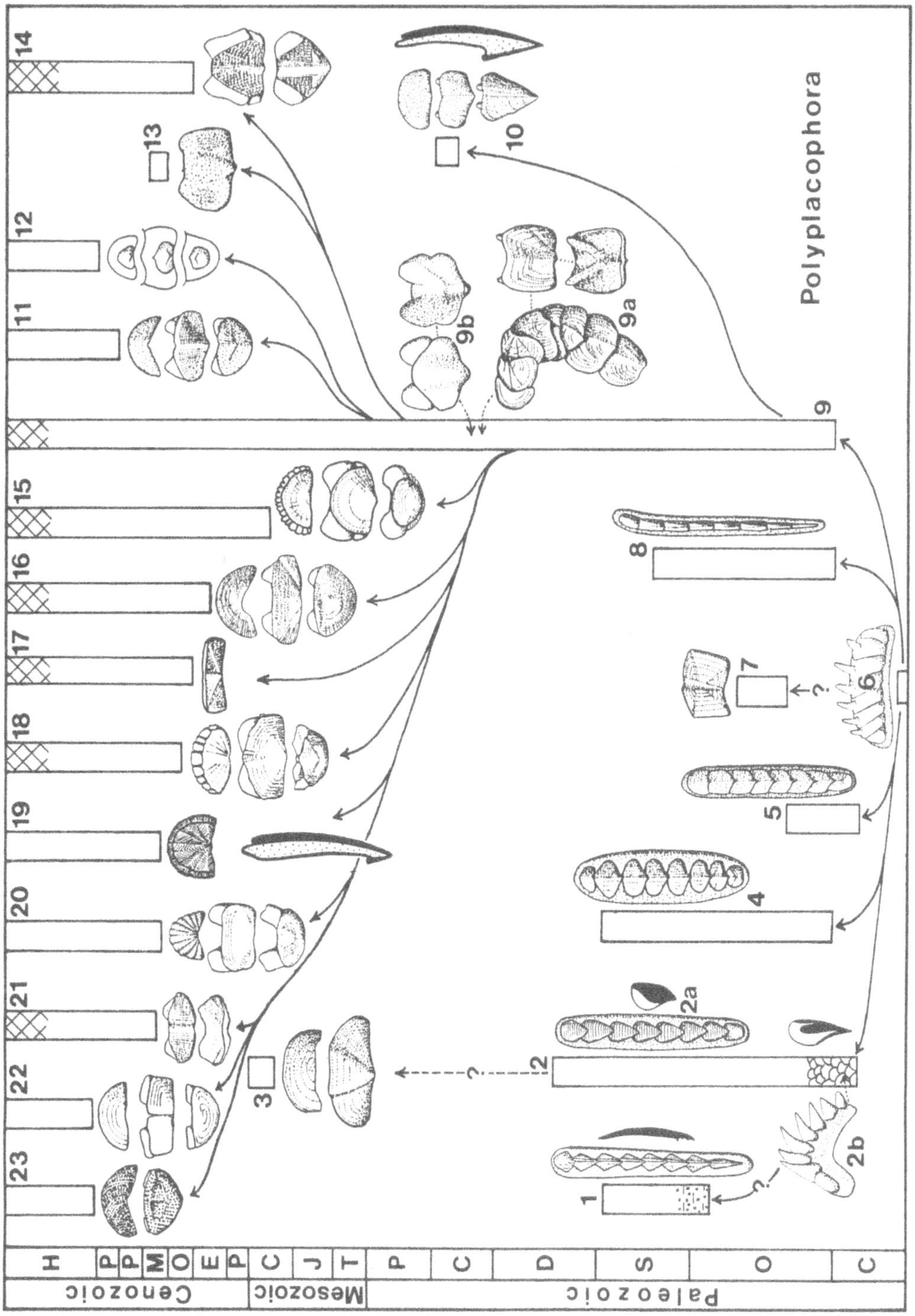

Debrock, Hoare and Mapes, 1984). 11: Hanleyidae [*Hanleya hanleyi* (Bean)]. 12: Choriplacidae [*Choriplax grayi* (Adams and Angas)]. 13. Afossochitonidae (*Afossochiton cudmorei* Ashby) 14: Acanthochitonidae (*Acanthochitona* sp.). 15: Chitonidae [*Acanthopleura spinosa* (Bruguière)]. 16. Ischnochitonidae [*Ischnochiton textilis* (Gray)]. 17. Callochitonidae [*Callochiton laevis* (Montagu)]. 18: Molpaliidae [*Plaxiphora aurata* (Spalowsky)]. 19: Schizochitonidae [*Schizochiton incisus* (Sowerby)]. 20. Callistoplacidae [*Nuttallina californica* (Nuttall in Reeve)]. 21. Chaetopleuridae (*Chaetopleura gemma* Dall). 22: Schizoplacidae [*Schizoplax brandtii* (Middendorff)]. 23: Subterenochitonidae [*Subterenochiton gabrieli* (Hull)]. The figures of chitons and chiton plates are redrawn from A.G. Smith (1960), Bergenhayn (1960), Runnegar *et al.* (1979), Yü (1984), and Debrock, Hoare and Mapes (1984). Phylogenetic relationships are based largely on order of appearance in the fossil record and the taxonomy by Smith (1960), with the tentative addition of the Phosphatoloricata by Bischoff (1981).

CLASS SCAPHOPODA (Plates 126-127)

Cayeux (1916) described "feuilletée" structure in a Mesozoic scaphopod, but this term merely indicates a layered appearance in the shell, not necessarily foliated structure as defined by most later authors (see Glossary, Volume I). Clark and West (1984) described calcitic foliated structure in an Upper Carboniferous *Plagioglypta*. This observation may be important for interpreting the affinities of *Plagioglypta*, because all other scaphopods for which data are available are entirely aragonitic.

Among modern scaphopods, members of the order Gadiloida (*Cadulus* and its allies) are aragonitic and porcelaneous, but details of their shell microstructure remain unknown. Members of the order Dentalioida are uniformly aragonitic (with the possible exception of *Plagioglypta*), with prismatic, CL, CCL and homogeneous structures (Bøggild, 1930; Fantinet, 1959; Haas, 1972b). Recent *Fissidentalium vernedei* (Sowerby) has an outer layer of fine CCL structure (Plate 126A) which grades inward into branching CL structure (Plates 126B, 127A,B). The inner shell layer is mostly homogeneous, with equidimensional to slightly elongate granules 0.3 - 0.7 microns in maximum dimension (Plate 127C). Haas (1972b) illustrated for this species an outer layer of fine, irregular simple prisms grading inward into crossed acicular to irregular CCL structure (the latter is limited to an early ontogenetic stage), a middle layer of branching CL structure, and an inner layer of crossed acicular structure intercalated with finely prismatic sublayers. *Coccodentalium* differs from *Fissidentalium* in having irregular spherulitic prismatic structure in its exterior tubercles (Haas, 1972b). Haas (1972b) mentioned that the microstructure of a recrystallized, Middle Devonian scaphopod resembles modern species.

CLASS GASTROPODA (Text-Fig. 2; Plates 128-134)

The following review of archaeogastropod, docoglossan and paragastropod shell microstructure is organized around a preliminary synthesis of the taxonomic systems of Knight *et al.* (1960), Linsley and Kier (1984), Salvini-Plawen and Haszprunar (1987), and Lindberg (1988). As noted by Hickman (1988), the concept of Archaeogastropoda has been challenged in recent years as a monophyletic taxon and as an evolutionary grade. A thorough discussion of archaeogastropod evolution lies beyond the scope of this chapter. However, several recent papers on gastropod systematics are presently cited which have a direct bearing on higher-level classification in this class. The reader is referred to Haszprunar (1988) and Hickman (1988) for further information on gastropod evolutionary systematics, and for references to the extensive literature on this subject.

Only one taxonomic revision is proposed in the present section, *i.e.*, transference of the family Holopeidae from the Platyceratoidea to the Pleurotomarioidea. Such a change is compatible with similarities in shell morphology and shell microstructure between the Holopeidae and Eotomariidae, and with the significant mineralogical and microstructural differences between holopeids and platyceratids.

This review excludes most archaeogastropod families for which shell microstructural data are unavailable. The reader may consult the index to the master bibliography in Volume I for many additional references on gastropod shell microstructure.

ORDER ARCHAEOGASTROPODA Thiele, 1925

Gastropod-like conispiral shells are known from the early Cambrian, *e.g.*, *Pelagiella* Matthews and *Aldanella* Vostokova (Pojeta, 1980, p. 71; Runnegar, 1981, 1983), but their affinities remain uncertain. Runnegar (1983) suggested that *Aldanella* is ancestral to the suborder Pleurotomariina, which appears in the Late Cambrian. Runnegar separated *Aldanella* from *Pelagiella* and *Costipelagiella*, which he transferred to the Monoplacophora. Linsley and Kier (1984) placed *Pelagiella* and tentatively also *Aldanella* in their new class Paragastropoda. MacKinnon (1985) grouped *Pelagiella* and *Aldanella* in the gastropod order Pelagielloida.

Nyers (1987) suggested that the phosphatic early Cambrian *Thorslundella* is intermediate between cyclomyan monoplacophorans and gastropods. It has symmetrical, paired muscle scars and its shell consists of alternating horizontal laminae of homogeneous francolite and organic layers. Like the Ordovician - Silurian chiton *Cobcrephora, Thorslundella* should be analyzed further to verify the original nature of its calcium phosphate.

I. Suborder Bellerophontina Ulrich and Scofield, 1897.

A. Superfamily Bellerophontoidea M'Coy, 1851. Plate 128.

Knight (1952), Knight *et al.* (1960), Yochelson (1967, 1975), and Peel (1972, 1976) summarized the evidence for gastropod affinities of the Bellerophontoidea. Runnegar and Jell (1976) placed *Bellerophon, Sinuites* and related genera together with *Sinuitopsis, Multifariites* and *Protowenella* in the order Bellerophontoida within the Monoplacophora. They believed that these taxa are monoplacophorans despite their gastropod-like apertural slit, because they have symmetrical, paired muscle scars. Runnegar and Jell (1976) suggested that the anisostrophic Early Cambrian *Aldanella* and other pelagielloideans, rather than bellerophonts, gave rise to later gastropods. Knight (1952) and Yochelson (1975) believed that pelagielloideans are unrelated to anisostrophic gastropods.

Berg-Madsen and Peel (1978) indicated that the group of bellerophontiform molluscs includes both untorted monoplacophorans (*Multifariites* and *Sinuitopsis*) and true gastropods (*Bellerophon* and *Sinuites*). Peel (1980) subsequently indicated that *Sinuites* is a monoplacophoran.

Bellerophontiform molluscs, including the Bellerophontidae and Sinuitidae, appear to be uniformly aragonitic and porcelaneous, with prismatic, CL and CCL shell layers, but without a trace of nacreous structure.

1. Family Bellerophontidae M'Coy, 1851. Late Cambrian - Early Triassic.

***Bellerophon* Montfort, 1808.** Bøggild (1930, p. 299) reported that an Ordovician *Bellerophon* consists of foliated calcite. This diagnosis was doubted by Batten (1972, p. 31), who suggested that Bøggild's specimen was recrystallized, and, hence, originally aragonitic. Other bellerophonts which Bøggild examined were described by him as probably originally aragonitic. Stehli (1956) reported a thin calcitic layer on the exterior of Upper Carboniferous bellerophonts from the Kendrick Shale of Kentucky and the Buckhorn Asphalt of Oklahoma. Present observations of the Kendrick Shale specimens indicate that this calcitic layer is a diagenetic crust.

Rollins (1967, Pl. 10, Fig. 3) illustrated a Kendrick Shale *Bellerophon* (*Bellerophon*) with an inner layer of cone CCL (?) structure, and possibly with an outer layer of linear CL structure. MacClintock (1967, p. 103) described aragonitic cone CCL structure in the inner shell layer of Upper Carboniferous *Bellerophon* (*Pharkidonotus*) *percarinatus* Conrad, 1842, from Oklahoma and Pennsylvania. He described the thin, outer shell layer as optically homogeneous. The present specimen of *Bellerophon percarinatus* from the Kendrick Shale (UNC 5442, Kentucky) has an outer layer of predominantly cone CCL structure, plus an inner layer of predominantly irregular CCL structure intercalated with thin bands of irregular simple prisms.

Bellerophon (*Pharkidonotus*) *labioreflexus* Sturgeon (Upper Carboniferous, Ohio) has outer and inner shell layers with linear CL (?) structure, and a middle layer with cone CCL structure (Rollins, 1967, Pl. 10, Fig. 4).

***Knightites* Moore, 1941.** Upper Carboniferous *Knightites* (*Cymatospira*) *montfortianus* (Norwood and Pratten, 1855) (Pennsylvania) may have some cone CCL structure (Rollins, 1967, Pl. 7, Fig. 8; Pl. 9, Fig. 3).

***Patellilabia* Knight, 1945.** Possible cone CCL structure was illustrated by Rollins (1967, Pl. 10, Fig. 5) for Middle Devonian *Patellilabia* (*Phragmosphaera*) *lyra* (Hall) from the Hamilton Group of New York.

***Retispira* Knight, 1945.** Middle Devonian *Retispira leda* (Hall) (Hamilton Group of New York) has some cone CCL structure (Rollins, 1967, Pl. 9, Fig. 6; Rollins *et al.*, 1971, Fig. 3E). Present observations of this species (UNC 7367) indicate an outer CL or cone CCL layer, plus an inner, fine CCL to irregular CCL layer, with sublayers of irregular simple prisms.

2. Family Sinuitidae Dall in Zittel-Eastman, 1913. Early Late Cambrian - Early Triassic.

According to Peel (1980) and Runnegar (1981), the muscle attachment scars of *Sinuites* indicate that it is a monoplacophoran. Runnegar and Jell (1976, p. 121) included in the

Sinuitidae the genera placed here by Knight *et al.* (1960), plus three genera which Knight *et al.* classified as cyrtolitids (*Trigyra*, *Cloudia* and *Strepsodiscus*), and a few more recently named genera.

***Euphemites* Warthin, 1930.** MacClintock (1967) reported prismatic and CL structures in an Upper Carboniferous *Euphemites* from Ohio, Oklahoma and Texas. According to MacClintock (1967), descriptions of nacreous structure in *Euphemites* by Weller (1930) and Moore (1941) are erroneous. Rollins (1967, Pl. 8, Figs. 1-2) illustrated concentric, linear to branching CL structure in Upper Carboniferous *Euphemites nodocarinatus* (Hall, 1858) from Illinois. He also illustrated an optically homogeneous shell layer exterior to a branching to linear, concentric, CL layer in an Upper Carboniferous *Euphemites* sp. from Pennsylvania (Rollins, 1967, Pl. 8, Fig. 3)

Batten (1972, Fig. 29) showed that Upper Carboniferous *Euphemites vittatus* has alternating bands of vertical, irregular simple prisms and irregular CCL structure in its inner shell layer.

***Ptomatis* Clarke, 1899.** Rollins (1966, Fig. 4C) described for Middle Devonian *Ptomatis patulus* (Hall, 1843) from New York a coarse "prismatic" structure near the aperture with "prism" diameters over 100 microns. Rollins (1967) indicated that this structure is apparently CCL. His Pl. 10, Fig. 2 suggests the presence of linear CL and cone CCL structures.

Present observations of Middle Devonian *Ptomatis patulus* from the Hamilton Group of New York (UNC 5439, 5451) indicate an originally aragonitic shell with a thin, outer layer of fine, nearly vertical irregular simple prisms; a thick, middle layer of cone CCL structure; and, better developed in later whorls, an inner layer of fine CCL to very faintly CL (?) structure. The parietal shelf consists of irregular CCL structure.

3. Family uncertain. Middle Devonian. Plate 128.

***Praematuratropis* Rollins *et al.*, 1971.** Judging from descriptions by Rollins (1967) and *Rollins et al.* (1971), Middle Devonian *Praematuratropis ovatus* (Hamilton Group, New York) has an inductural layer of concentric, linear CL structure, plus CCL structure in its main shell layers (Rollins, 1967, Pl. 10, Fig. 1; Pl. 9, Figs. 1,2,4; Rollins *et al.*, 1971, Fig. 8E and p. 145). Present observations of this species (UNC 5452, 5453) confirm the presence of linear CL structure in the inductura (Plate 128A), and indicate that the underlying shell layer is fine CCL to irregular CCL, grading inward into a thick, middle layer of cone CCL structure (Plate 128B), and an inner layer of irregular CCL to cone CCL structure. There is also a thin, irregular simple prismatic, innermost shell layer.

II. Suborder Pleurotomariina Cox and Knight, 1960.

Knight *et al.* (1960, p. I198) suggested that the superfamily Pleurotomarioidea evolved directly from the Bellerophontoidea, and then gave rise to other archaeogastropod groups. Runnegar (1983) believed that the suborder Pleurotomariina evolved from the Early Cambrian *Aldanella* (family Aldanellidae), which he said was already distinct from the suborder Macluritina.

Salvini-Plawen (1980) and Salvini-Plawen and Haszprunar (1987) united the superfamilies Fissurelloidea, Pleurotomarioidea ana Trochoidea in the suborder Vetigastropoda, based on comparative anatomy of modern species.

Knight *et al.* (1960, p. I197) indicated that shells in this suborder have an outer calcitic layer and an inner aragonitic layer that is nacreous except in the patelliform genera. It is now known that the shell is entirely aragonitic in some members of this suborder, and that many Pleurotomariina combine nacreous and CL and/or CCL structures in the same shell.

Text-fig. 2 (facing page). Shell microstructural data for modern and fossil Gastropoda (left) and Paragastropoda (right), excluding the suborder Bellerophontina. The symbols indicate the predominant microstructure(s) of the middle and inner layers of the adult, post-juvenile shell. Prismatic structures are not shown except for *Pelagiella*, which is apparently entirely prismatic. Double-wide columns (see Holopeidae and Trochidae) indicate that two microstructural grades occur within the group. Legend: **Thin, horizontal lines** = aragonitic nacre. **Thick, horizontal lines** = calcitic semi-nacre. **Small x** = aragonitic CL and/or CCL.

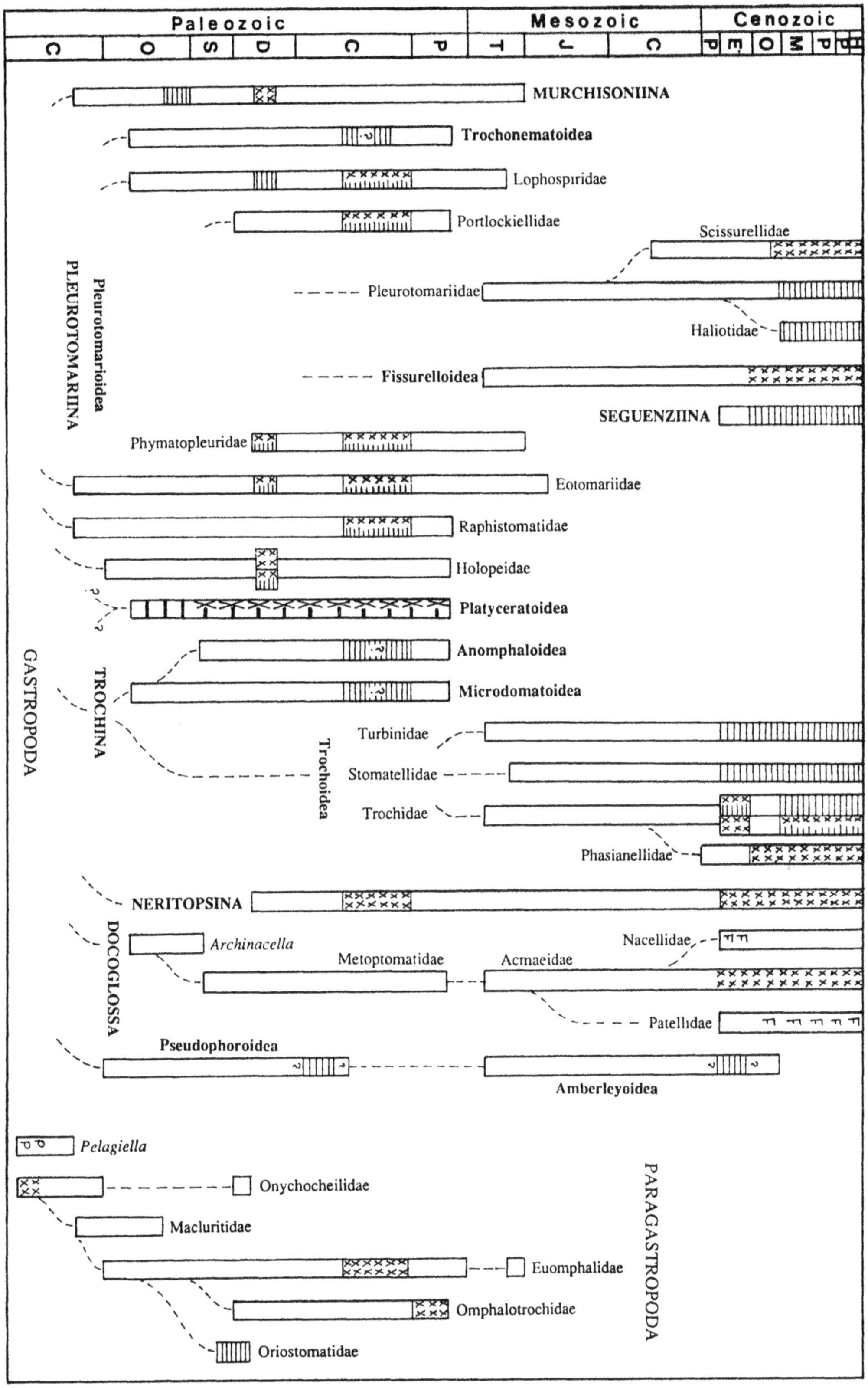

Large X = calcitic crossed semi-foliated and complex crossed semi-foliated. **F** = calcitic foliated with or without associated aragonitic CL structure. **P** = aragonitic prismatic.

A. Superfamily Trochonematoidea Zittel, 1895. Middle Ordovician - Middle Permian.

Knight *et al.* (1960, p. I224) suggested that these nacreous gastropods evolved from the Lophospiridae (superfamily Pleurotomarioidea), and that, "turbiniform and other low-spired caenogastropods were probably derived from such archaeogastropod groups as the Trochonematacea..." (*ibid.*, p. I291).

B. Superfamily Fissurelloidea Fleming, 1822.

McLean (1984) argued that the Fissurelloidea descended directly from bellerophontiform gastropods, independently of the Pleurotomarioidea and Trochoidea, judging from the presence of a tubulated, crossed lamellar shell and a mantle slit in both groups. According to Salvini-Plawen (1980) and Salvini-Plawen and Haszprunar (1987) these features may have evolved convergently in bellerophonts and vetigastropods, because comparative anatomy of modern species suggests that the Fissurelloidea are very closely related to the Pleurotomarioidea and Trochoidea. Descent of the Fissurelloidea from one of these two superfamilies is compatible with the presence of CL and CCL structures in some representatives of all three groups.

1. Family Fissurellidae Fleming, 1822. Middle Triassic - Recent.

***Fissurella* Bruguière, 1789.** In Recent *Fissurella nubecula* Linné, 1758, the protoconch consists of outer and inner layers of aragonitic, fine simple prisms and a middle layer of horizontal, elongate rods (Batten, 1975). In the later juvenile shell, the middle layer changes to "type 2 crossed lamellar", *i.e.*, rod-type CL structure intergrading with intersected crossed acicular and/or dissected crossed prismatic structure. The adult shell consists of an outer layer of simple CL structure and an inner layer of irregular CCL structure.

Bøggild (1930, p. 70) observed only aragonitic CL and CCL structures in fissurellids, except for Recent *Fissurella crassa.* The latter species has a thick, dark brown, calcitic, prismatic outer shell layer.

***Diodora* Gray, 1821 and *Emarginula* Lamarck, 1801.** Recent *Diodora* and *Emarginula* have a protoconch which differs slightly in microstructure from *Fissurella.* However, adult shells of these three genera are microstructurally similar (Batten, 1975).

C. Superfamily Pleurotomarioidea Swainson, 1840. Plates 129-130.

Salvini-Plawen and Haszprunar (1987) indicated that the superfamily Pleurotomarioidea is probably paraphyletic. Many Paleozoic and some Recent representatives of this superfamily resemble some Seguenziidae (suborder Sequenziina) and Trochidae (suborder Trochina) in combining aragonitic nacreous and CL and/or CCL structures in the same shell. Upper Carboniferous *Worthenia* (family Lophospiridae) and *Glabrocingulum* (family Eotomariidae) have well-developed aragonitic CL structure and nacre in shell layers separated only by a thin layer of irregular simple prisms. However, nacreous structure predominates in the adult shell of most modern pleurotomarioideans, sometimes to the exclusion, or near exclusion, of aragonitic crossed structures (*e.g.*, in many Pleurotomariidae and Haliotidae). Aragonitic crossed structures occur without nacreous structure in the modern Scissurellidae and in some Paleozoic Holopeidae.

Available evidence suggests that both aragonitic CCL and nacreous structures evolved early in the Pleurotomarioidea, but that CCL structure is primitively restricted to the larval and early juvenile shell. Species with largely nacreous adult shells probably evolved into species with largely or entirely aragonitic CL and/or CCL adult shells through paedomorphic elimination of the adult nacreous structure.

1. Family Sinuopeidae Wenz, 1938. Late Cambrian - Middle Permian.

This family is unknown microstructurally.

2. Family Pleurotomariidae Swainson, 1840. Middle Triassic - Recent. Plate 129A-D.

***Pleurotomaria* Defrance, 1826 (including *Entemnotrochus*, *Mikadotrochus* and *Perotrochus*).** As described by Erben and Krampitz (1972), *Pleurotomaria* deposits its nacreous middle shell layer after the first juvenile whorl. The outer and inner layers of the protoconch change microstructurally during ontogeny. The outer layer changes from "irregularly complex-prismatic" (actually irregular CCL) to "crossed acicular" (actually crossed acicular plus very fine, rod-type CL), to irregular spherulitic prismatic with some intersection of the elongate structural subunits, and finally to irregular spherulitic prismatic without intersecting crystallites. The inner shell layer of the protoconch varies from vertical, regular simple prismatic to irregular simple prismatic, intersected crossed platy, and intersected crossed blocky.

The adult shell of Recent *Perotrochus adansonianus* (Crosse and Fischer, 1861) (UNC 5430) has a very thin, outer layer of nearly vertical to slightly reclined prisms. This layer varies from irregular spherulitic prismatic to irregular simple prismatic except for its middle part, which shows some incipient fine CCL structure (Plate 129B). The middle shell layer is sheet nacreous to columnar nacreous (Plate 129D). The nacreous layer is intercalated with thin bands of vertical, irregular simple prisms.

3. Family Haliotidae Rafinesque, 1815. Miocene - Recent (*fide* Hickman, 1984). Plate 129E-F.

The Haliotidae share a number of primitive features with the Pleurotomariidae, and they may have evolved from that family (Hickman, 1984, p. 229).

***Haliotis* Linné, 1758.** Bøggild (1930, p. 70) noted that *Haliotis* commonly has calcite and aragonite in its outer shell layer. Mutvei *et al.* (1985) showed that this outer layer may be entirely aragonitic (in *Haliotis asinina* and *H. glabra*) or it may have intimately mixed, yet discrete, aragonitic and calcitic portions (in *Haliotis* cf. *H. discus*, *H. fulgens*, *H. roei*, *H. tuberculata* and *H. rotundata*). Illustrations by Mutvei *et al.* (1985) indicate irregular spherulitic prismatic and other prismatic structures in this outer layer.

Present observations of Recent *Haliotis rufescens* Swainson (UNC 8803, California) indicate that the outer shell layer is largely or entirely calcitic and irregularly prismatic. The prisms show well-developed rhombohedral calcite cleavage planes (Plate 129E). The middle and inner shell layers are aragonitic and largely columnar nacreous (Plate 129F). The nacre tablets are usually about 8 microns wide and 0.4 microns thick.

4. Family Scissurellidae Gray, 1847. Late Cretaceous - Recent.

This family was placed in the superfamily Pleurotomarioidea by Knight *et al.* (1960). Batten (1975) argued that the Scissurellidae should be removed from the Pleurotomarioidea "despite nearly identical external morphology", and placed in the Fissurelloidea. At the time scissurellids appeared, the only possible pleurotomarian ancestors were haliotids and pleurotomariids, and both families appear to be farther removed from scissurellids than from fissurellids. Haliotids can also be excluded from ancestry to scissurellids because of their late (Miocene) appearance in the fossil record. According to Salvini-Plawen and Haszprunar (1987), Batten's (1975) comparison of the morphology of embryonic stages of fissurellids and scissurellids is invalid. Also, according to Bandel (1982, p. 44), the shell microstructure of adult scissurellids resembles the protoconch of fissurellids and various other Vetigastropoda. Salvini-Plawen and Haszprunar (1987, p. 757) concluded that present data "unequivocally unite the Scissurellidae with the Haliotidae and Pleurotomariidae", so they included all three families in their (admittedly) paraphyletic superfamily Pleurotomarioidea. Hickman (1984) indicated that scissurellids share a number of primitive features with the Pleurotomariidae.

Knight *et al.* (1960, p. I221) noted that scissurellids combine nacreous and porcelaneous structures in their shells. However, the presence of nacre in this family has yet to be verified by subsequent researchers. The occurrence of aragonitic CL and CCL structure in many Paleozoic Pleurotomarioidea strengthens the hypothesis of close affinity with the Scissurellidae.

***Scissurella* d'Orbigny, 1824.** Recent *Scissurella* (*Anatoma*) *crispata* Fleming, 1832, has a protoconch with outer and inner layers of aragonitic, simple prismatic structure. In the adult shell, it has a "type 2 crossed lamellar" layer between outer and inner prismatic layers (Batten, 1975). Batten's "type 2 crossed lamellar structure" is the present rod-type CL structure, locally intergrading with intersected crossed acicular and/or dissected crossed prismatic structure.

***Incisura* Hedley, 1904, and *Sinezona*, Finlay, 1927.** The ultrastructure of the protoconch of *Incisura* and *Sinezona* resembles that of *Scissurella* (see Batten, 1975, p. 11).

5. Family Portlockiellidae Batten, 1956. Early Devonian - Middle Permian.

***Shansiella* Yin, 1932.** Judging from illustrations and descriptions by Batten (1972), Upper Carboniferous *Shansiella carbonaria* (Norwood and Pratten, 1855) has, from exterior to interior, a thin, outer layer of planar, irregular spherulitic prismatic structure; a thin "platy" layer with vertical plates or tiles oriented about 45 degrees to internal growth lines; a thick, branching CL layer (Batten's "elongate complex prismatic layer"); a myostracal (?) layer of irregular simple prisms; a columnar nacreous layer; and an inner layer with bands of irregular CCL and cone CCL structure alternating with irregular simple prisms.

6. Family Raphistomatidae Koken, 1896. Late Cambrian - Middle Permian.

***Trepospira* Ulrich and Scofield, 1897.** Upper Carboniferous *Trepospira discoidalis* Newell, 1935 (UNC 3273) has an entirely aragonitic shell with a very thin, outer layer of diffuse CL (?) or fine CCL to irregular CCL structure, underlain by a thick, middle layer of columnar nacre with tablets 9 to 10 microns wide. The very thin, inner shell layer is irregular simple prismatic.

Carboniferous *Trepospira illinoiensis* (Worthen, 1884) has a (middle?) sheet nacreous layer exterior to the irregular simple prismatic myostracum, and a fine CCL ("spherolitic complex prismatic") layer interior to the myostracum (Batten, 1972, Fig. 27).

***Callistadia* Knight, 1945.** Upper Carboniferous *Callistadia spirallia* preserves color patterns on its shell, but the shell microstructure is unknown (Hoare and Sturgeon, 1978).

7. Family Eotomariidae Wenz, 1938. Late Cambrian - Early Jurassic. Plate 130.

Batten (1967) indicated that Cambrian *Taeniospira* probably gave rise to *Mourlonia*, and *Mourlonia* evolved into a number of other taxa, such as *Schwedagonia*.

***Bembexia* Oehlert, 1888.** Middle Devonian *Bembexia sulcomarginata* (Conrad, 1842) was reported by Rollins *et al.* (1971, Fig. 11) to consist of vertical prisms. Present observations of this species from the Middle Devonian of New York (UNC 5443, 5447, 5751) show an outer layer of cone CCL to irregular CCL structure grading inward into irregular simple prismatic to irregular CCL structure; a thick, middle layer of columnar nacre (Plate 130A.1) locally underlain by a thin band of irregular simple prisms (Plate 130A.2); and an inner layer of fine CCL structure (Plate 130A.3) grading inward into irregular to cone CCL and CL structure (Plate 130A.4, 130B). The latter may be covered interiorly by a thin layer of irregular simple prisms.

***Glabrocingulum* Thomas, 1940.** Upper Carboniferous *Glabrocingulum* (*Glabrocingulum*) *grayvillensis* (Norwood and Pratten, 1855) has an outermost "elongate complex prismatic layer" (Batten, 1972, Fig. 10). This is actually a CL layer in which the dip directions of the second-order lamellae are uniformly oblique rather than directly opposite. Present observations of Upper Carboniferous *Glabrocingulum* (*Ananias*) *nodocostatus* Hoare, 1961 (UNC 13784b, Kendrick Shale Member, Breathitt Formation, Kentucky) reveal an outer layer of aragonitic CL structure, a thin myostracal layer of irregular simple prisms, an underlying nacreous layer, and an innermost layer of irregular CCL structure.

8. Family Holopeidae Wenz, 1938. Early Ordovician - Middle Permian.

Knight *et al.* (1960, p. I238) united the Holopeidae and the Platyceratidae in the superfamily "Platyceratacea", citing their similar turbiniform shells and "nacreous" microstructure. However, Morris and Cleevely (1981) noted that *Holopea*'s shape also resembles the Cambrian - Ordovician genus *Sinuopea* (Family Sinuopeidae, superfamily Pleurotomarioidea). They speculated that turbiniform shells characterized the common ancestors of the Pleurotomarioidea, Trochoidea and Neritoidea.

Contrary to Knight *et al.* (1960), the Holopeidae differ both mineralogically and microstructurally from the Platyceratidae. The Holopeidae have aragonitic prismatic, nacreous, CL and CCL structures, whereas the Platyceratidae are entirely calcitic in their adult shell, with prismatic, semi-nacreous, crossed semi-foliated (CSF) and complex crossed semi-foliated (CCSF) structures. The Platyceratidae differ microstructurally from all other Paleozoic gastropods, although similar semi-nacreous and semi-foliated structures occur in certain brachiopods and corals (see *Glossary*, Volume I, and figures therein).

Middle Devonian *Gyronema lirata* (Hall, 1861) is morphologically and microstructurally similar to the Middle Devonian eotomariid *Bembexia sulcomarginata* (Conrad) (superfamily Pleurotomarioidea). Both taxa have columnar nacre between outer and inner porcelaneous shell layers, a turbiniform shell, similar radial ornament and, except for the labral slit in *Bembexia*, a nearly identical aperture. It may be noted that some Eotomariidae have only a slight labral slit (Knight *et al.*, 1960, p. I204-I207). These similarities suggest that the Holopeidae are more closely related to the Eotomariidae than to the Platyceratidae. The Holopeidae are, therefore, presently removed from the Platyceratoidea and placed near the Eotomariidae.

***Holopea* Hall, 1847.** Middle Devonian ?*Holopea hebe* (Hall, 1861) (Hamilton Group, New York; UNC 5448, 5745, 5746) was originally aragonitic with a thin, outer layer of irregular simple prisms or fibrous prisms; a thin, middle layer of irregular CCL structure grading inward into concentric CL structure; and an inner layer of irregular CCL structure.

***Gyronema* Ulrich in Ulrich and Scofield, 1897.** Middle Devonian *Gyronema lirata* (Hall, 1861) (Hamilton Group, New York, UNC 5446, 6650, 7623) was entirely aragonitic. The early whorls have a thin, outer layer of irregular simple prisms or fibrous prisms, underlain by a thicker layer of irregular CCL and minor cone CCL and fine CCL structure, in turn underlain by columnar nacre with tablets 7 to 10 microns wide. Below the nacreous layer is an innermost layer of fine CCL to irregular CCL structure, with a few thin bands of vertical, irregular simple prisms. In the later whorls, the outer prismatic and CCL layers are missing (probably through abrasion), and the nacreous layer and a thin, inner CCL layer comprise much of the shell thickness. The juvenile shell has proportionally more prismatic and crossed structures than the adult shell, which is largely nacreous.

9. Family Lophospiridae Wenz, 1938. Ordovician - Middle Triassic.

Knight *et al.* (1960, p. I209) placed *Ruedemannia* and *Worthenia* together in the subfamily Ruedemanniinae Knight, 1956. If these genera are related by direct descent, then the nacreous structure in adult *Ruedemannia* was partly replaced evolutionarily by aragonitic crossed structures in adult *Worthenia*.

***Ruedemannia* Foerste, 1914.** Middle Devonian *Ruedemannia trilix* (Hall, 1861) from New York (Hamilton Group, UNC 5438, 7621, 7622) was originally aragonitic with a very thin, outer layer of fine, irregular simple prisms; a middle columnar nacreous layer with bands of irregular simple prisms; and a thin, inner layer of fine, irregular simple prisms.

***Worthenia* deKoninck, 1883.** Upper Carboniferous *Worthenia tabulata* (Conrad, 1835) (UNC 3274) has a thin, outer layer of fine CCL (?) structure; a thick, middle layer of aragonitic branching CL structure with broad, diffuse first-order lamellae; and a thick, inner layer of columnar nacre with tablets about 12 microns wide. Batten (1972, Fig. 15) showed that the second-order lamellae in what is presently regarded as the the CL layer dip in two oblique directions instead of two opposite directions. He referred to this CL structure as "elongate complex prismatic". Batten (1972, Figs. 21, 24, 25) illustrated the rather poorly organized columnar nacre of this species.

10. Family Phymatopleuridae Batten, 1956. Middle Devonian - Upper Triassic.

***Glyptotomaria* Knight, 1945.** Middle Devonian *Glyptotomaria* (*Dictyotomaria*) *capillaria* (Conrad, 1842) from New York (UNC 5440) was originally aragonitic with a thin, outer layer of fine, irregular simple prisms; a thick, middle layer of columnar nacre; and a thin, inner layer of irregular simple prisms. The inner layer is locally transitional to irregular CCL, and it is better developed in the later whorls.

***Phymatopleura* Girty, 1939.** Carboniferous *Phymatopleura nodosus* (Girty, 1912) has an outer CCL layer and an underlying nacreous layer; the latter grades inward from sheet nacreous to columnar nacreous (Batten, 1972, Fig. 26; see his "complex prismatic" and nacreous shell layers).

D. ?Superfamily Neomphaloidea McLean, 1981.

1. Family Neomphalidae McLean, 1981. Recent.

***Neomphalus* McLean, 1981.** McLean (1981) placed *Neomphalus fretterae* McLean, 1981, in the suborder Euomphalina partly on the basis of its gill, epipodial tentacle, and radular structures. Batten (1984a) noted that the ultrastructure of the CL layer in *Neomphalus* resembles patellids and mesogastropods more than *Euomphalus*, and he added that *Neomphalus* differs from *Euomphalus* in its lack of a calcitic outer shell layer. Batten (1984a) concluded that *Neomphalus* is an advanced archaeogastropod or a primitive mesogastropod. Salvini-Plawen and Haszprunar (1987) noted similarities between *Neomphalus* and the Vetigastropoda (Pleurotomarioidea, Fissurelloidea and Trochoidea) in the structure of their oesophagus and trochoid embryonic shell. They placed *Neomphalus* close to but not necessarily within the Vetigastropoda.

As described by Batten (1984a), the protoconch of *Neomphalus fretterae* has a thin, outer layer of vertical (irregular?) simple prismatic structure, and an inner layer of intercalated irregular simple prismatic, fine CCL or irregular CCL, and dissected crossed prismatic structure. The adult shell has a thin, outer layer of asymmetric prismatic structure and a middle layer of cone CCL structure (Batten, 1984a, Fig. 8). The inner shell layer consists of concentric, linear CL structure. The shell has microtubules about 0.1 micron in diameter (*ibid.*, Fig. 12).

III. Suborder Trochina Cox and Knight, 1960.

According to Knight *et al.* (1960, p. I237) the inner shell layer in this order is aragonitic and nacreous, and in some instances the shell is entirely aragonitic. However, evidence summarized below indicates that the Platyceratidae were entirely calcitic in their adult shell.

A. Superfamily Platyceratoidea Hall, 1859. Plates 131-134.

The family Holopeidae is presently removed from this superfamily and placed in the superfamily Pleurotomarioidea near the Eotomariidae. Except for the absence a labral slit, the Holopeidae are morphologically, microstructurally and mineralogically more like the Eotomariidae than the Platyceratidae. Platyceratids have an entirely calcitic adult shell with prismatic, semi-nacreous, crossed semi-foliated (CSF), and complex crossed semi-foliated (CCSF) structures. These calcitic crossed structures appear to be unique in the Gastropoda. This superfamily is tentatively retained in the Trochina, following Knight *et al.* (1960), pending further analysis of their relationships with the suborders Pleurotomariina and Neritopsina.

1. Family Platyceratidae Hall, 1859. Middle Ordovician - Middle Permian, Upper Triassic.

Bandel (1988) has indicated the presence of platyceratids in the Triassic Cassiano Formation of northeastern Italy. The taxa were previously erroneously classified as capulids.

According to Batten (1984b), *Cyclonema* has shell features which suggest placement in the Neritidae. However, it is associated with crinoids, and its tendency for dwarfism and irregular shapes suggests coprophagous feeding on crinoids, like *Platyceras* (see also Thompson, 1970). Ulrich (in Ulrich and Scofield, 1897, p. 1046) suggested that *Cyclonema* evolved from *Gyronema* or *Holopea* (family Holopeidae, superfamily Pleurotomarioidea). Knight (1934, 1946) and Thompson (1970) favored derivation of *Cyclonema* from *Holopea*, with *Cyclonema* then evolving into *Platyceras*. Bowsher (1955) proposed that *Cyclonema* and *Naticonema* have close common ancestry, and that *Cyclonema* evolved into *Dyeria* (a subgenus of *Cyclonema*), whereas *Naticonema* evolved into *Platyceras*.

Despite their similar turbiniform shape, *Cyclonema* and the other platyceratids differ mineralogically and microstructurally from the Holopeidae. The Holopeidae are entirely aragonitic with prismatic, nacreous and/or CCL adult shell layers, whereas *Cyclonema* and *Platyceras* have largely, if not entirely, calcitic adult shells with prismatic, semi-nacreous, CSF and CCSF structures.

The shell microstructure and original mineralogy of the Platyceratidae have been variously interpreted by previous workers. Bøggild (1930, p. 72) indicated that some *Cyclonema* were aragonitic, whereas others had an outer calcitic shell layer. Knight *et al.* (1960, p. I239-I240) wrote that the outer shell layer is relatively thick and calcitic, and "the inner shell layers of the primitive genus *Cyclonema* are seemingly nacreous and aragonitic, but this layer appears to be lost in the more advanced *Platyceras*."

Present observations confirm the conclusion by Thompson (1970) that adult *Cyclonema* was originally largely calcitic. Shells of *Cyclonema* collected from several Ordovician localities invariably have an outer calcitic prismatic layer and an inner calcitic semi-nacreous layer. The latter resembles nacre in its general organization, but the tablets are much wider and more irregularly shaped, and preservational evidence clearly indicates that they were originally calcitic. Interestingly, the protoconch of *Cyclonema* is usually missing in otherwise well-preserved fossils. Thompson (1970) attributed this absence to the formation of a septum, and hence weakened attachment, between the protoconch and the later-formed shell. Absence of the protoconch may also reflect an original aragonitic mineralogy which is more susceptible to diagenetic destruction than the calcitic, adult shell.

Judging from the wide taxonomic distribution of aragonite among Paleozoic gastropods, it is probably safe to assume that the Platyceratidae evolved from at least partially aragonitic ancestors. The columnar semi-nacreous microstructure of *Cyclonema*, one of the earliest and morphologically least specialized platyceratids, probably evolved from ancestral columnar nacreous structure through a simple mineralogical transformation from aragonite to calcite. *Cyclonema*'s columnar semi-nacreous structure probably then evolved through slight reorganization into the crossed semi-foliated (CSF) and complex crossed semi-foliated (CCSF) structures of *Platyceras*. Semi-nacreous, CSF and CCSF structures have similar laminar, second-order structural elements; in semi-nacre these laminae are horizontal, whereas in CSF and CCSF structures they dip in two (CSF) or more (CCSF) directions. Some *Platyceras* combine semi-nacreous and semi-foliated structures in the same shell, whereas others have apparently entirely replaced their ancestral semi-nacreous structure with semi-foliated structures (see Plates 131 - 134). This evolutionary replacement recalls the analogous replacement of aragonitic nacre with aragonitic CL and CCL structures in the Bivalvia (see Carter, Volume I).

***Cyclonema* Hall, 1852.** Late Ordovician *Cyclonema* (*Cyclonema*) *humerosa* Ulrich, 1897, from Ohio (UNC 1284) has a calcitic, outer shell layer of fine to medium width (10 to 26 microns) regular to irregular simple prisms. Some of the prisms are optically homogeneous, with length-parallel extinction, but most are optically heterogeneous. The thicker, inner shell layer consists of calcitic, columnar semi-nacre. The contact between the outer prismatic and inner semi-nacreous shell layers is gradational, with some prism columns grading into the tablets of a semi-nacre column. Toward the shell interior, some semi-nacre columns disappear through geometric selection whereas others expand to diameters of 20 to 27 microns. These semi-nacre columns are two to three times wider than the nacre columns in *Haliotis rufescens* (cf. Plate 128B). Each semi-nacre column shows more or less length-parallel extinction independently of the adjacent columns. The columns sometimes maintain optical continuity with a prism in the outer shell layer.

Batten (1984b, Fig. 1A) illustrated a vertical fracture through the wall of Late Ordovician *Cyclonema bilix* (Conrad, 1842). The outer shell layer is irregular simple prismatic, and the middle shell layer has vertical columns with horizontal laminae. Batten suggested that the outer shell layer was originally calcitic, but he inferred, incorrectly, that the "stacked nacreous" middle layer was originally aragonitic. Batten also described a thin, innermost simple prismatic shell layer. Thompson (1970, p. 229) had earlier concluded, based on studies of specimens from several Ordovician localities, that all three shell layers in *Cyclonema bilix* were originally calcitic.

Present observations of Ordovician *Cyclonema* (*Cyclonema*) *bilix* (Conrad, 1842) (UNC 1286, Indiana) reveal originally calcitic microstructures similar to *C. humerosa* except that the outer shell layer is thicker and more irregularly prismatic. The shell has two main layers, an outer simple prismatic to irregular simple prismatic layer, and an inner columnar semi-nacreous layer.

***Platyceras* Conrad, 1840.** According to Batten (1984b, p. 1186-1187), Devonian *Platyceras* has a thin, outer shell layer of simple prisms and a thick, inner layer of "nacre", which he believed was originally aragonitic. Batten attributed preservation of the nacre to protection by the calcitic, outer shell layer. However, pterioid bivalves from the same Devonian localities collected by Batten also have a calcitic outer shell layer, yet their underlying nacreous layers are always recrystallized (Carter and Tevesz, 1978). All presently studied Silurian, Devonian and Carboniferous *Platyceras* preserve the microstructure of their middle and inner shell layers, whereas originally aragonitic shell layers in other gastropods, bivalves and cephalopods from the same localities are invariably recrystallized.

Batten (1984b) described that the "nacre" in *Platyceras spinigerum* Worthen, 1883, consists of tablets with needle-shaped subunits oriented parallel to the optic C-axis and concentrically arranged around the center of each tablet. He also noted that the "nacre" tablets show a brick-wall pattern in some vertical sections, but a columnar pattern in other vertical sections. This brick-wall pattern may represent a vertical section through a semi-nacreous layer, or a radial section through a CSF layer.

Batten (1984b) noted that *Platyceras* (*Platyostoma*) *lineata* has "nacre" tablets which are curved toward the exterior of the shell (his Fig. 1F). This bowing is typical of the elongate, gutter-shaped laminae which comprise the first-order lamellae in some platyceratid CSF structures (Plate 132A). The non-horizontal "nacreous" laminae noted by Batten (1984b) for *Platyceras* are the second-order lamellae of CSF structure as seen in non-radial, vertical sections. The second-order lamellae in CSF and CCSF structures consist of flat or gutter-shaped, elongated, irregular tablets or laminae, rather than euhedral blades. In this respect, CSF and CCSF structures differ from the crossed foliated and complex crossed foliated structures of patelloid gastropods and bivalves.

Adult *Platyceras* generally have three main shell layers: (1) an outer layer of vertical to reclined, regular simple prismatic to irregular simple prismatic structure; (2) a middle layer of concentric CSF structure; and (3) an inner layer of vertical, irregular simple prisms interstratified with semi-nacre and/or CSF and CCSF structures.

In Middle Devonian *Platyceras* (*Orthonychia*) *conicum* (Hall) (Williamsville, New York) the thin, outer shell layer is irregular simple prismatic to nearly homogeneous (Plate 131A.1). This layer grades inward into a thick, middle layer of concentric CSF in which the first-order lamellae appear in radial, vertical sections as more or less vertical columns of outwardly curved, second-order lamellae (Plate 132A). The second-order lamellae dip in two opposing directions about 15 degrees from the horizontal (Plates 131B, 132B). The inner shell layer is semi-nacreous (Plate 131C) and regular to irregular calcitic simple prismatic. The middle and inner shell layers contain sparse, vertical microtubules.

Middle Silurian *Platyceras* (*Platyceras*) *niagarense* (Hall, 1852) (UNC 1244, Waldron Shale, Indiana) has a thin, outer shell layer of irregular to regular simple prismatic structure (Plate 132C.1). The middle shell layer consists of concentric CSF structure with second-order lamellae dipping about 35 degrees from the depositional surface. The first-order lamellae are outwardly reclined and inwardly inclined (Plate 132C.2). In some specimens the innermost first-order lamellae become reclined again, giving this layer a zigzag appearance in radial, vertical sections. The thin, inner shell layer consists of semi-nacreous and fine, irregular simple prismatic structure.

Middle Devonian *Platyceras* (*Platyostoma*) *lineata* (Conrad, 1847) (UNC 1240, 1241, New York) has a zigzag, concentric CSF shell layer similar to the preceding species, except that the pattern appears more S-shaped than Z-shaped in radial sections, with the middle limb of the "S" predominating. The second-order lamellae dip about 12 to 20 degrees from the depositional surface. The thin, outer shell layer consists of vertical, regular simple prisms. There is a thin, local sublayer of calcitic semi-nacreous structure between the outer layer and the CSF layer. The thin, inner shell layer consists of semi-nacreous structure, possibly with traces of diffuse, radially oriented CSF structure.

In Middle Devonian *Platyceras argo* (UNC 1267, Williamsville, New York) the concentric CSF middle shell layer shows a double "S" zigzag pattern in radial, vertical sections. The second-order lamellae dip about 40 degrees from the depositional surface. The thin, outer shell layer consists of outwardly reclined and inwardly more vertical, regular to irregular simple prisms. The thick, inner shell layer is semi-nacreous.

In Middle Devonian *Platyceras* (*Platyceras*) *unguiforme* Hall (UNC 5432) from Morrisville, New York, the thin, outer shell layer is regular simple prismatic. This grades inward into the concentric CSF middle shell layer. The latter is relatively thick and shows diffuse first-order lamellae which are outwardly reclined and inwardly nearly vertical. Their second-order lamellae dip about 12 degrees from the horizontal. The middle shell layer grades inward into a moderately thick, horizontally laminated, semi-nacreous layer and then a thin layer of semi-nacreous to irregular simple prismatic structure.

In Middle Devonian *Platyceras thetis* Hall (UNC 1254, Moscow Shale, Yates Co., New York) the thin, outer shell layer consists of vertical, irregular to regular simple prisms. The thick, middle shell layer has concentric CSF structure with broad, poorly defined first-order lamellae, and with second-order lamellae dipping 10 to 25 degrees from the horizontal. The thin, inner shell layer consists of irregular simple prisms.

In Middle Devonian *Platyceras carinatum* (UNC 1266, Moscow Shale, Pavilion, New York) the thin, outer shell layer consists of irregular simple prisms (Plates 133A.1, D; 134A). The middle shell layer is CSF with wide first-order lamellae, and with second-order lamellae dipping about 15 degrees from the horizontal (Plates 133A.2, B.2, C; 134B,C). The second-order lamellae are generally flat, but they are rarely outwardly curved (*i.e.*, gutter-shaped). The CSF layer is underlain by a calcitic, simple prismatic layer (Plate 133A.3, B.3, E) and then by irregular CCSF (Plate 133A.4, B.4) and regular simple prismatic structure.

Lower Carboniferous *Platyceras* (*Orthonychia*) *aequilaterale* (UNC 2295, Keokuk Group, Crawfordsville, Indiana) has a thin, outer shell layer of slightly reclined, irregular to regular simple prismatic structure. This grades inward into calcitic semi-nacre, and then poorly defined, concentric CSF structure. The latter has wide first-order lamellae with second-order lamellae dipping about 9 degrees relative to the depositional surface.

2. Family Platyceratidae? Early Carboniferous.

Burchette and Riding (1977, Fig. 3) described certain Lower Carboniferous, open-coiled, vermiform gastropods with calcitic shells consisting of outer simple prismatic and inner "laminated" [semi-nacreous?] structures, sometimes with intercalated inner prismatic sublayers. Previous authors referred these fossils to the Annelida, *e.g.*, as *Serpula* cf. *advena* Salter and *Spirorbis helicteres* Salter, but Burchette and Riding (1977) indicated that their laminar microstructure differs from serpulids. The microstructure and helical coiling of the early shell suggest gastropod affinities. These fossils are older than the earliest vermetids, which range from ?Upper Cretaceous or Eocene to Recent, so it was concluded that they represent a new group of disjunct, suspension-feeding archaeogastropods.

Of the open-coiled, Paleozoic gastropods summarized by Yochelson (1971), only the Platyceratidae (where open-coiling is weakly developed in *Orthonychia*) are presently known to have a laminar, calcitic structure similar to these fossils. The specimens described by Burchette and Riding (1977) may, therefore, have evolved from the Platyceratidae.

B. Superfamily Microdomatoidea Wenz, 1938. Middle Ordovician - Middle Permian.

According to Knight *et al.* (1960, p. I242) members of this superfamily are nacreous, and this group probably evolved from the Pleurotomarioidea.

C. Superfamily Anomphaloidea Wenz, 1938. Silurian - Middle Permian.

According to Knight *et al.* (1960, p. I243) the inner shell layers in this group are "seemingly nacreous".

D. Superfamily Trochoidea Rafinesque, 1815.

According to Knight *et al.* (1960, p. I246, I271, I274) the inner shell layer in this superfamily may be nacreous or, in the Cyclostrematidae and Phasianellidae, it may be porcelaneous.

1. Family Trochidae Rafinesque, 1815. Triassic - Recent.

***Angaria* Röding, 1798.** According to Bøggild (1930, p. 301), Eocene, Miocene and Recent "*Delphinula*" (= *Angaria*?) generally have an irregularly prismatic outer layer, a nacreous middle layer, and an optically homogeneous inner layer. However, Eocene "*Delphinula*" *callifera* has an outer CL layer and a middle nacreous layer, whereas Eocene "*Delphinula*" *marginata* has no nacre but outer concentric CL and inner radial CL layers.

***Cittarium* Philippi, 1847.** Recent *Cittarium pica* (Linnaeus) has an outermost calcitic "distal" layer consisting of regular simple prisms with transverse striations and imbricated rhombohedral crystals (Erben, 1971). Below this is a layer of aragonitic irregular spherulitic prismatic structure which, in the adult shell, has non-intersecting crystallites. However, in the juvenile, post-protoconch shell, the prismatic structural subunits intersect along the prism boundaries. The inner, aragonitic columnar nacreous layer is associated with a thin layer of intersected crossed blocky structure in the early juvenile shell.

***Trochus* Linné, 1758.** As described by Bøggild (1930, p. 302), Recent *Trochus* has a prismatic outer shell layer, a nacreous middle shell layer, and sometimes also an optically homogeneous innermost layer.

***Umbonium* Link, 1807.** According to Bøggild (1930, p. 303), Recent *Umbonium* "*vestarium*" [= probably *Umbonium vestiarium* (Linné, 1758)] has a concentric CL outer layer, a nacreous middle layer, a more or less optically homogeneous inner layer, and a callus with indistinct, CL structure.

2. Family Stomatellidae Gray, 1840. Triassic - Recent.

***Stomatella* Lamarck, 1816, and *Stomatia* Helbling, 1779.** According to Bøggild (1930, p. 301) Recent *Stomatella* and *Stomatia* have a finely prismatic outer layer and a nacreous inner layer.

3. Family Turbinidae Rafinesque, 1815. Middle Triassic - Recent.

***Turbo* Linné, 1758.** Bøggild (1930, p. 301) noted that Tertiary and Recent species of *Turbo* have an outer layer which is irregularly prismatic or optically grained, and an inner nacreous layer.

4. Family Phasianellidae Swainson, 1840. Paleocene - Recent.

Bøggild (1930, p. 301) described three shell layers in Recent *Phasianella bulloides*: (1) outer optically complex prismatic, (2) middle concentric CL with a zigzag pattern of first-order lamellae as seen in radial, vertical sections, and (3) inner prismatic. Knight *et al.* (1960, p. I274) indicated that shells in this family are entirely porcelaneous.

IV. Suborder Seguenziina Verrill, 1884.

A. Superfamily Sequenzioidea Keen, 1971. Eocene - Recent.

1. Family Seguenziidae Keen, 1971. Eocene - Recent.

Seguenzia Jeffreys, 1876. *Seguenzia* was included in the Trochidae by Thiele (1929-1935), Wenz (1938), Knight *et al.* (1960, p. I251) and Abbott (1974). Keen (1971) erected the family Seguenziidae, superfamily "Seguenziacea" for this taxon, which she noted is nacreous, like many archaeogastropods, but which has a taenioglossate radula, like some mesogastropods. Bandel (1979) noted similarities with the Trochidae in the radula and the sculpture of the protoconch, and he regarded the rod-like fecal pellets and nacreous microstructure of *Sequenzia* as indicative of archaeogastropod affinity. Salvini-Plawen and Haszprunar (1987) added that the nervous system also resembles archaeogastropods, but that the ctenidium, epipodium, excretory organ, and oesophagus exclude the Seguenziidae from the suborder Vetigastropoda (Pleurotomarioidea, Fissurelloidea and Trochoidea). Salvini-Plawen and Haszprunar (1987) placed the Seguenziidae in the archaeogastropod suborder Seguenziina.

In modern *Seguenzia* the protoconch is aragonitic and consists of an outer layer of vertical, irregular simple prisms ("acicular prismatic") underlain by thin layers of homogeneous ("granular"), fine CCL ("dendritic"), and then dissected crossed prismatic ("disected crossed acicular") structure (Bandel, 1979). The latter forms the entire outer shell layer in adult *Seguenzia monocingulata*, but in adult *Seguenzia megaloconcha* the adult outer shell layer is irregular spherulitic prismatic with intersecting, second-order prismatic crystallites at the borders of the spherulitic prisms (Bandel, 1979, p. 51). A columnar nacreous, middle shell layer is secreted after the second juvenile whorl. The nacre tablets are 11 - 12 microns wide. The inner layer of the adult shell is acicular fibrous prismatic in the glaze on the inner lip, and irregular simple prismatic ("blocky prismatic") on the inside of the inner whorls (Bandel, 1979).

V. Suborder Neritopsina Cox and Knight, 1960.

This group is generally considered to have branched early from the Archaeogastropoda (Knight *et al.*, 1960; Fretter, 1965), because of its primitive ctenidial structure (Salvini-Plawen and Haszprunar, 1987, p. 760). Some authors retain these gastropods in the Archaeogastropoda, whereas others place them in the separate order Neritomorpha.

Knight *et al.* (1960, p. I275) indicated that the outer shell layer here is calcitic, and the inner shell layers are aragonitic and "lamellar" (*i.e.*, crossed lamellar) and non-nacreous.

A. Superfamily Neritoidea Rafinesque, 1815.

1. Family Neritidae Rafinesque, 1815. Middle Triassic - Recent.

Bandel (1979, p. 56) pointed out that neritids differ from other archaeogastropods, including the Seguenziidae, in the sculpture of their embryonic shell.

***Nerita* Linné, 1758.** Bøggild (1930, p. 303) described that the calcitic outer shell layer consists of fine, undulating, irregular prisms. This layer is usually thick, except in Eocene *Nerita* (*Semineritina*) *mammaria* Lamarck, 1804. The middle shell layer is concentric CL and the inner layer is CCL.

***Neritina* Lamarck, 1816.** According to Bøggild (1930, p. 304) the calcitic outer shell layer in Eocene - Recent *Neritina* is thin and coarsely prismatic. The middle shell layer is aragonitic concentric CL in most species, but with outer radial CL and inner concentric CL sublayers in *Neritina philippinarum*. An innermost, thin CCL layer may also be present in *Neritina*.

2. Family Neritopsidae Gray, 1847. Middle Devonian - Recent.

***Naticopsis* M'Coy, 1844.** Upper Carboniferous *Naticopsis* (*Naticopsis*) *wortheniana* Knight, 1934 (Buckhorn Asphalt, Oklahoma) has a thin, outer calcitic layer and a thick, inner layer of CL aragonite (Squires, 1976).

***Neritopsis* Grateloup, 1832.** According to Bøggild (1930, p. 303) Miocene *Neritopsis moniliformis* has an outer, radially prismatic layer and an inner, concentric CL layer.

VI. Suborder Murchisoniina Cox and Knight, 1960.

Knight *et al.* (1960, p. I290) indicated that this suborder represents an anatomical and structural grade intermediate between archaeogastropods and caenogastropods.

A. Superfamily Murchisonioidea Koken, 1896.

Knight *et al.* (1960, p. I290) suggested that this superfamily evolved from the Pleurotomarioidea, and evolved into the caenogastropod superfamily Loxonematoidea. The apparent evolutionary change from nacre to CCL structure in Ordovician versus Devonian *Murchsonia* is compatible with this hypothesis.

1. Family Murchisoniidae Koken, 1896. ?Late Cambrian, Early Ordovician - Late Triassic.

***Murchisonia* d'Archiac and de Verneuil, 1841.** Mutvei (1983) described relict nacreous structure in a phosphatized *Murchsonia* from the Upper Ordovician of Graf, Iowa. The nacre tablets are rather wide (over 20 microns) with a radiating ultrastructure, but these large dimensions and the ultrastructure may partly reflect diagenesis.

Middle Devonian *Murchisonia* (*Murchisonia*) *micula* Hall, 1861, from the Hamilton Group of New York (UNC 5737, 5444, 5747) was entirely aragonitic with an outer layer of irregular CCL to cone CCL structure, a thin, middle layer of vertical, irregular simple prismatic structure, and an inner layer of irregular CCL structure.

ORDER DOCOGLOSSA Troschel, 1866

Knight *et al.* (1969, p. I231) placed the suborder Patellina (= Docoglossa) in the Archaeogastropoda. Salvini-Plawen and Haszprunar (1987, 1988) retained the Docoglossa in the Archaeogastropoda on the basis of the structure of the nervous system. However, they noted that the radular apparatus, osphradial fine structure, and protoconch suggest that docoglossans are an early offshoot of the Archaeogastropoda.

Golikov and Starobogatov (1975) removed patelloidean limpets from the Archaeogastropoda and returned them to the order Docoglossa. Lindberg (1986) and Lindberg and Hickman (1986) supported this decision, but Lindberg (1986) rejected the term Docoglossa on the basis that it describes an anatomical and functional grade which is not unique to the Patelloidea. Lindberg (1986) suggested replacing "Docoglossa" with "Patellogastropoda". Haszprunar (1988) and Lindberg (1986, 1988) agreed that these are anatomically primitive snails which diverged early from other gastropods.

The following taxonomic outline of the Docoglossa follows Lindberg (1988, p. 55), except for the present inclusion of the family Metoptomatidae, a possible ancestral stock (see Knight and others, 1960, p. I231). The Metoptomatidae were not discussed by Lindberg (1988).

Docoglossan shell microstructure has been studied by Dall (1871), Pilsbry (1891), Thiem (1917), Bøggild (1930), MacClintock (1967), and Lindberg (1978, *et seq.*). According to Lindberg (1988), judging from the associated comparative anatomy of modern species, the combination of foliated and aragonitic CL structure is primitive in this group, and certain anatomically advanced docoglossans have emphasized one of these two structures at the expense of the other. Thus, the shell microstructure of the Patellidae (foliated plus aragonitic CL) is primitive, whereas that of the Acmaeidae (aragonitic CL without foliated structure) is derived. Lindberg (1988) suggested that patelloid foliated structure evolved directly from nacreous structure, which he presumed was the ancestral condition in the Gastropoda.

On the other hand, if the Acmaeidae evolved considerably earlier than the Patellidae, as indicated by Knight *et al.* (1960), and if early Mesozoic acmaeids resembled Cenozoic acmaeids in lacking foliated structure, then aragonitic CL structure would appear to have evolved before foliated structure in the Docoglossa. By this hypothesis, docoglossan foliated structure more likely evolved from calcitic prismatic structure, as in certain pectinoid bivalves (Waller, 1972; see also Carter, Volume I), than directly from nacreous structure. Unfortunately, nothing is presently known about the shell microstructure of early Mesozoic docoglossans.

I. ?Family Metoptopmatidae Wenz, 1938. Middle Silurian - Middle Permian.

Knight *et al.* (1960, p. I231) suggested that this group evolved from early pleurotomarians or from bellerophonts. The group is unknown microstructurally.

II. Suborder Nacellina Lindberg, 1988.

A. Superfamily Acmaeoidea Forbes, 1850.

1. Family Acmaeidae Carpenter, 1857. Middle Triassic - Recent.

Yochelson (1986) suggested that the modern patelloids, and presumably the acmaeids, evolved from the Middle Ordovician - Early Silurian genus *Archinacella* Ulrich and Scofield. *Archinacella* was formerly regarded as a monoplacophoran, but it has a horseshoe-shaped muscle scar, like the Paleozoic Metoptomatidae and modern patelloids.

Bøggild (1930, p. 306) described the microstructure of several acmaeids, all of which are non-nacreous and have a thin to thick calcitic, outer shell layer. MacClintock (1967) indicated that acmaeids typically have "fibrillar" prismatic structure in a shell layer exterior to their pallial myostracum. The reader is referred to Bøggild (1930) and MacClintock (1967) for detailed descriptions of these shells. Lindberg (1978, *et seq.*) indicated the presence or absence of a calcitic outer layer, which may be simple prismatic, fibrous prismatic or "modified foliate". Lindberg (1978, 1979a,b, 1981) and Lindberg and Hickman (1982) have described the shell microstructure of *Collisella, Notoacmaea, Problacmaea* and *Rhodopetala.*

2. Family Lepetidae Dall, 1869. Miocene - Recent.

Shell microstructure has not been described for this group, but Lindberg's (1988) placement of this family in the Acmaeoidea implies the presence of aragonitic CL structure without foliated structure.

3. Family Lottiidae Gray, 1840. Pliocene - Recent.

According to Lindberg (1988, Fig. 1h), lottiids typically have an outer prismatic shell layer, aragonitic CL structure on both sides of the pallial myostracum, and a "fibrillar" (= fibrous prismatic?) layer between the exterior prismatic layer and the middle CL layer.

B. Superfamily Nacelloidea Thiele, 1891.

1. Family Nacellidae Thiele, 1929. Eocene - Recent.

Knight *et al.* (1960) regarded this group as a subfamily of the Patellidae, but Lindberg and Hickman (1986) placed it in a distinct family. Late Eocene *Cellana ampla* Lindberg and Hickman, 1986, has five shell layers, from exterior to interior: (1) "complex prismatic" (= composite prismatic?), (2) foliated, (3) radial CL (the ventral portion of this layer is sometimes "irregularly tabulate foliated"), (4) myostracum, and (5) radial CL and/or CCL. Lindberg (1988) noted that nacellids lack concentric CL structure.

III. Suborder Patellina von Ihering, 1876.

A. Superfamily Patelloidea Rafinesque, 1815.

1. Family Patellidae Rafinesque, 1815. ?Jurassic. Eocene - Recent.

Bøggild (1930), MacClintock (1967) and Lindberg (1988) described the microstructure of several patellids. These gastropods typically have both foliated and aragonitic CL structure in their shell, but they lack prismatic structure in their outer shell layer.

ORDER UNCERTAIN

A. Superfamily Pseudophoroidea S.A. Miller, 1889. Early Ordovician - Early Carboniferous.

According to Knight *et al.* (1960, p. I297), nacre is present in some members of this superfamily.

B. Superfamily Amberleyoidea Wenz, 1938. Middle Triassic - Oligocene.

1. Family Amberleyidae Wenz, 1938. Middle Triassic - Oligocene.

According to Knight *et al.* (1960, p. I303) amberleyids have nacreous shells.

CLASS PARAGASTROPODA

Linsley and Kier (1984) erected this class for anisostrophic molluscs which are apprarently untorted, judging from inferences of soft anatomy and functional morphology. Linsley and Kier indicated that this class is paraphyletic at the ordinal level, with the Orthostrophina evolving independently of the Hyperstrophina. They added that *Aldanella* (family Aldanellidae) is either a member of the Pelagielloidea, or it is not a mollusc, depending on the unresolved question of whether its aperture was tangential or radial. They agreed with Knight *et al.* (1960) that the Macluritidae gave rise to the Euomphaloidea.

ORDER ORTHOSTROPHINA Linsley and Kier, 1984.

As defined by Linsley and Kier (1984), this order includes the superfamily Pelagielloidea, with the families Pelagiellidae and, tentatively, the Aldanellidae. The Aldanellidae are unknown microstructurally.

A. Superfamily Pelagielloidea Knight, 1956.

1. Family Pelagiellidae Knight, 1956. Early Cambrian (Tommotian) - Middle Cambrian.

***Pelagiella* Matthew, 1895.** Early Cambrian *Pelagiella* has outer radially fibrous and inner concentrically fibrous shell layers which were probably originally aragonitic (Runnegar, 1983, p. 127). The fibers were nearly horizontal and not arranged into crossed lamellae. Early Cambrian *Pelagiella subangulata* (Tate) shows evidence for a predominantly horizontal, radial fibrous prismatic outer shell layer with fibers a few microns wide. The fibers in the inner shell layer are oblique to this direction (Runnegar, 1985, p. 252). In Middle Cambrian *Pelagiella deltoides* Runnegar and Jell, the structure of the outer shell layer is unknown, but the inner layer had flattened, elongate crystallites arranged oblique to the shell margins (Runnegar, 1985, p. 252).

***Costipelagiella* Horny, 1964.** Steinkerns of Middle Cambrian *Costipelagiella* cf. *C. zazvorkai* Horny reveal acicular, more or less radially oriented crystallites in the inner shell layer (MacKinnon (1985, p. 77, Fig. 10D)

ORDER HYPERSTROPHINA Linsley and Kier, 1984.

McLean (1981) removed the superfamily Euomphaloidea from the suborder Macluritina and placed it in the new suborder Euomphalina along with the superfamily Neomphaloidea. Batten (1984a) favored separating the Neomphaloidea from the Euomphaloidea and returning the latter to the Macluritina. Linsley and Kier (1984) tentatively placed the Euomphaloidea (including the Euomphalidae, Omphalotrochidae, Omphalocirridae and Oriostomatidae) in the Order Hyperstrophina, which also includes the Macluritoidea and Onychochiloidea.

According to Linsley and Kier (1984), shells in this order consist of outer calcitic and inner aragonitic, non-nacreous layers, except for the nacreous Oriostomatidae.

A. Superfamily Onychochiloidea Koken, 1925.

1. Family Onychochilidae Koken, 1925. Early to Late Cambrian; Early Devonian.

Knight *et al.* (1960) placed this family in the superfamily Macluritoidea. Ordovician representatives of this family have a calcitic outer shell layer (Wängberg-Eriksson, 1979). Linsley and Kier (1984) suggested that the Onychochilidae gave rise to the Clisospiridae and Macluritidae.

***Yuwenia* Runnegar, 1981.** Early Cambrian *Yuwenia bentleyi* Runnegar, 1981, was regarded by Runnegar (1983) as close to the ancestry of the Macluritoidea. *Yuwenia* had concentric CL structure with first-order lamellae about 5 microns wide (Runnegar, 1985, Fig. 7B). Obscure impressions near the shell margins suggest the presence of a vertical prismatic outer shell layer (*ibid.*, Fig. 7F).

2. Family Clisospiridae Miller, 1889. Early Ordovician - Middle Devonian.

Wängberg-Eriksson (1979) placed this family (including *Clisospira*, *Mimospira* and *Undospira*) in the superfamily Macluritoidea, but Linsley and Kier (1984) placed it in the Onychochiloidea.

***Mimospira* Koken, 1925.** Middle Ordovician *Mimospira tenuistriata* Wängberg-Eriksson, 1979, has a calcitic outer shell layer with reclined to nearly vertical prisms, and an inner, recrystallized layer which was, presumably, originally aragonitic (Wängberg-Eriksson, 1979. Fig. 8F,H).

B. Superfamily Macluritoidea Fischer, 1885.

1. Family Macluritidae Fischer, 1885. Late Cambrian - Ordovician.

Knight *et al.* (1960, p. I187) believed that the Macluritidae evolved from the Bellerophontoidea, independently of the Pleurotomarioidea. However, Linsley and Kier (1984) suggested that the Macluritidae evolved from the Onychochilidae.

Cox and Knight (1960) and Knight *et al.* (1960) indicated that the suborder Macluritina is characterized by an outer calcitic layer and an inner, non-nacreous, aragonitic layer. However, they were probably referring to the Euomphalidae, which are no longer placed in the Macluritidae. Shell microstructural data are unavailable for the taxa which Linsley and Kier (1984) placed in the Macluritoidea *sensu stricto*.

C. ?Superfamily Euomphaloidea de Koninck, 1881.

Knight *et al.* (1960) and Linsley and Kier (1984) suggested that this group evolved from the Macluritidae. Morris and Cleevely (1981) believed that the Euomphalidae, Trochonematidae, and Pseudophoroidea evolved from a common Cambrian ancestor. They also agreed with Sohl (1960) that the Mesozoic genera included in the Euomphaloidea by Knight *et al.* (1960) are misplaced [*e.g.*, *Vivianella*, *Paraviviana* and *Colpomphalus*, formerly of the Helicotomidae; *Discohelix*, *Nummocalcar* (*Nummocalcar*) and *N.* (*Platybasis*), formerly of the Euomphalidae]. Cretaceous *Weeksia* was placed in the family Weeksiidae by Sohl (1960), but was transferred to the Architectonicoidea by Morris and Cleevely (1981). Bandel (1982, 1988) and Bandel and Hemleben (1986) confirmed that several Mesozoic gastropods have been erroneously allied with the Euomphaloidea.

Knight *et al.* (1960, p. I189) indicated that this superfamily is characterized by an outer calcitic shell layer, and an inner lamellar (*i.e.*, crossed lamellar?), non-nacreous layer.

1. Family Euomphalidae de Koninck, 1881. Early Ordovician - Permian, ?Late Triassic.

Bøggild (1930, p. 301) indicated that euomphalids have a calcitic prismatic outer shell layer and an inner aragonitic layer.

***Straparollus* de Montfort, 1810.** Batten (1984a,b) described for Upper Carboniferous *Straparollus* (*Amphiscapha*) *catilloides* (Conrad, 1842) a calcitic outer layer of very fine, vertical fibrous prisms. Locally, this outer layer has the appearance of a fine CCL structure (Fig. 2B in Batten, 1984b, his "wavy aspect of the calcite prisms"). The middle shell layer consists of aragonitic, concentric, linear CL structure (Batten, 1984a, Fig. 17). The inner shell layer is calcitic and is "composed of prisms resembling myostracum." (Batten, 1984b, p. 1189). Batten noted that similar microstructures occur in *Euomphalus*.

Present observations of Permian *Straparollus (Amphiscapha) muricatus* (Knight) (UNC 14888, Beattie Limestone, Kansas) reveal a calcitic, irregular simple prismatic to fibrous prismatic outer shell layer. The inner shell layer is recrystallized.

2. Family Omphalotrochidae Knight, 1945. Devonian - Middle Permian.

According to Knight *et al.* (1960, p. I196) "this family is thought to have been derived from earlier euomphalids, possibly the close allies of *Centrifugus*."

***?Coronopsis* Waterhouse 1963.** This Permian genus was tentatively assigned to this family by Waterhouse (1963). However, it was not listed as a member of this family by Linsley and Kier (1984). *Coronopsis vagrans* Waterhouse, 1963, shows, in a shell now composed entirely of calcite, evidence for outer and inner shell layers, apparently with relict CL structure in the inner layer (Waterhouse, 1963, p. 109, Fig. 36).

3. Family Oriostomatidae Wenz, 1938. Late Silurian - Early Devonian.

According to Knight *et al.* (1960, p. I245) and Linsley and Kier (1984) the inner shell layer is nacreous in this group. Knight *et al.* (1960) placed this family in the Oriostomatoidea, but Linsley and Kier (1984) moved it to the Euomphaloidea.

EXPLANATION OF PLATES

PLATE 122. *Acanthopleura granulata* (Gmelin, 1791), Recent, north side of Florida Keys, Mile Marker 56; UNC 7370 (Figs. A,B) and UNC 13350 (Fig. C). A. Underside of terminal plate, showing the exterior surface of the tegmentum (above) and the outer part of the articulamentum (below). The tegmentum is granular with an irregular simple prismatic substructure. The articulamentum is finely concentric CL to crossed acicular. Preparation method 12. Bar scale = 50 microns. B. Closer view of the contact between the tegmentum and the articulamentum in Figure A. Bar scale = 10 microns. C. Radial, vertical fracture through the outer part of the tegmentum of a medial shell plate, showing transitional homogeneous/fine CCL structure. The arrow points toward the depositional surface. Preparation method 1. Bar scale = 5 microns.

PLATE 123. *Acanthopleura granulata* (Gmelin, 1791), Recent, north side of Florida Keys, Mile Marker 56; medial shell plate, UNC 13350. A. Crossed acicular (CA) to transitional CA/CL structure locally developed as a middle sublayer in the tegmentum, directly below the homogeneous/fine CCL outer sublayer in Plate 122C. Concentric, vertical fracture. Preparation method 1. Bar scale = 5 microns. B. Radial, vertical fracture through the outer tegmentum (upper right), the inner tegmentum (middle) and the articulamentum (lower left), showing irregular simple prismatic (ISP), transitional CA/CL, and composite prismatic (CP) structure in the tegmentum, plus the outer part of the CL layer in the articulamentum (CL). Radial, vertical, medial fracture. Preparation method 1. Bar scale = 10 microns. The arrows in A and B point toward the depositional surface.

PLATE 124. *Acanthopleura granulata* (Gmelin, 1791), Recent, north side of Florida Keys, Mile Marker 56; medial shell plate, UNC 13350. A,B. Rod-type concentric CL structure in the articulamentum as seen in in oblique fractures. Preparation method 1. Bar scales = 5 microns. Note the lack of well-defined second-order lamellae. The arrow in B points toward the depositonal surface.

PLATE 125. *Acanthopleura granulata* (Gmelin, 1791), Recent, north side of Florida Keys, Mile Marker 56; medial shell plate, UNC 13350. A. Oblique, nearly horizontal fracture through the rod-type, concentric CL structure of the articulamentum. Preparation method 1. Bar scale = 5 microns. B. Radial, mostly vertical fracture through the CL to CA structure (upper right and lower left) of the articulamentum, showing a predominantly homogeneous lens of myostracum (middle) with traces of vertical, irregular simple prismatic structure. Preparation method 1. Bar scale = 10 microns. The arrow points toward the depositional surface.

PLATE 126. *Fissidentalium vernedei* (Sowerby), Recent, Japan, UNC 8273. A. Radial, vertical fracture through the thin, fine CCL outer shell layer. Preparation method 1. Bar scale = 1 micron. B. Outer part of the concentric CL middle shell layer, immediately below the fine CCL layer in Figure A, showing irregular first-order lamellae. Preparation method 1. Bar scale = 10 microns. The arrows in A and B point toward the depositional surface.

PLATE 127. *Fissidentalium vernedei* (Sowerby), Recent, Japan, UNC 8273. A. Radial, vertical fracture through the inner part of the concentric CL middle shell layer, showing well-defined second-order lamellae and lath-like third-order lamellae. Preparation method 1. Bar scale = 10 microns. B. Closer view of a single first-order lamella showing the lath-like second-order lamellae. Preparation method 1. Bar scale = 1 micron. C. Radial, vertical fracture through the homogeneous inner shell layer near the posterior shell margin. Preparation method 1. Bar scale = 1 micron. The arrows in A and C point toward the depositional surface.

PLATE 128. *Praematuratropis ovatus* Rollins *et al.*, 1971, Middle Devonian, Marcellus Formation, New York, UNC 5452. A. Relict aragonitic CL structure of the inductura as seen in an oblique, vertical section. Preparation method 2. Bar scale = 10 microns. B. Relict aragonitic fine CCL to irregular CCL structure of the shell layer immediately below the inductura, grading inward into cone CCL structure (below). Preparation method 2. Bar scale = 10 microns. The arrows in A and B point toward the shell interior.

PLATE 129. A-D: *Perotrochus adansonianus* (Crosse and Fischer, 1861), Recent, Florida Straits, UNC 5430. Preparation method 1. Bar scales in A-D represent 5 microns. A-C. Aragonitic vertical to slightly reclined prismatic outer shell layer as seen in a radial, vertical fracture; the shell margin is toward the right. A. Outer part of outer layer, showing irregular simple prismatic and irregular spherulitic prismatic structure. B. Middle part of the outer layer, showing irregular simple prismatic and fine CCL structure. C. Inner part of the outer layer, showing irregular simple prismatic structure; the underlying nacreous layer appears at the bottom of the photograph. D. Rather poorly stacked columnar nacre in the middle shell layer. E-F: *Haliotis rufescens* Swainson, Recent, California, UNC 8803. E. Radial, vertical fracture through the middle of the calcitic, irregularly prismatic, outer shell layer. Each prism shows rhombohedral cleavage planes which reflect the calcitic mineralogy. The shell margin is toward the right. Preparation method 1. Bar scale = 20 microns. F. Radial, vertical fracture through the columnar nacreous middle shell layer. Preparation method 1. Bar scale = 1 micron. The arrows in A-F point toward the shell interior.

PLATE 130. *Bembexia sulcomarginata* (Conrad, 1842), Middle Devonian, Marcellus Formation, Morrisville, New York, UNC 5447. Preparation method 2. Middle and inner shell layers. A. Vertical, nearly concentric section through the columnar nacreous middle shell layer (1), a thin irregular simple prismatic layer (2), a fine CCL layer (3), and an underlying CL to cone CCL to irregular CCL layer (4). Bar scale = 10 microns. B. Closer view of the CL structure of the inner shell layer. Bar scale = 10 microns. Not shown here is the outer shell layer, exterior to the nacreous layer, which is cone CCL to irregular CCL. The arrows point toward the shell interior.

PLATE 131. *Platyceras* (*Orthonychia*) *conicum* (Hall), Middle Devonian, Onondaga Limestone, Williamsville, New York, UNC 1268. Preparation method 1. All shell layers retain their original calcitic mineralogy. A. Concentric, vertical fracture through the homogeneous to irregular simple prismatic outer shell layer (1) which grades (at 2) into the concentric crossed semi-foliated middle shell layer (3). Bar scale = 5 microns. B. Transverse, vertical fracture through the concentric crossed semi-foliated middle shell layer, showing second-order lamellae. Bar scale = 1 micron. C. Horizontal fracture through the semi-nacreous inner shell layer. Bar scale = 10 microns.

PLATE 132. A-B. *Platyceras* (*Orthonychia*) *conicum* (Hall), Middle Devonian, Onondaga Limestone, Williamsville, New York, UNC 1268. Preparation method 3. All shell layers retain their original calcitic mineralogy. A. Radial, vertical section through the concentric crossed semi-foliated middle shell layer showing poorly defined first-order lamellae. The second-order lamellae are gutter-shaped, bowing outward toward the shell exterior. Bar scale = 50 microns. B. Transverse, vertical section through the crossed semi-foliated shell layer in Fig. A, showing the two dip directions of the second-order lamellae. Bar scale = 50 microns. C. *Platyceras (Platyceras) niagarense* (Hall), Middle Silurian, Waldron Shale, Waldron, Indiana, UNC 1244. Radial, vertical section through the outer irregular simple prismatic to regular simple prismatic shell layer (1) and the concentric crossed semi-foliated middle shell layer (2), with outwardly reclined and inwardly inclined first-order lamellae. Bar scale = 100 microns. The thick arrows point toward the the shell interior, and the thin arrows point toward the shell margin.

PLATE 133. *Platyceras carinatum*, Middle Devonian, Moscow Shale, Pavilion, New York, UNC 1266. Preparation method 3. All bar scales = 100 microns. The thick arrows point toward the shell interior; the thin arrows point toward the shell margin. All shell layers retain their original calcitic mineralogy. A. Radial, vertical section showing the irregular simple prismatic outer shell layer (1), the concentric crossed semi-foliated middle shell layer (2), and an irregular simple prismatic (3) to irregular complex crossed semi-foliated (4) inner layer. B. Transverse, vertical section through the same shell layers as in Fig. A. Note the two dip directions of the second-order lamellae in layer 2. C. Horizontal section through layer 2 (concentric crossed semi-foliated) showing the faint outlines of the interdigitating first-order lamellae. D. Horizontal section through the prismatic outer shell layer (above) and the crossed semi-foliated middle shell layer (below). E. Horizontal section through the prismatic myostracal (?) shell layer (left) and the crossed semi-foliated shell layer (right).

PLATE 134. *Platyceras carinatum*, Middle Devonian, Moscow Shale, Pavilion, New York, UNC 1266. Preparation method 1. Bar scales = 5 microns. The thick arrows point toward the shell interior; the thin arrows point toward the shell margin. All shell layers retain their original calcitic mineralogy. A. Radial, vertical fracture showing the irregular simple prismatic outer shell layer (above) and the concentric crossed semi-foliated middle shell layer (below). B. Horizontal fracture through the concentric crossed semi-foliated middle shell layer, showing several second-order lamellae dipping toward the right. The shell margin is toward the bottom. C. Radial, vertical fracture through the irregular complex crossed semi-foliated inner shell layer, showing the varied dip directions of the second-order lamellae.

Part 4
Diagenetic Modifications of Shell Microstructure in the Terebratulida (Brachiopoda, Articulata)
Plates 135-140

Daniéle Gaspard

Université de Paris Sud, Département des Sciences de la Terre
Bât. 504, 91405, Orsay cedex, France
Contribution aux travaux de l'u.r.a. no. 157 du C.N.R.S., 21000 Dijon
Université de Bourgogne

Shell microstructural studies of Recent and fossil brachiopods shed light on the processes and products of biomineralization, and enlarge our field of descriptive characters. The various inorganic and organic components of unaltered terebratulid shells are described by Williams in Volume I and in Part 5 of this Atlas. The present chapter compares well-preserved and partially altered terebratulid shell microstructures to provide a sound basis for interpreting their *post mortem* diagenesis and fossilization.

Careful investigation of sections and depositional surfaces of terebratulid secondary and tertiary shell layers shows that their calcitic fibers and prisms are not internally homogeneous, although they appear as such crystallographically. In fact, their initial (juvenile) ends show minute calcitic seeds or granules embedded in a discontinuous organic matrix, the latter being visible in rows of fibers near the boundary between the primary and secondary shell layers (Plate 135C-F). This organic matrix disappears away from the boundary with the primary layer as the calcitic seeds become more numerous and coalesce in the secondary layer. These seeds have been interpreted as the elementary units of carbonate secretion (Gaspard, 1986). According to the scheme of calcification described by Williams (1971), they are secreted by the microvilli of the active surfaces of mantle epithelial cells. Near the boundary with the primary shell layer, the outlines or organic sheaths of the fibers appear before the fibers develop (Plate 135A,B). Biochemical analysis confirms the presence of an insoluble organic matrix around the fibers (Jope, 1971). Thus, terebratulid shell construction takes place at three distinct levels. At the elementary level, calcitic seeds are secreted (Plate 135). At the intermediate level, these seeds are grouped into internal growth bands (Plates 138A,B,D-F; 139D). At the highest level, the growth bands accumulate to build fibers in protein sheaths or a prismatic structural element.

For paleontologists, the study of Recent terebratulids facilitates recognition of diagenetic modifications in fossil shells. The early stages of diagenesis can sometimes be seen in modern shells degraded by microboring organisms and showing progressive modification of the original calcitic granules or seeds into spike-like or blocky crystallites (Plate 137E). Early diagenesis is shown in Pleistocene and Cretaceous terebratulid shells in Plates 136F and Plate 137F. In the same manner, internal growth banding in terebratulids may be well-preserved, partly preserved, or entirely obliterated according to the degree and nature of recrystallization (Plates 136D, 137F, 138C).

The protein sheaths which envelope calcitic fibers (Plate 136A) and their relationship with caeca (Plate 140E) are sometimes well-preserved in fossil shells, which more often are partly or entirely degraded by microboring organisms and bacteria (Plate 136B), or which have become silicified (Plates 137D, 136C). In the latter case, epigenesis is initiated through the formation of silica granules (Plate 136F; see also Gaspard, Volume I), which sometimes coalesce to form nodules, or which disappear, thereby resulting in regrouping of the fibers (Plate 136D). In some instances, flexible organic matrices and siliceous nodules occur side-by-side in the same part of the shell (Plate 136E), thereby indicating differential chemical modification of the shell, a phenomenon called "paleization" by Florkin (1969).

To better interpret microstructures in fossil shells, one must also bear in mind that the fibrous elements of the standard terebratulid secondary shell layer may be modified in size and shape in muscle fields, dental sockets, transverse bands, etc. (Plate 140F), as a consequence of variations in the secretory regime of the mantle epithelial cells.

Primary, granular shell layers are less complex than secondary, fibrous shell layers in the Terebratulida. The constituent granules in the primary layers may be rounded or acicular, and their oblique arrangements in one or two directions (Plate 140B) may be so strongly modified by recrystallization that they are no longer discernable (Plate 140A).

The fibrous and prismatic layers in terebratulid shells are crossed by the classical network of puncta, and in some cases also by a network of micropuncta, each of which is lined by mantle microcaeca (Plates 137A,B; 139E; see also Gaspard, 1973). Microcaeca are more susceptible to diagenesis than the mantle caeca. In the fossil state, they may be well-preserved, but they are more commonly poorly preserved in recrystallized fragments (Plate 137C), or entirely obliterated by diagenesis.

The caecal brush at the distal end of a punctum may be exceptionally well-preserved in fossil terebratulids (Plate 140C). However, traces of tubules in the perforate calcitic canopy of the caecal brush (Plate 140D) are not generally seen in fossil shells, where traces of the vesicular periostracum are rarely observed (Plate 139A-C).

Finally, mantle cell divisions in terebratulids are expressed on the depositional surface and in sections as dividing fibers or puncta (Plate 139E-F).

EXPLANATION OF PLATES

PLATE 135. A. Edge of the internal surface of a valve of *Dallina septigera* (Lovèn), Recent, Gulf of Gascogne, France (Thalassa cruise), showing a granular aspect. The boundary between the primary and secondary layers is marked by a polygonal network of organic sheaths around the juvenile fibers. Packaging of the coarsely agglomerated granules constitutes the first stage in edification of the fibers of the secondary shell layer. Preparation method 1. B. Internal surface of a valve of *Macandrevia africana* Coop., Recent, Angola Basin, SE Atlantic (Walda cruise), showing incomplete organic sheets (o.s.) in a granular layer. This expresses the junction of the primary layer (p.l.) with the secondary layer (s.l.) and corresponds to modification of the secretory regime of the anterior part of the outer epithelial cells. Preparation method 1. C. Edge of the internal surface of *Macandrevia africana* Coop. showing the primary layer (upper right corner) and terminal faces of fibers (fc) in the secondary layer with well-marked impressions of the corresponding outer epithelial cells. These impressions have a granular aspect. Preparation method 1. D. *Macandrevia africana* Coop., showing detail of the anterior edge of the terminal face of a fiber, showing calcitic granules or seeds disposed in radiating lines. The calcitic granules (c.g.) are embedded in a temporary organic matrix. Preparation method 1. E. Internal surface of *Terebratulina retusa* (Linné), Recent, near Corsica (Bracors cruise), Mediterranean Sea, showing fibers surrounded by their proteinic sheath and the slightly granular terminal face. Preparation method 6. F. Detail of the terminal face of a fiber of a juvenile specimen of *Gryphus vitreus* (Born), Recent, Mediterranean (Bracors cruise); compare the sizes of the calcitic seeds in the various species illustrated. Preparation method 1.

PLATE 136. A. Transverse section of a brachial valve of *Loriolithyris* sp., Lower Cretaceous, Djebel Megrez, Maroc, showing well-preserved protein sheaths in the secondary layer. Internal surface of the valve at the bottom of the micrograph. Preparation method 5. B. Transverse section of *Musculina sanctaecrucis* (Catzigras), Hauterivian, Auxerre, France, showing the degradation of remaining organic sheaths around the anvil-type fibers of the secondary layer. Preparation method 5. C. Section of *Sellithyris cenomanensis* Gaspard, Middle Cenomanian, Maine et Loire, France, showing organic sheets disposed as in a spiral staircase around a punctum. Preparation method 4. D. Transverse section of *Phaseolina phaseolina* (Lamarck), Upper Cenomanian, Velim, near Kolin, Czechoslovakia, showing loss of the anvil shape of the fibers by recrystallization following disappearance of the protein matrix. Preparation method 5. E. Section of *Phaseolina phaseolina* (Lamarck), Upper Cenomanian, Port des Barques, Charente Maritime, France, showing the primary layer (p.l.) underlain by the secondary layer (s.l.), within which occur flexible organic sheets (o.s.) and a diagenetic development of siliceous nodules (s.n.). Preparation method 4. F. Transverse section of *Phaseolina phaseolina* (Lamarck), Czechoslovakia, showing initial silicification in the form of granules at the sites of former organic sheaths. Preparation method 5.

PLATE 137. A. Transverse section of *Megerlia truncata* (Linné), Recent, near Corsica (Bracors cruise), Mediterranean, showing two micropunctae (mp) traversing stacked fibers. Each fiber crossing is shown by a ring along the micropunctum. Preparation method 5. B. Detail of the internal surface of *Dallithyris murrayi* Muir-wood, Recent, Maldive Islands, showing radiating pits around micropuncta (mp) crossing prisms and the lumen of a punctum. Preparation method 4. C. Section of *Sellithyris cenomanensis* Gaspard, Middle Cenomanian, Ile Madame, Charente Maritime, France, showing fragmentation of a micropunctum (mp) during diagenesis. Preparation method 4. D. Medial, longitudinal section of *Gemmarcula menardi* (Lamarck), Middle Cenomanian, Sarthe, France, showing part of the previous proteinic sheets around fibers in transverse section, preserved as a siliceous network. Preparation method 5. E. Transverse section of *Frenulina sanguinolenta* (Gmelin), Recent, off New Caledonia, showing early diagenesis of the shell in the form of blocky crystallites, after biodegradation by microboring organisms. Preparation method 5. F. Longitudinal section of a pedical valve of *Gryphus minor* (Philippi), Lower Pleistocene, Capa del Armi, Italia, showing recrystallized prisms with well-marked growth lines, and granules modified into spike-like forms. Preparation method 5.

PLATE 138. A. Longitudinal section of *Terebratulina hataiana* Coop., Recent, near the Philippines, showing daily growth lines that extend from the granular primary layer (p.l.) into the fibrous secondary layer (s.l.). Preparation method 4. B. Detail of the distal part of the terminal face of a fiber on the internal surface of *Gryphus vitreus* (Born), showing the stacking of diurnal growth lines. Preparation method 1. C. Section of *Sellithyris tornacensis* (d'Arch.), Tourtia of Tournai, Cretaceous, Montignies-sur-roc, Belgium, showing well-marked growth lines in certain fibers in spite of the progressive diagenesis. Preparation method 4. D. Longitudinal section of a shell of *Terebratulina retusa* (Linné) near the articulation; Recent, near Corsica, showing intermediate growth lines. Preparation method 5. E. Fractured margin of the brachial valve of *Campages furcifera* Hedley, Recent, off New Caledonia (Musorstom cruise), showing the granular primary layer, below, and broken fibers in transverse section with arched diurnal growth lines, marked by the calcitic seeds (elementary units of carbonate secretion). The internal surface of the valve is toward the upper part of the micrograph. Preparation method 8. F. Section of *Macandrevia africana* Coop., Recent, SE Atlantic, showing growth lines and disturbance lines. Preparation method 7.

PLATE 139. A. External surface of *Megerlia truncata* (Linné), Recent, near Corsica, Mediterranean Sea (Bracors cruise), showing the vesicular periostracum. Preparation method 10. B. External surface of a pedical valve of *Campages furcifera* Hed., showing the elongated vesicles of the periostracum arranged in well-defined rows perpendicular to the growth lines. Preparation method 1. C. External surface of *Sellithyris cenomanensis* Gaspard, Upper Cenomanian, Sarthe, France, showing traces of the vesicular periostracum on the primary layer crossed by two puncta (pu). Preparation method 1. D. Fragment of a pedical valve of *Campages furcifera* Hed., showing a fiber in longitudinal section, with its organic sheath removed, and exhibiting growth lines underlined by granules or calcitic seeds. Preparation method 9. E. Section of *Megerlia truncata* (Linné), showing a dividing punctum and micropuncta (mp) crossing the fibrous secondary layer. The head of one branch of the punctum is terminated by short tubules of the caecal brush in the primary layer, marked by disturbance growth lines. Preparation method 5. F. Internal surface near the margin of the brachial valve of *Campages furcifera* Hed., showing effects of cell division. Preparation method 8.

Plate 140. A. Section of *Loriolithyris valdensis* (Lor.), Valanginian, Arzier, Canton of Vaud, Switzerland, showing the primary layer (p.l.) with oblique acicular crystallites underlain by the secondary layer (s.l.) both crossed by large puncta and all strongly recrystallized. Preparation method 4. B. Section of *Sellithyris cenomanensis* Gaspard, Cenomanian, Briollay, Maine & Loire, France, showing the acicular crystallites of the primary layer arranged in a chevron pattern. Preparation method 4. C. Section of *Sellithyris tornacensis* (d'Arch.), Cenomanian, Montignies-sur-roc, Belgium, showing a well-preserved caecal brush at the distal end of a punctum. Preparation method 4. D. External surface of *Terebratulina hataiana* Coop., Recent, off the Philippines, showing a perforate calcite canopy covering the distal end of a punctum. Preparation method 1. E. Section of *Macandrevia africana* Coop., showing the distal part of a caecum and the organic sheaths of the fibers surrounding it. Preparation method 5. F. Modification of the shape of fibers and of their terminal faces on the transverse band of *Terebratulina retusa* (Linné), Recent, near Oban, Scotland. Preparation method 1.

Part 5
Brachiopoda and Bryozoa
Plates 141-156

Alwyn Williams

Department of Geology & Applied Geology
The University, Glasgow G12 8QQ, Scotland

This portion of the Atlas supplements the illustrations of lophophorate skeletal microstructure in Volume I. Plates 141-149 show shell and mantle calcification in articulate and inartculate brachiopods. Plates 150-156 illustrate various portions of mineralized bryozoan skeletons. Numerical codes for specimen preparation method are explained in the introduction to this volume. The index to the bibliography in Volume I may be consulted to locate additional illustrations of brachiopod shell microstructure.

EXPLANATION OF PLATES

Section 1. Brachiopoda.

PLATE 141. A. Internal surface at the anterior shell margin (top left corner) of *Gryphus vitreus* (Born), Recent, mid-Atlantic, south of Portugal, showing the junction between the granular primary and fibrous secondary layers with the beginnings of fibers (fe) especially around a punctum. Preparation method 12. B. External surface of *Waltonia inconspicua* (Sowerby), Recent, Lyttleton Harbour, New Zealand, showing the pitted primary shell of vertical, acicular crystallites divided into strips by radial channels (ch) (probably coincident with outer epithelial intercellular boundaries) beneath torn periostracum with collapsed vesicles (ve). Preparation method 11. C. Section of *Notosaria nigricans* (Sowerby), Recent, Lyttleton Harbour, New Zealand, showing the primary layer of oblique, sporadically banded, acicular crystallites and anvil-type fibers of the secondary layer; exterior of valve beyond the top edge of the micrograph. Preparation method 14. D. Section of valve of *Liothyrella neozelanica* Thomson, Recent, South Island, New Zealand, showing acicular crystallites of the primary layer in relation to the first-formed part of an anvil-type fiber of the secondary layer seen in longitudinal section; exterior of valve lies beyond the top edge of the micrograph. Preparation method 14. E. Detail of the internal surface of the antero-medial edge of *Liothyrella neozelanica* showing the junction between the primary layer of oblique, acicular crystallites (below) and the first-formed secondary anvil-type fibers (above). Preparation method 12. F. Edge of brachial valve of *Thecidellina barretti* (Davidson), Recent, Discovery Bay, Jamaica, showing the mixture of vertical or oblique acicular crystallites, granules (gc) and lenses (le) making up the primary layer. Preparation method 12.

PLATE 142. A. Edge of brachial valve of *Thecidellina barretti* showing oblique acicular crystallites in a tubercle of the primary layer. Preparation method 12. B. Section of a pedicle valve of *Thecidellina barretti* showing the attachment of the periostracum (pe) to the substrate (su) with overlying growth-banded primary shells of near vertical acicular crystallites. Preparation method 14. C. Internal surface of the primary shell of a brachial valve of *Thecidellina barretti* in rhombic arrays. Preparation method 12. D. Section of a brachial valve of *Thecidellina barretti* showing a transgression (tr) of growth bands (gw) in the primary shell of acicular crystallites. Preparation method 14. E. Antero-medial (toward the left margin) internal surface within the edge of the brachial valve of *Neocrania anomala* (Muller), Recent, Firth of Clyde, Scotland, showing the arrangement of acicular crystallites as spines to narrow laminar laths in the primary layer. Preparation method 2. F. Submedial section of the brachial valve of *Neocrania anomala* showing the junction between the primary layer with its acicular crystallites of calcite (left) and the laminae of the secondary layer (right).

PLATE 143. A. Internal surface near the edge of the brachial valve of *Neocrania anomala* (beyond top left edge of micrograph) showing the overlapping arrangement of laminae just within the primary-secondary junction. Preparation method 12. B. Antero-medial internal surface (towards left) of *Notosaria nigricans* showing stacking of the anvil-shaped fibers with their terminal faces subtended within a posterior rhombic angle. Preparation method 12. C. Internal surface of a valve (lateral edge beyond left edge of micrograph) of *Liothyrella neozelanica* showing the disposition of secondary fibers, some of which have truncated and exaggerated terminal faces (fc) and carry evidence of cell division (mi). Preparation method 12. D. Section of the secondary shell of *Terebratulina retusa* (Linné), Recent, Isle of Cumbrae, Scotland, showing fibers in characteristic anvil-like transverse and elongated oblique sections with the calcite pads (pd) of a growth regression in the lower right hand corner of the micrograph, which is also the direction of the internal surface of the valve. Preparation method 14. E. Secondary fibers in the tooth of a pedicle valve of *Thecidellina barretti* showing their acicular crystallite fine structure. Preparation method 12. F. Section of the middle part of the inner socket ridge of a brachial valve of *Thecidellina barretti*, showing its spherulitic acicular crystallite fine structure. Preparation method 12.

PLATE 144. A. Section of *Gryphus vitreus* showing the primary layer of vertical acicular crystallites (above) underlain by the secondary layer of anvil-type fibers seen in transverse, oblique and longitudinal section, passing into the tertiary layer of growth-banded, prismatic calcite. Puncta partly filled with resin occur in the top right and left of the micrograph. Preparation method 14. B. Internal surface of the tertiary layer of *Gryphus vitreus* showing the discrete nature of the prisms with non-aligned rhombohedral cleavage. Preparation method 12. C. Internal surface of *Liothyrella neozelanica* showing the transitional area between the secondary and tertiary layers characterized by malformed fibers with curved outlines of epithelial cells impressed as pitted grooves and ridges on a mosaic of expanded terminal faces. Preparation method 12. D. Internal surface of a brachial valve of *Neocrania anomala* showing evidence for the simultaneous growth of several laminae along screw dislocation edges. Preparation method 12. E. Section of the laminar secondary layer of *Neocrania anomala*; valve exterior is beyond the top of the micrograph. Preparation method 14. F. Section of a brachial valve of *Neocrania anomala* showing a lens of cleaved calcite (myotest) surrounded by laminae of the secondary layer; external surface of valve beyond the top left corner of the micrograph. Preparation method 14.

PLATE 145. A. Fracture section slightly oblique to the interior of *Gacella insolita* Williams, Upper Ordovician, Stinchar Limestone, Brochloch, Scotland, showing how successive laminae are composed of arrays of parallel-sided blades terminating in rhombohedral angles, and usually set at acute angles from one lamina to the next (cross-bladed). Preparation method 13. B. Fracture section slightly oblique to the interior of *Triplesia extans* Emmons, Upper Ordovician (Rockland), Lowville, New York, showing a succession of cross-bladed laminae composed of amalgamated laths. Preparation method 13. C. Internal surface of *Schellwienella aspis* (Smythe), Lower Carboniferous, Lower Limestone Group, Lennoxtown, Scotland, showing the overlapping arrangement of laminae with rhombohedral angles marking the ends of individual blades. Preparation method 13. D. Section of *Derbya* cf. *cymbula* Hall and Clarke, Lower Permian, Putnam, Texas, showing a laminar succession. Preparation method 4. E. Fracture section slightly oblique to the interior of *Strophomena oklahomensis* Cooper, Upper Ordovician (Bromide), Rock Crossing, Oklahoma, showing how the blades of one lamina leave traces of their lateral boundaries on contiguous laminae. Preparation method 13. F. Section of *Strophomena oklahomensis* Cooper showing a recrystallized primary layer (pl) with traces of acicular crystallites and a finely laminated secondary layer. Preparation method 14.

PLATE 146. A-B. Internal surfaces of *Liothyrella neozelanica* showing how secondary fibers at the margin of a muscle field begin to lose their characteristic terminal faces (A) through differential secretion and resorption and pass into radially disposed ridges separating linearly arranged pits (B). Preparation method 12. C. Internal surface of *Gacella insolita* showing linearly arranged myotest ridges. Preparation method 12. D. Internal surface of a brachial valve of *Lacazella mediterranea* (Risso), Recent, Monaco, France, showing the arrangement of pits and ridges in the myotest. Preparation method 12. E-F. Internal and vertical fracture surfaces of a brachial valve of *Neocrania anomala* showing the raised subcircular ridges of the myotest and the vertical disposition of acicular crystallites in the ridges and the polygonal mounds. Preparation methods 12 and 13, respectively.

PLATE 147. A. Internal surface just within the junction of the primary (below) and secondary layers of *Liothyrella neozelanica* showing the growth of the fibers in relation to a punctum floored by a canopy (ca). Preparation method 12. B. Internal surface just within the junction of the primary (bottom right) and secondary layers of *Waltonia inconspicua* showing the perforated canopy of punctum. Preparation method 12. C-D. Skeletal traces of a brush seen as perforations of an external surface and as canals in a fracture surface of the canopy of a punctum penetrating the brachial valve of *Lacazella mediterranea*. E. Fracture surface slightly inclined to the internal surface of *Liothyrella neozelanica* showing the disposition of fibers of the secondary layer relative to an array of puncta. Preparation method 13. F. Section of the primary layer of the brachial valve of *Neocrania anomala* showing the branches and tubules of a punctum filled with resin; external surface of the valve along the top margin of micrograph. Preparation method 14.

PLATE 148. A. Section of the secondary laminar layer of a brachial valve of *Leptodus* cf. *richtofeni* Kayser, Permian, Sosio Beds, Sicily, showing a taleola (ta). Preparation method 14. B. Section of the secondary fibrous layer of *Sowerbyella variabilis* Cooper, Upper Ordovician (Bromide), Rock Crossing, Oklahoma, showing a taleola (ta). Preparation method 14. C-D. Spicule (C) from the mantle of *Terebratulina retusa* (Linné), Recent, Isle of Cumbrae, Scotland, and an enlargement (D) showing the acicular crystallites with rhombohedral cleavage beneath an external homogeneous skin. Preparation method 12. E. Section of *Notosaria nigricans* showing the calcite pads (pd) associated with a mantle regression separating an older secondary fibrous layer (on the left) from a younger primary layer (pl); valve exterior is beyond the left margin of the micrograph. Preparation method 14. F. Internal surface of a brachial valve of *Lacazella mediterranea* showing discrete rhombohedral growths. Preparation method 12.

PLATE 149. A. Vertical and fracture section of *Lingula anatina* Lamarck, Recent, beach detritus, Singapore, showing a typical shell succession of alternating organic and mineral layers seen as electron-dense and electron-light bands, respectively. Preparation methods 12 and 14. B. Fracture section slightly oblique to the internal surface of *Lingula anatina* showing an organic layer (ol) on a layer of apatite with conchoidally fractured surfaces and a vertical cleavage which imparts a strong lineation (ln) on the interface with the organic layer, which is also pitted by fine puncta (pa). Preparation method 13. C. Surface of a layer of apatite in *Lingula anatina* showing two lineations, one of which may be cleavage, the other traces of the overlying organic layer; oval outlines of two puncta (pu) are also visible. Preparation method 13. D. Ridges and pits impressed on successive layers of apatite in a left valve of *Lingula anatina*, which represent aligned constituents of interleaved organic layers or of the plasmalemmas of outer epithelium. Preparation method 13. E. Hexagonally packed ridges on a layer of apatite in *Lingula anatina*, representing outlines of epithelial cells with puncta (pu). Preparation method 13. F. Casts of outer epithelium on successive layers of apatite at the internal surface of *Lingulella hespera* Williams and Curry, Lower Ordovician, Tourmakeady Limestone, Ireland. Preparation method 12.

Section 2. Bryozoa.

PLATE 150. A. Fracture section of a lateral wall of *Membranipora membranacea* (Linné), Recent, Portaferry, Northern Ireland, showing the periostracum (pe) and lenses of vertical acicular crystallites (ac) constituting the primary layer. Preparation method 13. B. Internal surface of a transverse wall of *Membranipora membranacea* showing the scarp-like rhombic fronts developed in successive layers of aggregated, vertical crystallites. Preparation method 12. C. External surface of a lateral wall of *Electra pilosa* (Linné), Recent, Portaferry, Northern Ireland, showing the granular texture of crystallites of the primary layer seen end-on. Preparation method 12. D. Internal surface of a zoeecial wall of *Cupuladria biporosa* Canu and Bassler, Pointe-Noire, Ghana, showing the interleaved disposition of gently inclined aragonite prisms (ap) of the primary layer. Preparation method 12. E. Section of the basal part of the pad of *Cupuladria biporosa* showing an organic partition extending through the primary shell of acicular crystallites; the basal surface is to the right. Preparation method 14. F. Detail of part of the section illustrated in E showing the spherulitic disposition of the acicular crystallites.

PLATE 151. A. Fracture section of the walls of a zooecial chamber of *Cellaria fistulosa* (Linné), Recent, Portaferry, Northern Ireland, showing the laminar secondary layer (sl) and the prismatic tertiary layer (tl). Preparation method 13. B. Internal surface of a zooecial chamber of *Cellaria fistulosa* showing the stacking of acicular crystallites in the prismatic tertiary layer. Preparation method 12. C. Transverse section of the cryptocyst of *Cellaria fistulosa* showing the symmetrical nature of its skeletal succession with secondary (sl) and tertiary (tl) layers developed on either side of the medial primary layer (pl). Preparation method 14. D. Undersurface of the base of a colony of *Celleporella pumicosa* Hincks, Recent, Strangford, Northern Ireland, showing the disposition of inclined tabular crystallites. Preparation method 12. E. Internal surface of a frontal wall of *Celleporella hyalina* (Linné), Recent, Plymouth, England, showing overlapping laminae of the secondary layer. Preparation method 12. F. Internal surface of a lateral wall of *Celleporella hyalina* showing a muscle scar. Preparation method 12.

PLATE 152. A. Fracture section of the basal layer of *Sertella carinata* (MacGillivray), Recent, Murray Island, Australia, showing alternating bands of differently disposed, acicular crystallites of calcite. Preparation method 13. B. Section of the junction of organic partitions of adjacent zooecial tubes of *Iodictyum sanguineum* (Ortmann), Recent, Misaki, Japan, showing the spherulitic arrangement of acicular crystallites in the primary layer in relation to the partitions. Preparation method 14. C. Section of the basal layer of *Sertella carinata* showing a tightly folded organic partition in the primary layer; zooecial chambers and frontal surface lie beyond the top of the micrograph. Preparation method 14. D. Section of the zooecia of *Sertella carinata* showing the skeletal succession of organic partition (op), primary (pl) and secondary (sl) layers between adjacent zooecial chambers. Preparation method 14. E. External surface of a zooecium of *Crisidia cornuta* (Linné), Recent, Ardkeen, Northern Ireland, showing strips of acicular crystallites of the primary layer, with growth banding (gw) and the external opening of a punctum (pu). Preparation method 12. F. Internal view of a punctum penetrating the laminae of the secondary layer of *Crisidia cornuta*. Preparation method 12.

PLATE 153. A. Internal surface of a zooecium of *Crisidia cornuta* showing the spirally growing, overlapping tablets of the laminar secondary layer. Preparation method 12. B. Internal surface of a zooecium of *Crisidia cornuta* showing the keeled blades with growth banding (gw) in the laminar secondary layer. Preparation method 12. C. Internal surface of a zooecium of *Plagioecia patina* (Lamarck), Recent, Killyleagh, Northern Ireland, showing right- and left-hand screw dislocations of tablets in the laminar secondary layer. Preparation method 12. D. Section of zooecial walls of *Crisidia cornuta* showing a laminar secondary layer composed of tablets (tb) and keeled blades (kb). Preparation method 14. E. External surface of a zooecial wall of *Plagioecia patina* showing the star-shaped distal opening of a punctum surrounded by acicular crystallites of the primary layer. Preparation method 12. F. Section of a zooecial wall of *Plagioecia patina* showing the skeletal succession of primary (pl) and secondary (sl) layers. Preparation method 14.

PLATE 154. A. External surface of a zooecium of "*Entalophora*" sp., Recent, Atlantic, showing the generally inclined, spherulitically disposed, acicular crystallites of the primary layer and the distal opening of a punctum (pu). Preparation method 12. B-C. Internal zooecial surface of "*Entalophora*" sp., showing growth-banded blades of the laminar secondary layer, some with keels and branches. Preparation method 12. D. Section of the zooecial wall of "*Entalophora*" sp., showing a laminar secondary layer with both keeled and flat blades. Preparation method 14. E. Section of the zooecial wall of "*Entalophora*" sp. showing the skeletal succession of primary (pl) and secondary (sl) shell. Preparation method 14. F. Section of the base of a colony of *Lichenopora radiata* (Audouin), Recent, Plymouth, England, showing the succession of a primary layer (pl) of acicular crystallites and a secondary laminar layer (sl) of keeled blades and tablets penetrated by taleolae (ta). Preparation method 14.

PLATE 155. A. Internal surface of a zooecium of *Lichenopora radiata* showing tubercles (te) with cores of granular calcite forming pseudopuncta (pn) within the laminar secondary layer. Preparation method 12. B. Section of the basal wall of *Lichenopora radiata* showing a pseudopunctum (pn) of granular calcite within the laminar succession of the secondary layer; internal surface lies beyond the top right corner of micrograph. Preparation method 14. C. External surface of a zooecial branch of *Hornera frondiculata* (Lamouroux), Recent, Malta, showing overlapping blades and tablets of the secondary layer. Preparation method 12. D. Section of a branch of *Hornera frondiculata* showing arrays of tablets and blades making up the laminae of the secondary layer. Preparation method 14. E. Internal surface of a zooecium of *Hornera frondiculata* showing tabular laminae of the secondary layer in attitudes of spiral growth. Preparation method 12. F. Internal surface of a zooecium of *Heteropora* sp., Recent, Antarctic, showing discrete blades set at acute angles to one another making up the laminae of the secondary layer. Preparation method 12.

PLATE 156. A. Section of zooecial wall of *Heteropora* sp. showing the predominance of keeled plates in the laminar succession of the secondary layer. Preparation method 14. B. Internal surface of a zooecium of *Heteropora* sp. showing that the overlapping laminae of the secondary layer are composed of laterally amalgamating as well as keeled blades, all with strong growth banding. Preparation method 12. C. Gently inclined fracture section of a zooecial wall of *Heteropora* sp. showing successive laminae of the secondary layer consisting of laterally amalgamated laths in cross-bladed arrays. Preparation method 13. D. Internal surface of a zooecial wall of *Heteropora* sp. showing a mural pore (ml) and hook-like calcitic projections arising from the floor of the secondary layer which is paved with narrow, discrete blades. Preparation method 12. E. Section of the frontal wall of a zooecium of *Polypora corticosa* Ulrich, Mississippian, Glen Dean Limestone, Eckerty, Indiana, showing "stylets" (ta; compare taleolae) deflecting laminar secondary shell; obverse surface beyond the top of the micrograph. Preparation method 14. F. Section of a zooecial wall of *Constellaria constellata* (Van Cleve), Ordovician, Maysville Group, Cincinnati, Ohio, showing a symmetrical skeletal succession of granular, medial primary (pl) and flanking laminar secondary layers. Preparation method 14.

Part 6
Arthropoda (Crustacea and Trilobita)
Plates 157-161

John E. Dalingwater

Department of Environmental Biology
University of Manchester, M13 9PL, England

Harry Mutvei

Sektionen för Paleozoologi, Naturhistoriska Riksmuseet
104 05 Stockholm, Sweden

Introduction

The five plates in this section show a bias which reflects the research interests of the authors: decapod crustaceans and trilobites. All the material figured, except one piece, has been especially prepared for this atlas. Indeed, preparations of the Cambrian trilobite *Ellipsocephalus* show details not previously published: in particular, the clearly laminate nature of an (apatitic) outer layer and the apparent penetration of this layer by calcitic crystallites from the principal layer of the cuticle. These and other aspects will be more fully analyzed in a future publication, but are included here to add a current flavor to the outline of trilobite cuticle structure given in the arthropod chapter in Volume I. The other plates are intended to provide further illustration and illumination to the information presented in Volume I.

The scale bars in Plates 157-161 are labelled in microns. In the plate explanations, "vertical" indicates perpendicular to the plane of the cuticle, and "transverse" indicates perpendicular to the long axis of the body or limb.

Guide to the Literature. The reasons for selecting particular species for study have often been arbitrary for many arthropod groups, and the findings from these species have been considered applicable to the group as a whole. Therefore, it seems appropriate for us to refer only to higher taxa in this section. We further restrict ourselves by including only the most recent significant contributions for each major group; these, naturally, give leads into important areas of the earlier literature. Additional references are listed in the master bibliography and index in Volume I. **Grade Arthropoda** (general works): Neville (1975), Hackman (1984). **Phylum Trilobita**: Miller and Clarkson (1980), Mutvei (1981), Wilmot and Fallick (1989). **Phylum Crustacea, Class Ostracoda**: Rosenfeld (1979), Bate and Sheppard (1982), Sohn and Kornicker (1988); **Class Malacostraca**: Giraud-Guille (1984a,b), Roer and Dillaman (1984), Greenaway (1985), Compère and Goffinet (1987).

EXPLANATION OF PLATES

PLATE 157. Transverse, vertical breaks of great chela protopodite cuticle of the Recent crayfish *Austropotamobius pallipes* (Lereboullet). Preparation method 1. A. Complete thickness of the cuticle, showing the epicuticle (Ep), exocuticle (Ex), calcified zone (CZ), and membranous layer (ML). B. Detail of the exocuticle. C. Detail of the calcified zone. D. Detail of the innermost part of the calcified zone and the membranous layer.

PLATE 158. Vertical sections of abdominal cuticle of the Recent crab *Carcinus maenas* (Linné). Preparation method 17. A. Exocuticle lamina units. B. Detail of exocuticular microfibrillar arrangement. C. Calcified zone lamina unit. D. Macrofibers of the calcified zone consisting of microfibrils. E. Inner part of the membranous layer. F. Membranous layer microfibrils.

PLATE 159. A. Transverse vertical break of cephalic cuticle of the trilobite *Bumastus* sp. from the Upper Silurian of Gotland, showing the central laminate zone (CLZ); preparation no. 080988-1. Preparation method 15. B. Transverse vertical break of cephalic cuticle of the trilobite *Asaphus* (*Asaphus*) *raniceps* Dalman from the Lower Ordovician of Öland, Sweden; preparation no. 191274-0. OLZ = outer laminate zone; CLZ = central laminate zone; ILZ = inner laminate zone. Preparation method 3.

PLATE 160. Vertical "sections" of the cuticle of the trilobite *Ellipsocephalus polytomus* Linnarsson, from the Middle Cambrian of Öland, Sweden, showing the apatitic outer layer (AOL); preparation no. Ar 46218b-i. Preparation method 16. A. Complete thickness of the cuticle. B. Detail of lamina units in the apatitic outer layer.

PLATE 161. Further details of the cuticle of the Middle Cambrian trilobite *Ellipsocephalus*, showing the apatitic outer layer (AOL) and the principal layer (PL); preparation no. Ar 46218b-i. Preparation method 16. A. Outer parts of vertical canals (after etching but before washing these canals were seen to extend through the complete thickness of the cuticle). B. Calcitic crystals apparently penetrating the apatitic outer layer.

Part 7

Corals and Coralline Sponges

Plates 162-169

Jobst Wendt

Institut und Museum für Geologie und Paläontologie
Universität Tübingen, Sigwartstraße 10, D-7400
Tübingen 1, West Germany

The following plates supplement the illustrations of skeletal microstructure for corals and coralline sponges in Volume I.

EXPLANATION OF PLATES

PLATE 162. Paleozoic tabulate corals. A. Longitudinal section through the walls (epitheca) of three adjacent corallites of *Thamnopora cristata* (Blumenbach), showing calcitic orthogonal microstructure; dark line in center of walls is due to granular structure and recrystallization. Upper Devonian (Frasnian), Menorca, Spain; thin section, crossed nicols. B. Same specimen as Fig. A; transverse fracture through epitheca with calcitic orthogonal microstructure; GPIT 7964/39908. C. Transverse fracture through epitheca of *Syringopora ramulosa* (Goldfuss), with calcitic orthogonal microstructure. Carboniferous, Bacharia, Russia, GPIT 7967/39922. D. Same specimen as Fig. C, showing clinogonal microstructure in transverse fracture through epitheca, GPIT 7967/39920.

PLATE 163. Paleozoic tabulate (A,B) and rugose (C,D) corals. A. Calcitic granular structure of epitheca in *Favosites rariporus* Frech; coarse cement void-filling crystals at lower margin. Upper Devonian (Frasnian), Menorca (Spain). GPIT 7986/39905. B. Epitheca and septum (upper right) of *Tetradium fibratum* Safford; note coarse calcite crystals replacing aragonitic skeleton of unknown microstructure. The Tetradiidae are the only tabulates known to make aragonitic skeletons. Ordovician, Rowena, Kentucky (USA), thin section, plane polarized light. C. Orthogonal microstructure of septum of *Ipciphyllum arnouldi* Termier. Coarse void-filling cement crystals at lower right. Uppermost Permian, Djebel Tebaga, Tunisia; GPIT 7994/41541. D. Close-up of Fig. C; note coarsening and local fusion of individual fibers due to diagenetic grain growth and/or intergranular cementation; GPIT 7994/41542.

PLATE 164. Paleozoic rugose corals. A. Transverse fracture through septum of undetermined phaceloid rugose coral with calcitic granular microstructure in the center passing into orthogonal microstructure toward the septal flanks. Uppermost Permian, Djebel Tebaga, Tunisia, GPIT 7993/41554. B. Close-up of Fig. A, showing same diagenetic effects as Plate 163, Fig. D; GPIT 7993/41555. C. Transverse thin section through septum of *Numidiaphyllum gillianun* Flügel; note irregular mosaic of secondary calcite crystals replacing original aragonitic skeleton with relict orthogonal microstructure preserved at right margin (arrows). This is the only known aragonitic rugose coral. Uppermost Permian, Djebel Tebaga, Tunisia; plane polarized light. D. Same as Fig. C, crossed nicols.

PLATE 165. Scleractinian corals. A. Longitudinal fracture through three septa of *Margarosmilia septanectens* (Loretz) showing original aragonitic clinogonal microstructure; Cassian Formation, lower Carnian, Seelandalpe near Cortina d'Ampezzo, Dolomites, Italy; GPIT 7943/39993. B. Same specimen as Fig. A; oblique view of outer surface of epitheca composed of orthogonally arranged parallel aragonite crystals. GPIT 7943/39998. C. Transverse fracture through septum of *Fungia* sp. composed of compound trabeculae. Recent, Pacific; GPIT 7944/40002. D. Fiber fascicles (fasciculi) on septal flank of *Fungia* sp., same specimen as Fig. C; GPIT 7944/40011. E. Probably originally calcitic fiber fascicles and granulations on septal flank of an undetermined caryophylliid. Upper Cretaceous, Maastrichtian, Maastricht, Holland. GPIT 3326/53511. F. Close-up of Fig. E; GPIT 3326/53508.

PLATE 166. Coralline Sponges. A. Calcitic granular microstructure of the archaeocyathid *Metaldetes profundus*; oblique fracture through wall and void-filling cement (lower right). Lower Cambrian, Wreck Reef, Labrador, Canada; GPIT 7990/41552. B. Close-up of Fig. A; note large crystal indicating local recrystallization. GPIT 7990/41553. C. Calcitic granular microstructure in the pharetronid sponge *Himatella milleporata* (Münster), coarse cement crystals at right margin. Cassian Formation, lower Carnian, S. Cassian, Dolomites, Italy; GPIT 2947/43236. D. Calcitic granular microstructure of vesicula in the sphinctozoan sponge *Cystothalamia slovenica* Senowbari-Daryan, consisting of randomly oriented equant crystals. Upper Triassic (Norian), Hydra Island, Greece; GPIT 5882/08341. E. Irregularly fibrous microstructure (aragonite) in an undetermined stromatoporoid (naturally weathered surface). Cassian Formation, lower Carnian, Seelandalpe near Cortina d'Ampezzo, Dolomites, Italy; GPIT 3135/46589. F. Irregularly fibrous microstructure in the cryptocoeliid demosponge *Vaceletia crypta* (Vacelet); longitudinal fracture through calotte; note that each fiber is an aragonite trilling. Recent, Myrmidon Reef, Great Barrier Reef, Australia, GPIT 5881/08361.

PLATE 167. Coralline Sponges. A. Inner surface of tube in the demosponge *Acanthochaetetes wellsi* Hartman and Goreau, showing irregularly arranged Mg-calcite fibers twisted around wall pores. Recent, Anae Island, Guam, Pacific; GPIT 3605/63483. B. Longitudinal fracture through tube wall of *Acanthochaetetes seunesi* Fischer, composed of irregularly arranged calcitic fibers. Upper Cretaceous, Cenomanian, Gotein, Basses-Pyrénées, France; GPIT 4930/05800. C. Surface of epitheca of the sphinctozoan *Thalamopora cribrosa* (Goldfuss) composed of calcitic fibers arranged orthogonally around inhalant pore. Upper Cretaceous, Cenomanian, Essen, West Germany; GPIT 3783/65244. D. Aragonitic orthogonal to clinogonal microstructure in pillar (longitudinal fracture) of stromatoporoid *Cassianostroma küpperi* Flügel; central cavity is caused by recent weathering. Cassian Formation, lower Carnian, Seelandalpe near Cortina d'Ampezzo, Dolomites, Italy; GPIT 3176/48190. E. Mg-calcitic clinogonal microstructure in massive pharetronid skeleton of *Murrayona phanolepis* Kirkpatrick, Recent, locality unknown; GPIT 02320/59665. F. Mg-calcitic spherulitic microstructure in massive pharetronid skeleton of *Petrobiona massiliana* Vacelet and Levi. Etched vertical section, Recent, Mediterranean; GPIT 2956/39295.

PLATE 168. Coralline Sponges. A. Transverse fracture through exhalant tube of the sphinctozoan sponge *Enoplocoelia armata* (Klipstein), showing primary skeleton of exhalant tube composed of clinogonally arranged aragonite fibers (top) and surface of tabula (secondary skeleton) composed of aragonite spherulites (bottom). Cassian Formation, lower Carnian, Seelandalpe near Cortina d'Ampezzo, Dolomites, Italy; GPIT 7937/41604. B. Close-up of Fig. A, showing aragonitic clinogonal microstructure of exhalant tube; GPIT 7937/41605. C. Close-up of Fig. A; surface of tabula composed of aragonite spherulites. Cavities in centers of spherulites are caused by recent weathering. GPIT 7937/41607. D. Longitudinal fracture through exhalant canal of the pharetronid *Sestrostomella robusta* Zittel, with small diactine spicule embedded in aragonitic spherulitic non-spicular skeleton. Cassian Formation, lower Carnian, Seelandalpe near Cortina d'Ampezzo, Dolomites, Italy; GPIT 3026/40754. E. Excentric growth of aragonitic spherulites in *Astrosclera willeyana* Lister, almost completely closing primary cavity of abandoned portion of skeleton. Recent, Europa Island, Strait of Mozambique, GPIT 3028/42548. F. Aragonitic clinogonal microstructure in vertical pillar of *Leiospongia verrucosa* (Münster), longitudinal fracture. Cassian Formation, lower Carnian, S. Cassian, Dolomites, Italy; GPIT 2957/42714.

PLATE 169. Coralline Sponges. A. Vertical fracture through peripheral portion of *Stromatospongia norae* Hartman, showing massive skeleton composed of subparallel aragonite fibers in which spicules of amorphous silica (acanthostyles) are embedded. Recent, Montego Bay, Jamaica; GPIT 3112/45343. B. Same specimen as Fig. A; vertical fracture through abandoned portion of skeleton composed of clinogonally arranged aragonite fibers with embedded siliceous spicule (acanthostyle); GPIT 3112/45349. C. Oblique view on surface of outer wall (epitheca) of *Ceratoporella nicholsoni* (Hickson) showing tips of orthogonally arranged aragonite fibers. Recent, Montego Bay, Jamaica; GPIT 3121/45492. D. Peripheral portion of *Ceratoporella nicholsoni* (Hickson) with margin of open pore. Spiny siliceous spicule (acanthostyle) embedded in clinogonally arranged aragonite fibers. Recent, Montego Bay, Jamaica; GPIT 3019/40721. E. Calcitic scales covering and partly embedded in the non-spicular pharetronid skeleton of *Murrayona phanolepis* Kirkpatrick. Recent, unknown provenance, GPIT 2320/59661. F. Group of siliceous microscleres (spirasters) and tip of megasclere (tylostyle) embedded in the soft tissue of *Acanthochaetetes wellsi* Hartman and Goreau. Anae Island, Guam, Pacific; GPIT 3605/63491.

Part 8
Echinodermata
Plates 170-175

Andrew B. Smith

Department of Palaeontology, British Museum (Natural History)
Cromwell Road, London SW7 5BD, England

The skeletal microstructures illustrated in the following six plates encompass the broad range of morphologies encountered in echinoderms. These plates illustrate the following skeletal fabrics: Three-dimensional meshworks: rectilinear (Plate 170); galleried (Plate 171); labyrinthic (Plate 172); microperforate (Plate 173A,B); laminar (Plate 173C,D); fascicular (Plate 173E,F). Single layers: retiform (Plate 175A,B); perforate (Plate 175C,D); imperforate (Plate 175E,F). Definitions for these structures are given in the *Glossary* in Volume I.

In addition to these qualitative variations, stereom may also vary considerably between species according to the quantitative features of coarseness, trabecular thickness, and porosity. Stereom coarseness can be quantified using mean pore diameter (A), *i.e.* coarse (A > 25 microns), medium (A = 10-25 microns) and fine (A < 10 microns). Trabecular thickness (t) is measured at the narrowest point between adjacent pores. Mean pore diameter (A) and mean trabecular thickness (t) are both measured from SEM micrographs of surfaces (Text-figure 1). Where pores are oval or irregular in outline, the maximum pore diameter is measured ("A" in Text-figure 1). Surface porosity of stereom is calculated by overlaying a grid on the SEM micrograph of a stereom surface and point-counting pore space against trabeculae.

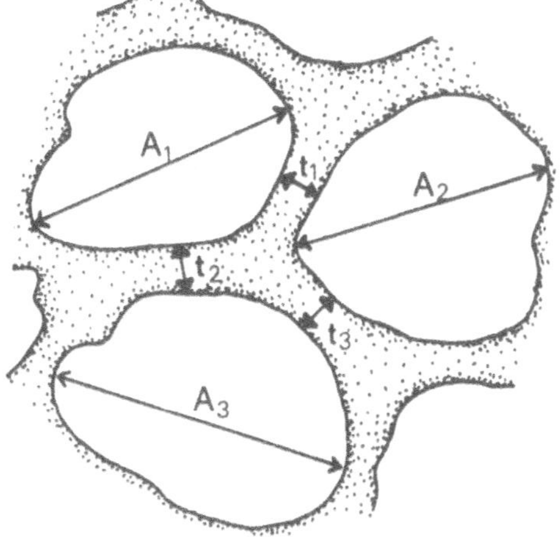

Text-Figure 1. Measurements of pore diameter (A) and trabecular thickness (t) in echinoderm stereom.

Skeletal composition. All stereom surfaces illustrated are calcitic. Plate 174, Figures A-B also show attaching muscle fibers and collagen bundles, respectively. The molpadiid granules in Plate 174, Figures D-F consist of ferrous phosphate.

Geological age of specimens. All specimens are Recent except where indicated otherwise.

Scale bars. All scale bars are labelled in microns.

Sample preparation methods. All figures in this section are scanning electron micrographs of gold/palladium-coated surfaces. The soft tissue structures in Plate 174, Figures A and B were partially digested in dilute sodium hypochlorite and then critical-point dried.

Guide to the literature. For further surveys of echinoderm stereom the following works should be consulted: **Echinoids:** Jensen (1972), Smith (1980). **Ophiuroids:** Macurda (1976). **Crinoids:** Macurda and Meyer (1975, 1976), Macurda and Roux (1981), Roux (1977a,b). **Holothurians:** Stricker (1985). Additional references are listed in the master bibliography and subject index in Volume I.

Acknowledgement. I am very grateful to David Pawson for supplying me with dermal granules of the holothurian *Molpadia*.

EXPLANATION OF PLATES

PLATE 170. Rectilinear stereom. A. *Eucidaris metularia* (echinoid): outer plate surface of interambulacral plate (epithelial coated surface). B. *Cidaris cidaris* (echinoid): interambulacral plate cross-section. C. *Poriocidaris purpurata* (echinoid): outer surface of interambulacral plate (areole of primary tubercle).

PLATE 171. Galleried stereom. A. *Tripneustes gratilla* (echinoid): outer plate surface of the boss of a primary interambulacral tubercle (collagen insertion area). B. *Ophiura albida* (ophiuroid): outer plate surface of the ligament insertion area on a vertebra. C. *Desmocrinus brevis* (crinoid): cross-section through a columnal (ligament insertion area). D. *Psammechinus miliaris* (echinoid): cross-section through an interambulacral plate - middle plate layer (collagen sutural fibers).

PLATE 172. Labyrinthic stereom. A. *Brisingella coronata* (asteroid): outer surface of adambulacral plate showing the sharp boundary between coarse labyrinthic stereom that makes up the shaft of the ossicle and the very much finer labyrinthic stereom forming the ambulacral muscle attachment area. B. *Ophiocoma erinaceus* (ophiuroid): outer surface of muscle attachment flange on proximal vertebra. C. *Calveriosoma hystrix* (echinoid): outer interambulacral plate layer (epithelial-coated surface). D. *Tripneustes gratilla* (echinoid): cross-section of interambulacral plate - inner plate layer. E. *Neocrinus blakei* (crinoid): outer surface of ambulacral lappet plate (epithelial-coated surface). F. *Echinus esculentus* (echinoid): outer surface of aboral face of auricle (internal mesodermal lining).

PLATE 173. Fascicular, microperforate, laminar and labyrinthic stereom. A-B. Microperforate stereom in *Echinolampas crassa* (echinoid): interambulacral plate cross-section - inner plate layer. C. Laminar (Lm) and labyrinthic (Lb) stereom in *Eupatagus hastingsi* (echinoid, Middle Eocene, Barton, Hampshire, U.K.): interambulacral plate cross-section with inner plate surface at the top. Arrows indicate boundary between outer plate surface and face of cross-section. D. Laminar (Lm) and labyrinthic (Lb) stereom in *Paramaretia peloria* (echinoid): cross-section of interambulacral plate - middle plate layer. E. Fascicular (F) and labyrinthic (Lb) stereom in *Astropecten irregularis* (asteroid): outer surface of ambulacral ossicle at approximately mid-length. F. Fascicular stereom of *Echinocardium cordatum* (echinoid): cross-section of inner layer of an interambulacral plate.

PLATE 174. A. Intervertebral muscle fibers attaching to a fine retiform stereom layer on the vertebra of *Ophiothrix fragilis* (ophiuroid). Note how the muscle fibers splay out as they reach the stereom and are wrapped around the fine surface trabeculae. Preparation method 20. B. Spine catch apparatus (collagen) inserting into galleried stereom forming the boss of a primary interambulacral tubercle of *Psammechinus miliaris* (echinoid). Note how the collagen bundles penetrate into the galleries of the stereom. Preparation method 20. C. Ambulacral spine of *Brisingella coronata* (asteroid) in cross-section; fractured trabeculae showing concentric growth pattern. D-F. Ferrous phosphate dermal granule of *Molpadia oolitica* (holothurian). D. Single dermal granule; note the characteristic irregular accretionary structure. E. Higher magnification of D, showing the subspherical subunits which comprise the dermal granule. F. Higher magnification of E, showing that the subspherical subunits themselves consist of much smaller subunits.

PLATE 175. Retiform, perforate, imperforate and labyrinthic stereom. A-B. Retiform stereom layer (RL) forming the muscle attachment platform in a tubercle of *Brissopsis lyrifera* (echinoid). A. Cross-section of plate; note the dense, perforate stereom layer (PL) lying just below the retiform layer, underlain by labyrinthic stereom (Lb). The arrow indicates the boundary between the outer plate surface and the face of the cross-section. B. Outer plate surface with underlying coarse labyrinthic stereom on the top left and bottom right. C-D. Perforate stereom layers. C. Cross-section of a ventral arm plate of *Ophiocoma erinaceus* (ophiuroid), showing a perforate sterom layer (PL) and an underlying labyrinthic stereom layer (Lb); exterior surface is toward the left; arrows mark the boundary between the outer plate surface and the face of the cross-section. D. Internal surface of an interambulacral plate of *Stomopneustes variolaris* (echinoid). E-F. Imperforate stereom (I). E. Cross-section through an interambulacral plate of *Echinoneus cyclostomus* (echinoid) showing one glassy tubercle (external surface at top). L = labyrinthic stereom. F. Plastron tubercle of *Brissopsis lyrifera* (echinoid) showing imperforate (I) mamelon and platform (articulation surfaces) surrounded by two rings of fine, labyrinthic stereom (Lb), the inner for collagen attachment, the outer for muscle attachment.

Part 9
Vertebrate Dental Structures
Plates 176-192

Sandra J. Carlson

Department of Geology,
University of California, Davis, California, 95616

Introduction

The accompanying plates supplement the discussion of vertebrate dental microstructures in Volume I. The scale bars are either 5, 10, 100, or 500 microns long, as indicated in the plate explanations. The following orientational terms are used in the plate explanations: Cross-section: in a plane perpendicular to the sagittal plane. Longitudinal section: in a plane parallel with the sagittal plane. Sagittal plane: the plane parallel with the long axis of the tooth and passing through the middle of the labial and lingual sides of the tooth. Tangential section: in a plane tangential to an exterior surface of the tooth. The abbreviations in this section are explained in the introduction to the Atlas.

Species examined. This list follows the taxonomic order of the classification of vertebrates by Carroll (1988).

1. *Carcharinus* sp. (modern shark, Elasmobranchii, Carcarhinoidea), Recent, California; UCD Department of Zoology, teaching collection. Plate 176A-D.
2. *Procarcharodon angustidens* (Elasmobranchii, Lamnoidea), Lower Miocene, Pungo River Formation, Aurora, North Carolina; UNC 13351. Plate 176E-F.
3. *Sebastes* sp. (Actinopterygii, Scorpaeniformes), Recent, caught by M. Patterson near the Bodega Marine Laboratory, Northern California. Plate 177A-C.
4. *Acrocheilus latus* (Actinopterygii, Cypriniformes), Pliocene, Glenns Ferry Formation, Snake River Plain, Idaho; UMMP uncatalogued. Plate 177D-F.
5. *Ambystoma tigrinum* (tiger salamander, "Lissamphibia", Urodela), Recent, New Mexico; research collection of B. Shaffer; UCD Department of Zoology. Plate 178A-D.
6. *Melanosaurus* sp. (Squamata, Lacertilia), Eocene, Willwood Formation, Big Horn Basin, Wyoming; UMMP uncatalogued. Plate 178E-F.
7. *Ptychozoon lionatum* (flying gecko, Squamata, Gekkonidae), Recent, Thailand; research collection of B. Shaffer; UCD Department of Zoology. Plate 179A-C.
8. *Masticophis ?flagellum* (whip snake, Squamata, Serpentes), Recent, northern California; research collection of B. Shaffer, UCD Department of Zoology. Plate 179D-F.
9. *Sphenodon punctatus* (tuatara, Sphenodonta, Sphenodontidae), Recent, New Zealand; research collection of D.C. Fisher, UMMP. Plate 180A-F.
10. *Phytosaur* sp. (Archosauria, Thecodontia), Triassic, locality unknown; UMMP vertebrate collection. Plate 181A.
11. *Crocodylus ?acutus* (Crocodylia, Crocodylidae), Recent, locality unknown; UMMP vertebrate collection. Plate 181C-F.

12. *Leidyosuchus* sp. (Crocodylia, Crocodylidae), Eocene, Willwood Formation, Big Horn Basin, Wyoming; UMMP vertebrate collection. Plate 181B.
13. *Alligator mississippiensis* (Crocodylia, Alligatoridae), Recent, locality unknown; UMMP vertebrate collection. Plate 182A-D.
14. *Allognathosuchus* sp. (Crocodylia, Alligatoridae), Eocene, Willwood Formation, Big Horn Basin, Wyoming. Plate 182E-F.
15. *Morganucodon* sp. ("*Eozostrodon*", Prototheria, Triconodonta), Triassic, Wales; UCMP 134819. Plate 183A-B.
16. *Obdurodon insignis* (Prototheria, Monotremata), Miocene, Etadunna Formation, South Australia; AMNH 97228. Plate 183C-G.
17. *Ptilodus cedrus* (Multituberculata, Ptilodontoidea), Eocene, Cedar Point Quarry, Wyoming; premolar (UMMP 63112) and uncatalogued incisor. Plate 184A-B.
18. *Taeniolabis taöensis* (Multituberculata, Taeniolabidoidea), Paleocene, Puercan, Barrel Springs, New Mexico; AMNH 12924. Plates 185A-F, 186A-C.
19. *Catopsalis joyneri* (Multituberculata, Taeniolabidoidea), Cretaceous, Bug Creek, Anthills, Montana; UMMP uncatalogued. Plate 186E.
20. *Sphenopsalis nobilis* (Multituberculata, Taeniolabidoidea), Upper Paleocene, Shabarakh, Usu, Mongolia; AMNH 21719. Plate 186G.
21. *Lambdopsalis bulla* (Multituberculata, Taeniolabidoidea), Paleocene, China; IVPP uncatalogued. Plate 186J.
22. *Eucosmodon primus* (Multituberculata, Taeniolabidoidea), Paleocene, Puercan, Tsosie Rincon, New Mexico; AMNH 59999. Plate 186H.
23. *Stygimys kuszmauli* (Multituberculata, Taeniolabidoidea), Cretaceous, Bug Creek, Anthills, Montana; UMMP uncatalogued. Plate 186D.
24. *Neoliotomus ultimus* (Multituberculata, Taeniolabidoidea), Eocene, Wasatchian, Big Horn Basin, Wyoming; incisor (UMMP 63294) and premolar (UMMP 65144). Plate 184C-F.
25. *Kryptobaatar dashzevegi* (Multituberculata, Taeniolabidoidea), Lower Cretaceous, Bayn Dzak, Mongolia; ZPAL MGM 1/37. Plate 186F.
26. *Microcosmodon rosei* (Multituberculata; Taeniolabidoidea), Eocene, Clarkforkian, Big Horn Basin, Wyoming; UMMP 71549. Plate 184G.
27. *Didelphis virginianus* (Marsupialia, Didelphoidea), Recent, northern California; UCD Department of Zoology, teaching collection. Plate 187A-C.
28. *Canis familiaris* (domestic dog, Carnivora, Arctoidea), Recent, northern California; UCD Department of Zoology, teaching collection. Plate 187D-F.
29. *Felis domesticus* (Carnivora, Aeluroidea), Recent, northern California; UCD Department of Zoology, teaching collection. Plate 188A-F.
30. *Homo sapiens* (Primates, Hominoidea), Recent, North Carolina; UNC 14886. Plate 189A-F, 192A.
31. *Mus musculus* (mouse, Rodentia, Muroidea), Recent, California; research collection, B. Shaffer; UCD Department of Zoology. Plate 190A-D.
32. *Lepus ?californicus* (rabbit, Lagomorpha, Leporidae), Recent, California; UCD Department of Zoology, teaching collection. Plate 190E-H.
33. *?Pliohippus* sp. (Perissodactyla, Equoidea), Mio-Pliocene, northern California, UCD Department of Zoology, teaching collection. Plate 191A-E.
34. *Coryphodon* sp. (Pantodonta, Coryphodontidae), Eocene, Willwood Formation, Bighorn Basin, Wyoming; UMMP uncatalogued. Plate 191F.
35. *Mammut americanum* (mastodon, Proboscidea, Mammutoidea), Pleistocene, Michigan; UMMP vertebrate collection. Plate 192C.
36. *Elephas maximus* (Proboscidea, Euelephantoidea), Recent, India; UMMP vertebrate collection. Plate 192F.
37. *Loxodonta africana* (Proboscidea, Euelephantoidea), Recent, Africa, UMMP vertebrate collection. Plate 192E.
38. *Mammuthus jeffersoni* (mammoth, Proboscidea, Euelephantoidea), Pleistocene, locality unknown; UMMP vertebrate collection. Plate 192B,D.

EXPLANATION OF PLATES

PLATE 176. Elasmobranchii. A. *Carcharinus* sp. enameloid layer over orthodentine. Note the complex enameloid ultrastructure, virtually identical to prismatic enamel. Despite the extraordinary structural similarity to prismatic enamel, these features are almost certainly convergent (non-homologous). Cross-section; tooth crown is toward the upper left. Preparation method 6. Scale bar = 100 microns. B. *Carcharinus* sp.; close-up of cross-section in 1A. Preparation method 6. Scale bar = 10 microns. C. *Carcharinus* sp.; development of "prism" decussation. Tangential section; tooth crown is toward the left. Preparation method 7. Scale bar = 10 microns. D. *Carcharinus* sp. Similarity to prismatic enamel is remarkable; compare with Plate 192C. Tangential section; tooth crown is toward left. Preparation method 7. Scale bar = 5 microns. E. *Procarcharodon angustidens*; structural complexity is apparent; similar to crossed lamellar microstructure found in some bivalved molluscs. Vertical section; tooth crown is toward the right. Preparation method 1. Scale bar = 100 microns. F. *Procarcharodon angustidens.* Crystallites within one layer are parallel to each other, and perpendicular to crystallites in adjacent layers. Tangential section; tooth crown is toward the top. Preparation method 1. Scale bar = 5 microns.

PLATE 177. Actinopterygii. A. *Sebastes* sp.; teeth are reflexed posteriorly and have a slight swelling at their apices. Teeth are scattered along the dentary bone, not arranged in single file; note the tooth bud emerging lingual to functional teeth, near lower left. Anterior of jaw is toward the left. Preparation method 5. Scale bar = 100 microns. B. *Sebastes* sp.; the pulp cavity (center) is surrounded by orthodentine, with a thin enameloid outer layer. Longitudinal section, preparation method 1. Scale bar = 100 microns. C. *Sebastes* sp. Base of tooth is about to be shed, illustrating the fabric of natural resorption. Preparation method 5. Scale bar = 100 microns. D. *Acrocheilus latus* pharyngeal tooth. Structurally complex enameloid layers over orthodentine and trabecular dentine. Longitudinal section; tooth crown is toward the upper right. Preparation method 6. Scale bar = 100 microns. E. *Acrocheilus latus.* Close-up of orthodentine; more dense peritubular dentine lines the the dentinal tubules, whereas less dense intertubular dentine comprises the bulk of the dentine fabric. Longitudinal section; tooth crown is toward the top. Preparation method 6. Scale bar = 5 microns. F. *Acrocheilus latus.* Close-up of trabecular dentine; spongy texture is apparent. Fossilized dentinal tubules are visible at lower right. Longitudinal section; tooth crown is toward the upper right. Preparation method 6. Scale bar = 5 microns.

PLATE 178. "Lissamphibia" and Squamata. A. *Ambystoma tigrinum*; note staggered rows of functional and erupting teeth. Anterior of jaw is toward the bottom; lingual side is toward the right. Preparation method 5. Scale bar = 100 microns. B. *Ambystoma tigrinum.* Close-up of tooth attached by ligament to cylindrical bony pedestal. Anterior is toward the left. Preparation method 5. Scale bar = 100 microns. C. *Ambystoma tigrinum*; close-up of individual tooth removed from its pedestal; note the bicuspid crown. Anterior is toward the viewer, lingual side is toward the right. Preparation method 5. Scale bar = 10 microns. D. *Ambystoma tigrinum.* Pulp cavity is surrounded by orthodentine, and covered by thin enamel layer. Longitudinal section; anterior is toward the left, lingual side is toward the viewer. Preparation method 1. Scale bar = 10 microns. E. *Melanosaurus* sp.; incremental lines in nonprismatic enamel follow contours of external tooth topography. Longitudinal section; tooth crown is toward the bottom. Preparation method 7. Scale bar = 5 microns. F. *Melanosaurus* sp.; incremental lines in nonprismatic enamel; note their angular relationship to the enamel-dentine junction, at left. Longitudinal section; tooth crown is toward the bottom. Preparation method 7. Scale bar = 5 microns.

PLATE 179. Squamata. A. *Ptychozoon lionatum*; note tooth bud, at left, emerging lingually to the functional teeth, arranged in single file along the dentary bone. The tooth crown has a shallow groove oriented antero-posteriorly. The anterior is toward the left; the lingual side is toward the viewer. Preparation method 5. Scale bar = 500 microns. B. *Ptychozoon lionatum* pleurodont dentition; the teeth are fused along the lingual side of of the dentary bone. The anterior is toward the viewer; the lingual side is toward the right. Preparation method 5. Scale bar = 100 microns. C. *Ptychozoon lionatum*; close-up of orthodentine, illustrating felt-like mat of apatite crystallites permeated with dentinal tubules. Longitudinal section; tooth crown is toward the top. Preparation method 1. Scale bar = 5 microns. D. *Masticophis ?flagellum*; dentinal tubules radiate from the pulp cavity, on left, and bifurcate toward the tooth surface. Longitudinal section; tooth crown is toward the top. Preparation method 1. Scale bar = 5 microns. E. *Masticophis ?flagellum*; teeth are recurved posteriorly. Anterior side is toward the lower right; lingual side is toward the viewer. Preparation method 5. Scale bar = 100 microns. F. *Masticophis ?flagellum.* Dentinal tubules radiate from the pulp cavity; a thin enamel layer covers the tooth. Longitudinal section, preparation method 1. Scale bar = 5 microns.

PLATE 180. Sphenodonta: *Sphenodon punctatus*. A. Thin enamel layer is present toward the top. Longitudinal section. Preparation method 6. Scale bar = 100 microns. B. Preprismatic enamel; each "preprism" is approximately 5 microns in diameter. Slightly oblique view of tangential section; the tooth crown is toward the top. Preparation method 7. Scale bar = 5 microns. C. Close-up of preprismatic structure; note that the crystallites are oriented toward the center of each "preprism." Tangential section; tooth crown is toward the top. Preparation method 7. Scale bar = 5 microns. D. Somewhat different appearance of enamel near location of 180C. Note the absense of prism sheaths; variation in crystallite orientation alone delineates boundaries between "preprisms." Tangential section; tooth crown is toward the top. Preparation method 7. Scale bar = 5 microns. E. Close-up of enamel layer in 180A. Columns of radiating crystallites, perpendicular to enamel-dentine junction, at bottom, comprise the microstructure of preprismatic enamel. Longitudinal section; tooth crown is toward the left. Preparation method 6. Scale bar = 5 microns. F. Close-up of a different location along enamel layer, showing columns of radiating crystallites and scallop-shaped incremental lines. Longitudinal section; tooth crown is toward the left. Preparation method 6. Scale bar = 5 microns.

PLATE 181. Archosauria. A. *Phytosaur* sp. Columns of radiating crystallites, typical of preprismatic enamel, are illustrated here. However, the width of the columns is several times greater than in *Sphenodon punctatus*. Longitudinal section; the tooth crown is toward the left. Preparation method 7. Scale bar = 10 microns. B. *Leidyosuchus* sp.; typical preprismatic "herringbone" pattern of crystallite orientation is apparent in some locations, as illustrated here, but this pattern is not sustained along the entire length of the enamel section. Longitudinal section; the tooth crown is toward the left; the tooth surface is toward the top. Preparation method 7. Scale bar = 10 microns. C. *Crocodylus ?acutus*; nonprismatic enamel. Crystallites are oriented parallel to one another and perpendicular to enamel-dentine junction throughout the enamel thickness. Horizontal, non-scalloped incremental lines are visible. Longitudinal section; the tooth crown is toward the right. Preparation method 7. Scale bar = 10 microns. D. *Crocodylus ?acutus*; nonprismatic enamel. Homogeneity of crystallite orientation is apparent. Tangential section; tooth crown is toward the right. Preparation method 7. Scale bar = 5 microns. E. *Crocodylus ?acutus*; trabecular dentine. Spongy, porous texture is apparent. Tangential section; tooth crown is toward the upper right. Preparation method 7. Scale bar = 10 microns. F. *Crocodylus ?acutus*. Close-up of 181E. Trabecular dentine shares structural similarities with echinodermal stereom. Tangential section; tooth crown is toward the left. Preparation method 7. Scale bar = 5 microns.

PLATE 182. Crocodylia. A. *Alligator mississippiensis*; enamel ultrastructure intermediate between nonprismatic and preprismatic. Crystallite orientation varies more than in nonprismatic enamel, but not as consistently or as clearly as in preprismatic enamel. Longitudinal section; the tooth crown is toward the right. Preparation method 7. Scale bar = 10 microns. B. *Alligator mississippiensis*; close-up of enamel structure. Longitudinal section; the tooth crown is toward the right. Preparation 7. Scale bar = 5 microns. C. *Alligator mississippiensis*; close-up of 182D, showing crystallite orientation. Tangential section; the tooth crown is toward right. Preparation method 7. Scale bar = 5 microns. D. *Alligator mississippiensis*. In this view, a periodic pattern of varying crystallite orientation is apparent. The "preprisms" have a negative rather than a positive topography, as in *Sphenodon punctatus*. Tangential section; tooth crown is toward the right. Preparation method 7. Scale bar = 10 microns. E. *Allognathosuchus* sp.; enamel appears nonprismatic; incremental lines are horizontal and not scalloped. Longitudinal section; tooth crown is toward the upper right. Preparation method 7. Scale bar = 5 microns. F. *Allognathosuchus* sp.; enamel appears preprismatic. The area illustrated in this photomicrograph is located very near the area illustrated in 182E, demonstrating the range of variability in microstructure within small areas of a tooth. Longitudinal section; tooth crown toward upper right. Preparation method 7. Scale bar = 5 microns.

PLATE 183. Prototheria. A. *Morganucodon* sp., premolar; natural fracture surface showing pulp cavity in center, surrounded by orthodentine, with a thin enamel layer. Longitudinal section. Preparation method 6. Scale bar = 100 microns. B. *Morganucodon* sp.; close-up of enamel layer. Crystallite convergence and divergence is apparent, but further preparation is necessary to resolve individual "preprisms." Longitudinal section; the tooth crown is toward the top. Preparation method 6. Scale bar = 5 microns. C.

Obdurodon insignis, molar, enamel layer on dentine. The anterior is toward the right; the lingual side is toward the viewer. Preparation method 5. Scale bar = 100 microns. D. *Obdurodon insignis*; Pattern 2 prismatic enamel. The prisms are arranged in columns, but individual prisms have round or oval, not arc-shaped, outlines. Tangential section; occlusal surface is toward the right. Preparation method 5. Scale bar = 5 microns. E. *Obdurodon insignis*; close-up of enamel, showing prisms with arc-shaped cross-sections. Tangential section; occlusal surface is toward the right. Preparation method 5. Scale bar = 5 microns. F. *Obdurodon insignis*; arrangement of Pattern 2 prisms breaks down; some prisms have circular outlines (near center), whereas others are arc-shaped (toward right). Tangential section; occlusal surface is toward the right. Preparation method 5. Scale bar = 5 microns.

PLATE 184. Multituberculata. A. *Ptilodus cedrus* premolar; prisms emerge from enamel-dentine junction at an angle; the angle changes abruptly approximately 2/3 of the way through the enamel layer. Longitudinal section; tooth crown is toward the left. Preparation method 7. Scale bar = 10 microns. B. *Ptilodus cedrus*, incisor; prism orientation changes abruptly 1/2 way through the enamel layer. Cross-section; tooth crown is toward the viewer. Preparation method 7. Scale bar = 5 microns. C. *Neoliotomus ultimus*, incisor; pattern 1 prismatic enamel. Prisms are arranged in columns. Tangential section; tooth crown is toward the right. Preparation method 7. Scale bar = 10 microns. D. *Neoliotomus ultimus*, premolar; close-up of 184F. Prismatic and interprismatic crystallites and prism sheaths are identifiable. Unusual "en echelon" prisms vary from circular to rhomb-shaped, in columns. Tangential section; tooth crown is toward the top. Preparation method 7. Scale bar = 5 microns. E. *Neoliotomus ultimus*, same as 184D; parallel orientation of prisms within enamel. Tangential section; tooth crown is toward the upper left. Preparation method 7. Scale bar = 10 microns. F. *Neoliotomus ultimus*, same as 184D; near region in 184E. Prism orientation changes in gradual curves. Tangential section; tooth crown is toward the top. Preparation method 7. Scale bar = 10 microns. G. *Microcosmodon rosei*, incisor; note changes in orientation of prisms within enamel layer. Oblique tangential section; enamel-dentine junction at bottom, tooth crown is toward the left. Preparation method 7. Scale bar = 10 microns.

PLATE 185. Multituberculata: *Taeniolabis taöensis*. A. Incisor showing constant prism orientation within enamel layer, at an angle to enamel-dentine junction. Enamel tubules are visible in lower right. Longitudinal section; tooth crown is toward the left. Preparation method 7. Scale bar = 100 microns. B. Close-up of 185A; prism sheaths and changing orientation of crystallites within prisms and interprismatic areas. Longitudinal section; tooth crown is toward the left. Preparation 7. Scale bar = 10 microns. C. Arc-shaped prisms are arranged in a fashion intermediate between Patterns 2 and 3. Prisms "dissipate" near the surface of the tooth; nonprismatic outer layer of enamel is visible in the upper left corner. Cross-section; tooth crown is toward the viewer. Preparation method 7. Scale bar = 100 microns. D. Close-up of enamel-dentine junction, showing enamel tubules and dentinal tubules. Cross-section; tooth crown is toward the viewer. Preparation method 7. Scale bar = 10 microns. E. Same surface as in 185C, at a different location. Note that the prisms are more irregularly shaped and arranged. Cross-section; tooth crown is toward the viewer. Preparation method 7. Scale bar = 100 microns. F. Close-up of 185E; pattern 3 prismatic enamel. Prismatic crystallites are perpendicular to the section, whereas interprismatic crystallites are parallel to the section. Cross-section; tooth crown is toward the viewer. Preparation method 7. Scale bar = 10 microns.

PLATE 186. Multituberculata. A. *Taeniolabis taöensis*; close-up of 185F, for comparison with 186B and C. Cross-section; tooth crown is toward the viewer. Preparation method 7. Scale bar = 5 microns. B. *Taeniolabis taöensis*; section deep within the enamel layer, near enamel-dentine junction. The prism sheaths appear less distinct, although the preparation technique is the same as in Plate 186C. The "seam" feature discussed by Lester and Hand (1987) is visible in the prism at the lower left. Tangential section, preparation method 7. Scale bar = 5 microns. C. *Taeniolabis taöensis*; well-developed prism sheaths, slightly incurved at base of arc. Note the irregular prism arrangement, intermediate between Patterns 2 and 3. Tangential section, preparation method 7. Scale bar = 10 microns. D. *Stygimys kuszmauli* molar; the prism crystallites lie parallel to the plane of section, the interprismatic crystallites lie perpendicular to the plane of section. Tangential section, preparation method 7. Scale bar = 5 microns. E. *Catopsalis joyneri* premolar; "seam" is visible, and enamel tubules are present. Tangential section.

Preparation method 7. Scale bar = 5 microns. F. *Kryptobaatar dashzevegi* premolar, showing unusual broad, flat arcs, unlike the more typical Pattern 3 enamel elsewhere on this section. Enamel tubules are visible. Tangential section, preparation method 7. Scale bar = 5 microns. G. *Sphenopsalis nobilis* molar; secondary "growths" over prism sheaths are attributed to preparation artifacts (this region is etched but not polished). Tangential section, preparation method 6. Scale bar = 5 microns. H. *Eucosmodon primus* incisor; semi-circular arcs. Tangential section. Preparation method 7. Scale bar = 5 microns. J. *Lambdopsalis bulla* molar; arc-shaped prisms; sheaths occasionally appear to form continuous circles around prisms. Tangential section. Preparation method 7. Scale bar = 5 microns.

PLATE 187. Marsupialia and Carnivora. A. *Didelphis virginianus* premolar; prismatic enamel in longitudinal section; the tooth crown is toward the left. Preparation method 6. Scale bar = 5 microns. B. *Didelphis virginianus.* Pattern 2 enamel microstructure. Tangential section; the tooth crown is toward the upper right. Preparation method 7. Scale bar = 5 microns. C. *Didelphis virginianus*; interprismatic crystallites are less apparent in this unground section. Enamel tubules are visible throughout the enamel fabric. Tangential section; the tooth crown is toward the upper left. Preparation method 6. Scale bar = 5 microns. D. *Canis familiaris* molar; hexagonal closest-packing of Pattern 1 enamel prisms. Note the constant thickness of interprismatic regions surrounding each prism, unlike multituberculate enamel. Tangential section; tooth crown is toward the upper left. Preparation method 7. Scale bar = 5 microns. E. *Canis familiaris*; "prism within a prism" fabric is apparent in tangential section. Prism decussation is clear in longitudinal section. Tangential section in upper half; longitudinal section in lower half. Preparation method 6. Scale bar = 10 microns. F. *Canis familiaris*; peritubular and intertubular dentine. Longitudinal section; tooth crown is toward the bottom. Preparation method 6. Scale bar = 10 microns.

PLATE 188. Carnivora: *Felis domesticus*. A. Molar illustrating the enamel layer lapping onto the dentine, and the microstructure of the most recently formed enamel. Longitudinal section; the tooth crown is toward the top. Preparation methods 6 and 7. Scale bar = 10 microns. B. Orthodentine microstructure as exposed in the pulp cavity. Tubules appear to radiate from central points throughout the dentine. The tooth crown is toward the top. Preparation method 6. Scale bar = 5 microns. C. Prism decussation is clearly illustrated and is most pronounced in the central 3/4 of the enamel layer. Longitudinal section; tooth crown is toward the left. Preparation method 7. Scale bar = 10 microns. D. "Seams" are present; arc-shaped prisms are arranged in Pattern 2 columns. Longitudinal section; tooth crown is toward the left. Preparation method 7. Scale bar = 5 microns. E. Arc-shaped Pattern 3 prisms. Tangential section; tooth crown is toward the left. Preparation 7. Scale bar = 5 microns. F. Circular Pattern 1 prisms, less than 1 millimeter from 188E. Tangential section; tooth crown is toward the left. Preparation method 7. Scale bar = 5 microns.

PLATE 189. Primates: *Homo sapiens*. A. Prism orientation within the enamel layer. Longitudinal section. Preparation method 1. Scale bar = 100 microns. B. Sinuous sheaths designate cross-striations within the enamel, purported to represent daily growth increments. Longitudinal section, preparation method 7. Scale bar = 5 microns. C. Arc-shaped Pattern 3 enamel microstructure. Tangential section, preparation method 7. Scale bar = 5 microns. D. Fracture section similar to ground and etched section in Plate 189B. View is perpendicular to Plate 189C. Longitudinal section, preparation method 1. Scale bar = 5 microns. E. Enamel-dentine junction. Longitudinal section. Preparation method 1. Scale bar = 100 microns. F. Close-up of Plate 189E, illustrating dentinal tubules and branching enamel tubules. Longitudinal section, preparation method 1. Scale bar = 5 microns.

PLATE 190. Rodentia and Lagomorpha. A. *Mus musculus* incisor in mandible, showing restricted band of enamel on orthodentine, with pulp cavity in center. Cross-section, preparation method 1. Scale bar = 500 microns. B. *Mus musculus*; prism orientation within enamel is consistent with the external topography of the gnawing edge of the ever-growing incisor. Longitudinal view, preparation method 6. Scale bar = 10 microns. C. *Mus musculus*; prisms bending in decussation. Longitudinal view, preparation method 6. Scale bar = 5 microns. D. *Mus musculus*; uniserial to "biserial" enamel structure. Longitudinal view; tooth crown is toward the left. Preparation method 7. Scale bar = 5 microns. E. *Lepus ?californicus*, incisor; close-up of Plate 190H. Pattern 2 enamel microstructure. Tangential section, preparation method 7. Scale bar = 5 microns. F. *Lepus ?californicus*; three distinct layers in enamel correspond to coordinated changes in orientation of prisms from enamel-dentine junction to tooth surface. Longitudinal section; tooth crown is toward the top. Preparation method 6. Scale bar = 10 microns. G. *Lepus ?californicus*; closer view of changes in prism orienation. Longitudinal section; tooth crown is toward the left. Preparation method 7. Scale bar = 5 microns. H. *Lepus ?californicus*; multiserial enamel structure. Tangential section, preparation method 7. Scale bar = 10 microns.

PLATE 191. Perissodactyla and Pantodonta. A. *?Pliohippus* sp. molar; Pattern 1 prismatic enamel. Tangential section, preparation method 7. Scale bar = 5 microns. B. *?Pliohippus* sp. Hunter-Shreger bands within enamel fabric. Longitudinal section, preparation method 7. Scale bar = 100 microns. C. *?Pliohippus* sp.; Pattern 2 prismatic enamel, immediately adjacent to 191A. Tangential section, preparation method 7. Scale bar = 5 microns. D. *?Pliohippus* sp.; Hunter-Shreger bands seen in a view perpendicular to Plate 191B. Tangential section, preparation method method 7. Scale bar = 100 microns. E. *?Pliohippus* sp.; Pattern 3 prismatic enamel, immediately adjacent to 191A and C. Tangential section, preparation method 7. Scale bar = 5 microns. F. *Coryphodon* sp.; Pattern 3 enamel, comparable to proboscidean and human enamel in Plate 192. Tangential section, preparation method 7. Scale bar = 10 microns.

PLATE 192. Primates and Proboscidea. A. *Homo sapiens*, same as in Plate 189; characteristic "keyhole" prism pattern with virtually no interprismatic enamel. Tangential section, preparation method 7. Scale bar = 5 microns. B. *Mammuthus jeffersoni* molar; close-up of Plate 192D. "Keyhole" Pattern 3 enamel microstructure, very similar to human enamel, as in 192A. Tangential section, preparation method 7. Scale bar = 5 microns. C. *Mammut americanum*, molar; Pattern 3 enamel microstructure. Tangential section, preparation method 7. Scale bar = 5 microns. D. *Mammuthus jeffersoni*. Domains of similarly oriented prisms can be distinguished within enamel. Tangential section, preparation method 7. Scale bar = 100 microns. E. *Loxodonta africana* molar; prism sheaths connect to form scalloped rows. Tangential section, preparation method 7. Scale bar = 5 microns. F. *Elephas maximus* molar; both circular and arc-shaped prisms visible in this section. Tangential section, preparation method 7. Scale bar = 5 microns.

Part 10
Conodontophorida
Plates 193-200

Jack C. Hall

Department of Earth Sciences
University of North Carolina at Wilmington
Wilmington, North Carolina, 28403-3297

Introduction

Observations of forty multi-element genera in the Conodontophorida suggest that their skeletal microstructure is more varied than previously documented. The present specimens come from Ordovician through Triassic localities in North America, Europe and Asia, and they represent all major conodont superfamilies (Appendix 1). Most of the microstructures illustrated in this section are probably original. The studied specimens show little or no evidence of thermal alteration. In all specimens examined the Color Alteration Index is less than 2.0. However, the possibility remains that some specimens show slight recrystallization. Recrystallization is difficult to evaluate on a micromorphologic basis for conodonts because they have no living representatives. The present illustrations are based on scanning electron microscopy of natural and induced fracture surfaces coated with gold/palladium.

Crystallite form. Conodont apatite crystallites occur in four forms: (1) elongate, hexagonal prisms with a dipryamid (Text-figure 1; Plate 193:5), (2) elongate, hexagonal prisms with a pyramid and basal pinacoid (Text-figure 2; Plate 193:5,6), (3) short, platy, hexagonal prisms with a pryamid and basal pinacoid (Text-figure 3, Plate 193:7), and (4) amorphous, cryptocrystalline masses similar to colophane (Text-figure 4, Plate 194:8). The elongate, rod-like crystallites range from less than 0.1 micron to 2.0 microns in diameter, with their length axis (parallel to their c-axis) two to over ten times their diameter. These elongate crystallites are generally less than 0.5 micron wide and 3 microns long. Elongate crystallites have previously been described for conodonts by Pietzner *et al.* (1968), Barnes *et al.* (1970, 1973), Lindström and Ziegler (1971) and Hall (1989). Two types of elongate crystallites have been recognized, sometimes within the same specimen (Text-figures 1,2). However, over 90% of elongate crystallites have pryamidal and pinacoidal faces. Short, platy crystallites have length dimensions along their c-axis shorter than their horizontal axis. They range from 0.3 to over 7.0 microns in diameter, and their length is commonly less than one-third their diameter. Short, platy crystallites commonly constitute the largest structural components in conodonts, with diameters ranging between 2.0 and 7.0 microns. Pietzner *et al.* (1968), Barnes *et al.* (1970), and Lindström and Ziegler (1971) mentioned but did not illustrate granular and flat or flaky structures which may be similar to the present short, platy crystallites.

Amorphous, cryptocrystalline masses in conodonts are roughly spherical, and they show no prismatic or pinacoidal faces even at magnifications exceding 20,000 diameters. They have maximum diameters of 2.0 to 3.0 microns. Cryptocrystalline masses are known to occur only in the "white matter" and basal fillings of conodonts. These masses are commonly botryoidal (Text-figure 4, Plates 194:8, 199:24,25, 200:26) but they occasionally show "corn flake" shapes (Plate 200:27). Pietzner *et al.* (1968) and Barnes *et al.* (1970) described similar structures.

In all specimens examined, no transitions were observed between short, platy crystallites and the two varieties of elongate crystallites, even though all three forms may occur in a single specimen. On the other hand, the margin of the basal cavity commonly shows transitions between well-defined prisms and amorphous, cryptocrystalline masses (Plate 200:28).

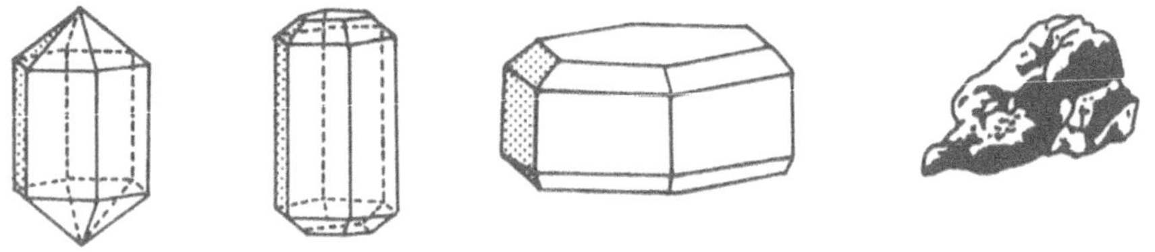

1 2 3 4

Text-figs. 1-4. Crystallite form in conodonts. 1. Elongate prism with dipyramid. 2. Elongate prism with pyramid and pinacoid. 3. Short, platy crystallite with pyramid and pinacoid. 4. Amorphous, cryptocrystalline mass.

Crystallite packing. Crystallite packing varies with the type of crystallite. Elongate crystallites may be packed tight and uniform along their prismatic and pinacoidal faces, or they may occur in bundles which are tightly bound parallel to the prismatic faces and loosely bound parallel to the pinacoidal faces, the latter reflecting offsets in the orientation of the crystallite bundles (Plates 194:9,10, 195:11-13). Elongate crystallites are generally uniformly packed or show slight to moderate offsets (Plates 194:9,10; 195:11).

Short, platy crystallites are commonly tightly packed parallel to their broad, pinacoidal faces, sometimes producing a massive appearance. With tight packing parallel to their prismatic faces, they may form rod-like arrangements due to similar tight packing parallel to their pinacoidal faces (Plate 193:7). Slight offsets parallel to the pinacoidal faces serve to distinguish individual crystallites. As with elongate crystallites, platy crystallites may be more loosely packed both parallel and perpendicular to their c-axis. The offsets of the pinacoidal faces may be so extreme as to produce a nearly open framework (Plates 196:14-16; 197:17-18). Most commonly, short, platy crystallites show slight offsets as in Plate 196:15. Offsets in platy or elongate crystallites may have produced what Pietzner *et al.* (1968) and Barnes *et al.* (1970) termed irregular or ill-defined rods.

Amorphous, cryptocrystalline masses may be loosely packed into porous clumps or they may be tightly packed into solid masses with a botryoidal appearance (Plates 199:24-25, 200:26). It is unknown whether the "corn flake" texture in the white matter or basal filling material in Plate 200:27 results from a unique crystallite form or distinctive packing of amorphous clumps. Cyptocrystalline masses usually exhibit a partially closed framework similar to that in Plate 199:25.

Crystallite orientation. In all specimens examined, both elongate and platy crystallites generally have their c-axis parallel or nearly parallel to the outer surface and long axis of the conodont element (Plates 197:19, 198:20). Similar alignment has been observed in a variety of conodont taxa (Hass and Lindberg, 1946; Rhodes and Wingard, 1957; Barnes *et al.*, 1970; Lindström and Ziegler, 1971; Barnes *et al.*, 1973; Barnes and Slack, 1975; Hassler and Hall, 1989). However, in a significant number of specimens, elongate crystallites are oreinted with their c-axis oblique to the outer surface of the conodont (Plate 198:22). In individual specimens, elongate crystallites have been observed with their c-axis parallel, oblique, and even perpendicular to the surface of the conodont (Plate 199:23). Variations in crystallite orientation from parallel have also been noted by Pietzner *et al.* (1968), although some of these may have been induced by their ultramicrotome sectioning techniques (see Barnes *et al.*, 1970; Lindström and Ziegler, 1971; Barnes *et al.*, 1973).

Unlike elongate crystallites, platy crystallites invariably have their c-axes oriented parallel (Plate 193:7) or perpendicular (Plate 198:21), but not oblique, to the outer surface of the conodont element.

Conclusions. Lindström (1964) suggested that the internal structure of conodonts may be taxonomically important. The present study indicates that conodonts are microstructurally more complex than previously realized. Much more data will, therefore, be required to produce a clear picture of the relationship between conodont ultrastructure and systematics.

Acknowledgements. I thank Dr. Stig M. Bergstrom, Director of the Orton Geological Museum and Professor of Geology at Ohio State University, for kindly providing many of the specimens used in this study. Mr. Scott E. Hassler of the University of North Carolina at Wilmington provided invaluable insights throughout this project. This research was partially supported by a Faculty Research and Development Grant from the University of North Carolina, for which I am grateful.

Appendix I. Species Examined

Examined specimens of the following species came from the United States unless indicated otherwise:

Aphelognathus kimmswickensis Sweet, Thompson and Satterfield, 1975, Middle Ordovician, Catheys Formation, Georgia.
Asphidognathus tuberculatus Walliser, 1964, Lower Silurian, Clinton Group, New York.
Baltoniodus gerdae (Bergstrom, 1971a), Middle Ordovician, Chickamauga Formation, Alabama.
Belodina compressa (Branson and Mehl, 1933), Middle Ordovician, Eggleston Formation, Virginia.
Belodella robusta Ethington and Clark, 1982, Middle Ordovician, Kenosh Formation, Nevada.
Carniodus carnulus Walliser, 1964, Lower Silurian, Lockport Group, New York.
Curtognathus typus Branson and Mehl, 1933, Middle Ordovician, Pond Spring Formation, Georgia.
Decoriconus fragilis (Branson and Mehl, 1933), Lower Silurian, Clinton Group, New York.
Drepanoistodus suberectus (Branson and Mehl, 1933), Middle Ordovician, Ben Hur Formation, Virginia.
Ellsonia gradata Sweet, 1970, Lower Triassic, Mianwali Formation, Pakistan.
Ellsonia robusta Sweet, 1970, Lower Triassic, Mianwali Formation, Pakistan.
Ellsonia triassica Muller, 1956, Lower Triassic, Mianwali Formation, Pakistan.
Erismodus radicans Hinde, 1879, Middle Ordovician, Ridley Formation, Georgia.
Falcodus alatoides Rexroad and Burton, 1961, Upper Mississippian, Maxville Group, Ohio.
Histiodella serrata Harris, 1962, Middle Ordovician, Kenosh Formation, Nevada.
Icriodella superba Rhodes, 1953, Middle Ordovician, Chickamauga Formation, Alabama.
Juanognathus variabilis Serpagli, 1974, Middle Ordovician, Kenosh Formation, Nevada.
Jumodontus granda Cooper, 1981, Middle Ordovician, Kenosh Formation, Nevada.
Kockelella ranuliformis Walliser, 1964, Lower Silurian, Lockport Group, New York.
Microozarkodina marathonensis (Bradshaw, 1969), Middle Ordovician, Kenosh Formation, Nevada.
Multioistodus compressus Harris and Harris, 1965, Middle Ordovician, Kenosh Formation, Nevada.
Neogondolella postserrata Behnken, 1975, Upper Permian, Bell Canyon Formation, Texas.
Neospathodus cristagalli (Huckriede, 1958), Lower Triassic, Mianwali Formation, Pakistan.
Neospathodus dieneri Sweet, 1970, Lower Triassic, Mianwali Formation, Pakistan.
Oepikodus communis (Ethington and Clark, 1964), Middle Ordovician, Kenosh Formation, Nevada.
Oistodus multicorrungnathus Harris, 1962, Middle Ordovician, Kenosh Formation, Nevada.
Oulodus oregonia (Branson, Mehl and Branson, 1971), Middle Ordovician, Chickamauga Formation, Tennessee.
Ozarkodina polinclinata (Nicoll and Rexroad, 1969), Lower Silurian, Clinton Group, New York.
Panderodus gracilis (Branson and Mehl, 1933), Middle Ordovician, Planilimbata Formation, Sweden.
Paraprioniodus costatus (Mound, 1965), Middle Ordovician, Kenosh Formation, Nevada.
Phragmodus undatus Branson and Mehl, 1933, Middle Ordovician, Dalecarlicus Formation, Sweden.
Plectodina aculeata (Stauffer, 1930), Middle Ordovician, Dalecarlicus Formation, Sweden.

Protopanderodus gradatus Serpagli, 1974, Middle Ordovician, Kenosh Formation, Nevada.
Pterocontiodus cryptodens (Mound, 1965), Middle Ordovician, Kenosh Formation, Nevada.
Pterospathodus amorphognathoides Walliser, 1964, Lower Silurian, Clinton Group, New York.
Pygodus anserinus Lamont and Lindström, 1957, Middle Ordovician, Little Oak Formation, Alabama.
Rhipidognathus symmetricus (Bergstrom and Sweet, 1966), Middle Ordovician, Hermitage Formation, Tennessee.
Scandodus sinnuosis Mound, 1965, Middle Ordovician, Kenosh Formation, Nevada.
Walliserodus sancticlairi Cooper, 1980, Lower Silurian, Lockport Group, New York.
Xaniognathus abstractus (Clark and Ethington, 1962), Upper Permian, Bell Canyon Formation, Texas.

EXPLANATION OF PLATES

PLATE 193. 5. Elongate, prismatic crystallites with dipyramid (A) and with pryamid and pinacoid (B); Sb element of *Xanignathus abstractus*. Scale bar = 5 microns. 6. Elongate, prismatic crystallites with pyramid and pinacoid; M element of *Kockella ranuliformis*. Scale bar = 5 microns. 7. Short, platy, prismatic crystallite with pinacoids and minor pyramidal development; P element of *Baltoniodus gerdae*. Scale bar = 20 microns.

PLATE 194. 8. Amorphous, cryptocrystalline masses in the basal filling of the M element of *Aphelognathus kimmswickensis*. Scale bar = 5 microns. 9. Elongate crystallites showing uniform packing along the prismatic and pinacoidal faces in the microcoelodontiform element of *Curtognathus typus*. Scale bar = 4 microns. 10. Elongate crystallites showing slight offsets in packing along the prismatic and pinacoidal faces; Pa element of *Phragmodus undatus*. Scale bar = 4 microns.

PLATE 195. 11. Elongate crystallites showing moderate offsets in packing along the prismatic and pinacoidal faces; Pa element of *Xanignathus abstractus*. Scale bar = 20 microns. 12. Elongate crystallites showing major offsets in packing along the prismatic and pinacoidal faces in The grandiform element of *Belodina compressa*. Scale bar = 6 microns. 13. Bundles of elongate crystallites showing offsets in packing along the prismatic and pinacoidal faces in the compressiform element of *Belodina compressa*. Scale bar = 3 microns.

PLATE 196. 14. Short, platy crystallites showing uniform packing along the prismatic and pinacoidal faces; Pb element of *Ozarkodina polinclinata*. Scale bar = 5 microns. 15. Short, platy crystallites showing slight offsets in packing along the prismatic and pinacoidal faces; M element of *Oulodus oregonia*. Scale bar = 3 microns. 16. Short, platy crystallites showing moderate offsets in packing along the prismatic and pinacoidal faces in the Pa element of *Icriodella superba*. Scale bar = 3 microns.

PLATE 197. 17. Short, platy crystallites showing major offsets in packing along the prismatic and pinacoidal faces in the Pb element of *Plectodina aculeata*. Scale bar = 5 microns. 18. Short, platy crystallites showing almost total offset along the prismatic and pinacoidal faces in the Sb element of *Plectodina aculeata*. Scale bar = 5 microns. 19. Elongate crystallites with the c-axes oriented parallel to the outer surface of the Sb element in *Erismodus radicans*. The arrow indicates the outer surface of the element. Scale bar = 4 microns.

PLATE 198. 20. Elongate crystallites with the c-axes oriented parallel to the outer surface of the Pb element in *Phragmodus undatus*. Note that the elongate prisms form the individual striae. Scale bar = 4 microns. 21. Short, platy crystallites with the c-axes oriented perpendicular to the outer surface of the Pb element of *Oulodus oregonia*. The arrow indicates the outer surface of the element. Scale bar = 5 microns. 22. Elongate crystallites with the c-axes oriented obliquely to the outer surface in the denticle of the grandiform element of *Belodina compressa*. The arrow indicates the outer surface of the element. Scale bar = 6 microns.

PLATE 199. 23. Bundles of elongate crystallites with the c-axes oriented at various oblique angles to the outer surface of compressiform element of *Belodina compressa.* The arrow indicates the outer surface of the element. Scale bar = 3 microns. 24. Amorphous, cryptocrystalline masses showing an open framework in the basal filling of the Sc element of *Oulodus oregonia.* Scale bar = 3 microns. 25. Amorphous, cryptocrystalline masses showing a partially closed framework in the basal filling of a gracilliform element of *Panderodus gracilis.* Scale bar = 4 microns.

PLATE 200. 26. Amorphous, cryptocrystalline masses showing a closed or solid framework in the basal filling of the homocurvatiform element of *Drepanoistodus suberectus.* Scale bar = 3 microns. 27. Amorphous, cryptocrystalline mass showing "corn flake" texture in the basal filling of the compressiform element of *Belodina compressa.* Scale bar = 10 microns. 28. Transition from elongate crystallites to amorphous, cryptocrystalline masses in the basal cavity of the tetraprioniodontiform element of *Multioistodus compressus.* Scale bar = 15 microns.

Part 11

References Cited

Alvarez, L.W., Alvarez, W., Asaro, F., *et al.*, 1980, Extraterrestrial cause for the Cretaceous Tertiary extinction. *Science*, v. 208, no. 4448, p. 1095-1108.

Andalib, F., 1972, Mineralogy and preservation of siphuncles in Jurassic cephalopods: *Neues Jahrbuch für Geologie und Paläontologie, Abhandlungen*, v. 140, p. 33-48.

Bandel, Klaus, 1979, The nacreous layer in the shells of the gastropod family Seguenziidae and its taxonomic significance: *Biomineralization Research Reports*, v. 10, p. 49-61.

Bandel, Klaus, 1982, Morphologie und Bildung der Frühontogenetische Gehäuse bei conchiferan Mollusken: *Facies*, v. 7, p. 1-198, 6 pls.

Bandel, Klaus, 1988, Early ontogenetic shell and shell structure as aids to unravel gastropod phylogeny and evolution: *Malacological Review*, 1988, Supplement 4, p. 267-272.

Bandel, Klaus, and Spaeth, Christian, 1984, Beobachtungen am rezenten *Nautilus*. *Mitteilungen aus dem Geologisch-Paläontologischen Institut der Universität Hamburg*, v. 54, p. 9-26.

Bandel, Klaus, Lehman, N.H., and Waage, K.M., 1982, Micro-ornament on early whorls of Mesozoic ammonites: implications for early ontogeny. *Journal of Paleontology*, v. 56, no. 2, p. 386-391.

Barnes, C.R., and Slack, D.J., 1975, Conodont ultrastructure: the subfamily Acanthodontidae: *Life Sciences Contributions, Royal Ontario Museum*, No. 106, p. 1-21, 6 figs.

Barnes, C.R., Sass, D.B., and Monroe, E.A., 1970, Preliminary studies of the ultrastructure of selected Ordovician conodonts: *Life Sciences Contributions, Royal Ontario Museum*, No. 76, p. 1-24, 10 pls.

Barnes, C.R., Sass, D.B., and Monroe, E.A., 1973, Ultrastructure of some Ordovician conodonts, p. 1-30 *in*: Rhodes, F.H.T., ed., *Conodont Paleozoology*: *Geological Society of America Special Paper*, no. 141, 296 p.

Bate, R.H., and Sheppard, L.M., 1982, The shell structure of *Halocypnis inflata* (Dana, 1849), p. 25-50 *in*: Bate, R.H., Robinson, E., and Sheppard, L.M., eds., *Fossil and Recent Ostracods*: E.H. Chichester, for the British Micropaleontological Society, 493 p.

Batten, R.L., 1967, Thoughts on the genus *Ptychomphalina* Fischer, 1887, and the family Eotomariidae Wenz, 1938: *Journal of Paleontology*, v. 41, no. 1, p. 262-265.

Batten, R.L., 1972, The ultrastructure of five common Pennsylvanian gastropod species of eastern United States: *American Museum Novitates*, No. 2501, p. 1-42.

Batten, R.L., 1975, The Scissurellidae - are they neotenously derived fissurellids? *American Museum Novitates*, No. 2567, p. 1-34.

Batten, R.L., 1984a, Shell structure of the Galapagos rift limpet *Neomphalus fretterae* McLean, 1981, with notes on muscle scars and insertions: *American Museum Novitates*, No. 2776, p. 1-13.

Batten, R.L., 1984b, The calcitic wall in the Paleozoic families Euomphalidae and Platyceratidae (Archaeogastropoda): *Journal of Paleontology*, v. 58, no. 5, p. 1186-1192.

Bayer, Ulf, 1978, Constructional morphology of ammonite septa. *Neues Jahrbuch für Geologie und Paläontologie*, Abhandlung B, v. 157, no. 1/2, p. 150-155.

Bergenhayn, J.R.M., 1930, Kurze Bemerkungen zur Kenntnis der Schalenstruktur und systematik der Loricaten: *Kungliga Svenska Vetenskapakademiens Handlingar*, ser. 3, v. 9, no. 3, p. 1-54, 10 pls.

Bergenhayn, J.R.M., 1960, Cambrian and Ordovician loricates from North America: *Journal of Paleontology*, v. 34, no. 1, p. 168-178.

Berg-Madsen, Vivianne, and Peel, J.S., 1978, Middle Cambrian monoplacophorans from Bornholm and Australia, and the systematic position of the bellerophontiform molluscs: *Lethaia*, v. 11, p. 113-125.

Bernat, Michel, 1975, Les isotopes de l'uranium et du thorium et les terres rares dans l'environment marin. *Cahiers ORSTROM, Série Géologie*, v. 7, p. 65-83.

Birkelund, T., and Hansen, H.J., 1974, Shell ultrastructures of some Maastrichtian Ammonoidea and Conoidea and their taxonomic implications. *Biologiske Skrifter*, v. 20, no. 6, p. 1-34, 16 pls.

Bischoff, G.C.O., 1981, *Cobcrephora* n.g., representative of a new polyplacophoran order Phosphatoloricata with calciumphosphatic shells: *Senckenbergiana Lethaea*, v. 61, no. 3, p. 173-215.

Blind, Wolfram, 1967, Die Wetterau; Structurelement und Lebensraum. *Natur und Museum (Senckenbergische Natur Forschende Gesellschaft, Bericht)*, v. 97, no. 2, p. 45-52.

Bøggild, O.B., 1930, The shell structure of the Mollusks: *Danske Videnskabernes Selskabs Skrifter. Naturvidenskabelig og Mathematisk Afdeling*, ser. 9, vol. 2, p. 231-326, 15 pls.

Bowsher, A.L., 1955, Origin and adaptation of platyceratid gastropods: *University of Kansas Paleontological Contributions, Mollusca*, Article 5, p. 1-11, 2 pls.

Bulykh, P.Y., 1983, *Kolongites* - a new heliolitid genus. *Paleontologicheskii Zhurnal*, 1983(1), p. 39-44. See English translation in *Paleontological Journal*, v. 17, no. 1, p. 35-41.

Burchette, T.P., and Riding, R., 1977, Attached vermiform gastropods in Carboniferous marginal marine stromatolites and biostromes: *Lethaia*, v. 10, p. 17-28.

Carroll, Robert, 1988, *Vertebrate Paleontology and Evolution*: New York, W.H. Freeman and Co., 698 p.

Carter, J.G., and Aller, R.C., 1975, Calcification in the bivalve periostracum: *Lethaia*, v. 8, p. 315-320.

Carter, J.G., and Clark, G.R. II, 1985, Classification and phylogenetic significance of molluscan shell microstructure, p. 50-71. *In*: Bottjer, D.J., Hickman, C.S., and Ward, P.D., eds., *Mollusks. Notes for a Short Course*. University of Tennessee Department of Geological Sciences, Studies in Geology 13, edited by T.W. Broadhead. The Paleontological Society.

Carter, J.G., and Tevesz, M.J.S., 1978, Shell microstructure of a Middle Devonian (Hamilton Group) bivalve fauna from central New York: *Journal of Paleontology*, v. 52, no. 4, p. 859-880.

Cayeux, Lucien, 1916, Introduction a l'étude pétrographique des roches sédimentaires: *Mémoire pour Servir à l'Explication de la Carte Géologique Détaillée de la France*: Ministere des Trauvaux Publics, Paris, Imprimerie Nationale, 2 vols., 534 p., 56 pls.

Clark, G.R., II, and West, R.R., 1984, Microstructure and morphology of Pennsylvanian specimens of the "scaphopod" *Plagioglypta*: *Geological Society of America Abstracts with Programs*, v. 16, p. 472.

Clarke, M.R., 1978, The cephalopod statolith - an introduction to its form. *Journal of the Marine Biological Association of the United Kingdom*, v. 58, no. 3, p. 701-712.

Compère, Philippe, and Goffinet, Gerhard, 1987, Ultrastructural shape and three-dimensional organization of the intracuticular canal system in the mineralized cuticle of the green crab *Carcinus maenas*: *Tissue and Cell*, v. 19, p. 839-857.

Couvreur, M., 1929, Anatomie microscopique des cérames des chitons: *Archives d'Anatomie Microscopique, Morphologie, et Expérimentale*, v. 25, p. 433-444, 1 pl.

Cox, L.R., *et al.*, 1969, Bivalvia, Part N, Mollusca 6, Volumes 1 and 2, *in*: Moore, R.C., ed., *Treatise on Invertebrate Paleontology*: Geological Society of America and University of Kansas Press, Lawrence, 952 p.

Dall, W.H., 1871, Preliminary sketch of a natural arrangement of the order Docoglossa. *Proceedings of the Boston Society of Natural History*, v. 14, p. 49-55.

de Baar, H.J.W., Bacon, M.P., Brewer, P.G., and Bruland, K. W., 1985, Rare earth elements in the Pacific and Atlantic Oceans. *Geochimica et Cosmochimica Acta*, v. 49, no. 9, p. 1943-1959.

Debrock, M.D., Hoare, R.D., and Mapes, R.H., 1984, Pennsylvanian (Desmoinesian) Polyplacophora (Mollusca) from Texas: *Journal of Paleontology*, v. 58, no. 4, p. 1117-1135.

Decker, 1969: cited incorrectly in Kingsley chapter (Volume I); should be "Decker, Morrill and Lennarz (1969)".

Decker, G.L., Morrill, J.B., and Lennarz, W.J., 1987, Characterization of sea urchin primary mesenchyme cells and spicules during biomineralization *in vitro*. *Development*, v. 101, p. 297-312.

Dullo, W.-C., and Bandel, Klaus, 1988, Diagenesis in molluscan shells: a case study from cephalopods, p. 719-729. *In*: Wiedman, J., and Kullmann, J., eds., *2nd International Cephalopod Symposium. Cephalopods Present and Past.* O.H. Schindewolf Symposium, Tubingen, 1985. E. Schweizerbart'sche Verlagsbuchhandlung, Stuttgart, p. 1-765.

Elderfield, H., and Greaves, M.J., 1982, The rare earth elements in seawater. *Nature*, v. 296, no. 5854, p. 214-219.

Erben, H.K., 1971, Anorganische und organische Schalenkomponenten bei *Cittarium pica* L. (Archaeogastropoda): *Biomineralization Research Reports*, v. 3, p. 51-64, 4 pls.

Erben, H.K., and Krampitz, G., 1972, Ultrastruktur und Aminosäuren-Verhältnisse in den Schalen der rezenten Pleurotomariidae (Gastropoda): *Biomineralization Research Reports*, v. 6, p. 12-31, 6 pls.

Fantinet, Daniel, 1959, Contribution à l'étude des Scaphopodes fossiles de l'Afrique du Nord: *Publications du Service de la Carte Géologique de l'Algérie, new series, Paléontologie, Mémoire no. 1*, p. 1-112, 12 pls.

Florkin, Marcel, 1969, Fossil shell "conchiolin" and other preserved biopolymers, p. 498-520 *in*: Eglington, Geoffrey, and Murphy, M.T.J., eds., *Organic Geochemistry*: *Methods and Results*: Springer-Verlag, Berlin, New York, 828 p.

Gaspard, Daniéle, 1986, Aspects figurés de la biominéralisation unités de base de la sécrétion carbonatée chez des Terebratulida actuels, *in*: Racheboeuf, P.R., and Emig, Christian, eds., *Biostratigraphie du Paléozoique 4, Les Brachiopodes Fossiles et Actuels.* 500 p., 197 figs., 30 pls., see p. 77-83.

Gibbins, 1969: cited incorrectly in Kingsley chapter (Volume I); should be "Gibbins, Tilney and Porter (1969)".

Gibbins, J.R., Tilney, L.G., and Porter, K.R., 1969, Microtubules in the formation and development of the primary mesenchyme in *Arbacia punctulata.* I. The distribution of microtubules. *Journal of Cell Biology*, v. 41, p. 201-226.

Giraud-Guille, Marie-Madeleine, 1984a, Fine structure of the chitin-protein system in the crab cuticle: *Tissue and Cell*, v. 16, p. 75-82.

Giraud-Guille, Marie Madeleine, 1984b, Calcification initiation sites in the crab cuticle: the interprismatic septa. An ultrastructural cytochemical study: *Cell and Tissue Research*, v. 236, p. 413-420.

Goldberg, W.M., 1988, Chemistry, histochemistry and microscopy of the organic matrix of spicules from a gorgonian coral. Relationship to Alcian blue staining and calcium binding. *Histochemistry*, v. 89, no. 2, p. 163-170.

Goldberg, W.M. and Benayahu, Y., 1987, Spicule formation in the gorgonian coral *Pseudoplexaura flagellosa.* 1: Demonstration of intracellular and extracellular growth and the effect of ruthenium red during decalcification. *Bulletin of Marine Science*, v. 40, p. 287-303.

Golikov, A.N., and Starobogatov, Y.I., 1975, Systematics of prosobranch gastropods: *Malacologia*, v. 15, p. 185-232.

Greenaway, Peter, 1985, Calcium balance and moulting in the Crustacea: *Biological Reviews of the Cambridge Philosophical Society*, v. 60, p. 425-454.

Haas, Winfried, 1972a, Untersuchungen über die Mikro- und Ultrastruktur der Polyplacophorenschale: *Biomineralization Research Reports*, v. 5, p. 1-52, 18 pls.

Haas, Winfried, 1972b, Micro- and ultrastructure of Recent and fossil Scaphopoda: *Proceedings of the 24th International Geological Congress*, Section 4, p. 15-19.

Haas, Winfried, 1976, Observations on the shell and mantle of the Placophora, p. 389-402 *in*: Watabe, Norimitsu, and Wilbur, K.M., eds., *The Mechanisms of Mineralization in the Invertebrates and Plants.* Columbia, South Carolina, University of South Carolina Press, 461 p.

Haas, Winfried, 1981, Evolution of calcareous hardparts in primitive molluscs: *Malacologia*, v. 21, p. 403-518.

Haas, Winfried, and Kriesten, Klaus, 1974, Studien über das Mantelepithel von *Lepidochitona cinerea* (L.) (Placophora): *Biomineralization Research Reports*, v. 7, p. 100-109.

Haas, Winfried, and Kriesten, Klaus, 1975, Studien über das Perinotumepithel und die Bildung von Kalkstacheln von *Lepidochitona cinerea* (L.) (Placophora): *Biomineralization Research Reports*, v. 8, p. 92-107.

Haas, Winfried, and Kriesten, Klaus, 1975, Studien über das Epithel und die kalkigen Hartgebilde des Perinotums bei *Acanthopleura granulata* (Gmelin) (Placophora): *Biomineralization Research Reports*, v. 9, p. 11-27, 8 pls.

Hackman, H.R., 1984, Arthropoda. Cuticle: biochemistry, p. 583-610 *in*: Bereiter-Hahn, Jürgen, Matoltsy, A.G., and Richards, K.S., eds., *Biology of the Integument. 1. Invertebrates.* Berlin, Heidelberg, New York, Tokyo; Springer-Verlag, 841 p.

Hall, J.C., 1989, The form and packing of crystallites in conodonts: *Geological Society of America Abstracts with Programs*, v. 21, no. 3, p. 19.

Haas, W.H., and Lindberg, M.L., 1946, Orientation of crystal units in conodonts: *Journal of Paleontology*, v. 20, p. 501-504.

Harmer, S.F., 1902, On the morphology of the Cheilostomata. *Quarterly Journal of Microscopical Science*, v. 46, p. 263-450, pls. 15-18.

Hartman, W.D., 1979, A new sclerosponge from the Bahamas and its relationship to Mesozoic stromatoporoids. *Colloques Internationaux du Centre National de la Recherche Scientifique*, no. 291, p. 467-474.

Hassler, S.E., and Hall, J.C., 1989, Variations in crystallite orientation in the lamellae of the Conodontophorida: *Geological Society of America Abstracts with Programs*, v. 21, no. 3, p. 20-21.

Haszprunar, Gerhard, 1988, A preliminary phylogenetic analysis of the streptoneurous gastropods: *Malacological Review*, 1988, Supplement 4, p. 7-16.

Hewitt, R.A., Lazell, B.H., and Moorhouse, S.J., 1983, An introduction to the inorganic components of cephalopod shells. *Neues Jahrbuch für Geologie und Paläontologie, Abhandlungen*, v. 165, no. 3, p. 331-361.

Hickman, C.S., 1984, *Pleurotomaria*: pedigreed perseverence? p. 225-231 *in*: Eldredge, Niles, and Stanley, S.M., *Living Fossils*, Springer Verlag, New York, Berlin.

Hickman, C.S., 1988, Archaeogastropod evolution, phylogeny and systematics: a re-evaluation: *Malacological Review*, 1988, Supplement 4, p. 17-34.

Hyman, L.H, 1967, *The Invertebrates, Volume 6, Mollusca I*: McGraw-Hill, New York, 792 p.

Jensen, M., 1972, The ultrastructure of the echinoid test: *Sarsia*, v. 48, p. 39-48.

Jeyasuria, P., and Lewis, J.C., 1987, Mechanical properties of the axial skeleton in gorgonians. *Coral Reefs*, v. 5, no. 4, 213-219.

Jones, W.C., 1978, The microstructure and genesis of sponge biominerals. *Colloque International du CNRS*, v. 291, p. 425-447.

Jones, W.C., and Ledger, P.W., 1986, The effect of diamox and various combinations of calcium on spicule secretion in the calcareous sponge *Sycon ciliatum*. *Comparative Biochemistry and Physiology*, A, v. 84, no. 1, p. 149-158.

Jope, Margaret, 1971, Constituents of brachiopod shells, *in*: Florkin, Marcel, and Stotz, E.H., eds., *Comprehensive Biochemistry, v. 26, part C. Extracellular and Supporting Structures, (continued)*. Amsterdam, Elsevier, p. 749-784.

Kabanov, G.K., 1967, [Skeleton of belemnoids. Morphology and biological analysis]. *Trudy Paleontological Institute*, v. 114, p. 1-100. [In Russian]

Keen, A.M., 1971, *Sea Shells of Tropical West America*, 2nd ed. Stanford University Press, 1064 p.

Kennett, J.A.B., and Dalingwater, J.E., 1986, A Progress report on using aspects of reproduction as criteria for distinguishing two closely related British *Tegenaria* species, p. 139-142. *In*: Eberhard, W.G., Lubin, Y.D., and Robinson, B.C., *Proceedings of the Ninth International Congress of Arachnology, Panama, 1983*. Smithsonian Institution Press, Washington, D.C., 333 p.

Kingsley, R.J., and Watabe, Norimitsu, 1987, Role of carbonic anhydrase in calcification in the gorgonian *Leptogorgia virgulata*. *Journal of Experimental Zoology*, v. 241, no. 2, p. 171-180.

Knight, J.B., 1931, The gastropods of the St. Louis, Missouri, Pennsylvanian outlier: the Subulitidae. *Journal of Paleontology*, v. 5, no. 3, p. 177-229, pls. 21-27. {proposes the term "inductura" for sheet-like parietal deposits}

Knight, J.B., 1934, The gastropods of the St. Louis, Missouri, Pennsylvanian outlier: VII. The Euomphalidae and Platyceratidae: *Journal of Paleontology*, v. 8, no. 2, p. 139-166, pls. 20-26.

Knight, J.B., 1946, [Review of] Les différences ches les gastéropodes capuliformes; organization des Platyceratidae, by Genevieve Delpey, 1940: *Geological Magazine*, v. 83, p. 280-284.

Knight, J.B., 1952, Primitive fossil gastropods and their bearing on gastropod classification: *Smithsonian Miscellaneous Collections*, v. 117, no. 13 (Publ. 4092), 56 p., 2 pls.

Knight, J.B., Cox, L.R., Keen, A.M., et al., 1960, *Treatise on Invertebrate Paleontology, Part 1, Mollusca 1*. Geological Society of America and University of Kansas Press, Lawrence, Kansas, xxiii + 351 p.

Larwood, G.P. and Taylor, P.D., 1979, Early structural and ecological diversification of the Bryozoa. *Systematics Association Special Volume No. 12*, p. 209-234.

Laubenfels, M.W. de, 1955, Sponges of Onotoa. *Pacific Science*, v. 9, no. 2, p. 137-143.

Laubenfels, M.W. de, 1956, Preliminary discussion of the sponges of Brazil. *Contribuicoes avulsas do Instituto Oceanographico Sao Paulo, Oceanogr. Biol. No. 1*, p. 1-4.

Lecompte, Marius, 1956, Stromatoporoidea, p. F107-F144. *In*: Bayer, F.M., Boschma, Hilbrand, Harrington, H.J., *et al.*, Part F, Coelenterata. *In*: Moore, R.C., ed., *Treatise on Invertebrate Paleontology*. Geological Society of America and University of Kansas Press, Lawrence, 498 p.
Ledger, P.W., and Jones, W.C., 1977, Spicule formation in the calcareous sponge *Sycon ciliatum*. *Cell and Tissue Research*, v. 181, p. 553-567.
Lehmann, U., 1967, Ammoniten mit Kieferapparat und radula aus Lias-Geschieben. *Paläontologische Zeitschrift*, v. 41, p. 38-45.
Lindberg, D.R., 1981, Rhodopetalinae, a new subfamily of Acmaeidae from the boreal Pacific: anatomy and systematics: *Malacologia*, v. 20, p. 291-305.
Lindberg, D.R., 1986, Radular evolution in the Patellogastropoda: *American Malacological Bulletin*, v. 4, p. 115.
Lindberg, D.R., 1988, The Patellogastropoda: *Malacological Review*, 1988, Supplement 4, p. 35-63.
Lindström, Maurits, 1964, *Conodonts*: Elsevier, Amsterdam, 196 p.
Lindström, Maurits, and Ziegler, Willi, 1971, Feinstrukturelle Untersuchungen an Conodonten 1. Die Überfamilie Panderodontacea: *Geologica et Palaeontologica*, v. 5, p. 9-17.
Linsley, R.M., and Kier, W.M., 1984, The Paragastropoda: a proposal for a new class of Paleozoic Mollusca: *Malacologia*, v. 25, no. 1, p. 241-254.
MacClintock, Copeland, 1967, Shell structure of patelloid and bellerophontoid gastropods (Mollusca): *Yale University Peabody Museum, Bulletin*, v. 22, p. 1-140.
MacKinnon, D.I., 1971, Perforate canopies to canals in the shells of fossil Brachiopoda. *Lethaia*, v. 4, p. 321-325.
MacKinnon, D.I., 1985, New Zealand late Middle Cambrian molluscs and the origin of Rostroconchia and Bivalvia: *Alcheringa*, v. 9, p. 65-81.
Macurda, D.B., 1976, Skeletal modifications related to food capture and feeding behavior of the basket star *Astrophyton*: *Paleobiology*, v. 2, p. 1-7.
Macurda, D.B., and Meyer, D.L., 1975, The microstructure of the crinoid endoskeleton: *University of Kansas Paleontological Contributions*, Paper 74, p. 1-22, 30 pls.
Macurda, D.B., and Meyer, D.L., 1976, The morphology and life habits of the abyssal crinoid *Bathycrinus aldrichianus* Wyville Thomson and its paleontological implications: *Journal of Paleontology*, v. 50, p. 647-667.
Macurda, D.B., and Roux, M., 1981, The skeletal morphology of the isocrinid crinoids *Annacrinus wyvillethomsoni* and *Diplocrinus maclearanus*: *Contributions from the Museum of Paleontology, University of Michigan*, v. 25, no. 9, p. 169-219.
McLean, J.H., 1981, The Galapagos Rift limpet *Neomphalus*: relevance to understanding the evolution of a major Paleozoic-Mesozoic radiation: *Malacologia*, v. 21, p. 291-336.
McLean, J.H., 1984, A case for derivation of the Fissurellidae from the Bellerophontacea: *Malacologia*, v. 25, p. 3-20.
Miller, John, and Clarkson, E.N.K., 1980, The post-ecdysial development of the cuticle and the eye of Devonian trilobite *Phacops rana milleri* Stewart, 1927: *Philosophical Transactions of the Royal Society of London*, ser. B, v. 288, p. 461-480, 7 pls.
Moore, R.C., 1941, Upper Pennsylvanian gastropods from Kansas: *State Geological Survey of Kansas, Bulletin 38, 1941 Reports of Studies, Part 4*, p. 121-164, pls. 1-3.
Morris, N.J., and Clevely, R.J., 1981, *Phanerotinus cristatus* (Phillips) and the nature of euomphalacean gastropods: *Bulletin of the British Museum (Natural History), Geology*, v. 35, p. 195-212.
Müller, A.H., 1960, *Lehrbuch der Paläozoologie, Band II, Invertebraten, Teil 2, Mollusca - Arthropoda 1*, xii + 448 p.
Mutvei, Harry, 1981, Exoskeletal structure in the Ordovician trilobite *Flexicalymene*: *Lethaia*, v. 14, p. 225-234.
Mutvei, Harry; Dauphin, Yannicke, and Cuif, Jean-Pierre, 1985, Observations sur l'organisation de la couche externe du test des *Haliotis* (Gastropoda): un cas exceptionnel de variabilité minéralogique et microstructurale: *Bulletin du Muséum Nationale d'Histoire Naturelle, Paris*, ser. 4, v. 7, sect. A, no. 1, p. 73-91.
Naef, Adolf, 1921-1928, Die Cephalopoden: Fauna e Flora di Golfo di Napoli. *Stazione Zool. di Napoli, Mon. 35*, pt. 1, v. 1-2. [p. 1-148, pls. 1-19 of v. 1 appeared in 1921; p. 149-863 of v. 1 appeared in 1923; v. 2, 364 p. and 37 pls., appeared in 1928]
Neville, A.C., 1975, *Biology of the Arthropod Cuticle*. Berlin, Heidelberg, New York, Springer-Verlag, 448 p.
Nyers, A.J., 1987, *Thorslundella*: a proposed early Cambrian protogastropod that secreted a phosphatic shell due to environmental constraints: *Neues Jahrbuch für Geologie und Paläontologie, Abhandlungen*, v. 174, no. 2, p. 171-192.

Orton, J.H., 1937, Oyster biology and oyster-culture. Buckland Lectures for 1935. Edward Arnold & Co., London. 211 p.

Peel, J.S., 1972, Observations on some Lower Paleozoic tremanotiform Bellerophontacea (Gastropoda) from North america: *Palaeontology*, v. 15, p. 412-422.

Peel, J.S., 1976, Musculature and systematic position of *Megalomphala taenia* (Bellerophontacea, Gastropoda) from the Silurian of Gotland: *Bulletin of the Geological Society of Denmark*, v. 23, p. 231-264.

Peel, J.S., 1980, A new Silurian retractile monoplacophoran and the origin of the gastropods: *Proceedings of the Geologists Association*, v. 91, p. 91-97.

Pietzner, Horst; Vahl, Johanna; Werner, Hans; and Ziegler, Willi, 1968, Zur chemischen Zusammensetzung und Mikromorphologie der Conodonten: *Palaeontographica*, v. 128, p. 115-152, pls. 18-27.

Pilsbry, H.A., 1891, *Manual of Conchology Volume XIII. Acmaeidae, Lepetidae, Patellidae, Titiscaniidae*. Philadelphia, Pennsylvania, 195 p., 74 pls.

Pojeta, J. Jr., 1980, Molluscan phylogeny: *Tulane Studies in Geology and Paleontology*, v. 16, p. 55-80.

Prenant, M., 1925, Contribution a l'étude II. Sur les conditions de formation des spicules de Didemnidae. Bulletin Biologique de la France et de la Belgique, v. 59, p. 403-435.

Quenstedt, F.A., 1845-1849, *Petrefactenkunde Deutschlands. Band 1, Die Cephalopoden*, 581 p., atlas with 36 pls, Fues (Tübingen).

Quenstedt, F.A., 1852, *Handbuch der Petrefactenkunde, 2nd ed.*, v + 792 p., atlas 62 pls., Laupp (Tübingen).

Quenstedt, F.A., 1858, *Der Jura*. Tübingen, 842 p.

Rhodes, F.H.T., and Wingard, P.S., 1957, Chemical composition, microstructure and affinities of the Neurodontiformes: *Journal of Paleontology*, v. 31, p. 448-454.

Roer, R.D., and Dillaman, Richard, 1984, The structure and calcification of the crustacean cuticle: *American Zoologist*, v. 24, p. 893-909.

Rollins, H.B., 1966, Morphological observations on the bellerophont *Ptomatis patulus* (Hall) (Gastropoda, Bellerophontacea): *American Museum Novitates*, No. 2242, p. 1-7.

Rollins, H.B., 1967, *The phylogeny and functional morphology of the Knightitinae, Carinaropsinae, and Praematuratropidae (Gastropoda; Bellerophontacea)*: Doctoral Dissertation, Columbia University, p. 1-182.

Rollins, H.B., Eldredge, Niles, and Spiller, Judith, 1971, Gastropoda and Monoplacophora of the Solsville Member (Middle Devonian, Marcellus Formation) in the Chenango Valley, New York State: *Bulletin of the American Museum of Natural History*, v. 144, no. 2, p. 129-170.

Rosenfeld, Amnon, 1979, Structure and secretion of the carapace in some living ostracodes: *Lethaia*, v. 12, p. 353-360.

Roux, Michel, 1977a, Les Bourgueticrinina (Crinoidea) recueillis par la "Thalassa" dans le Golfe de Gascogne: anatomie comparée des pédoncules et systématique: *Bulletin du Museum National d'Histoire Naturelle, Sect. A, Zoologie, Biologie, et Geologie Animales*, v. 296, p. 25-84.

Roux, Michel, 1977b, The stalk-joints of Recent Isocrinidae (Crinoidea): *Bulletin of the British Museum (Natural History), Zoology*, v. 32, p. 45-64, 20 figs.

Ruedemann, Rudolph, 1905, The structure of some primitive cephalopods. *Albany, New York State Education Department, Museum Bulletin*, no. 80, p. 296-341.

Runnegar, Bruce, 1981, Muscle scars, shell form and torsion in Cambrian and Ordovician molluscs: *Lethaia*, v. 14, p. 311-322.

Runnegar, Bruce, 1983, Molluscan phylogeny revisited: *Memoirs of the Association of Australasian Paleontologists*, v. 1, p. 121-144.

Runnegar, Bruce, 1985, Shell microstructures of Cambrian molluscs replicated by phosphate: *Alcheringa*, v. 9, p. 245-257.

Runnegar, Bruce, and Jell, P.A., 1976, Australian Middle Cambrian molluscs and their bearing on early molluscan evolution: *Alcheringa*, v. 1, p. 109-138.

Runnegar, Bruce, Pojeta, J. Jr., Taylor, M.E., and Collins, D., 1979, New species of the Cambrian and Ordovician chitons *Matthevia* and *Chelodes* from Wisconsin and Queensland: evidence for the early history of polyplacophoran mollusks: *Journal of Paleontology*, v. 53, no. 6, p. 1374-1394.

Salvini-Plawen, L.v., 1980, A reconsideration of systematics in the Mollusca (phylogeny and higher classification): *Malacologia*, v. 19, p. 249-278.

Salvini-Plawen, L.v., 1981, On the origin and evolution of the Mollusca: *Atti dei Convegni Lincei*, v. 49, p. 235-293.

Salvini-Plawen, L.v., 1985, Early evolution and the primitive groups, p. 59-150 *in*: Trueman, E.R., and Clark, M., eds., *The Mollusca, Vol. 10, Molluscan Evolution*, Academic Press, London.

Salvini-Plawen, L.v., and Haszprunar, G., 1987, The Vetigastropoda and the systematics of streptoneurous Gastropoda (Mollusca): *Journal of Zoology, London*, v. 211, p. 747-770.

Scherer, M., 1986, 3. Paleozoic reefs. Diagenesis of aragonitic sponges from Permian patch reefs of southern Tunisia, p. 291-310. *In*: Schroeder, J.H., and Purser, B.H., eds., *Reef Diagensis*. Springer-Verlag, Berlin, Heidelberg, etc., i-viii + 455 p.

Schindewolf, O.H., 1941, Zur Kenntnis der Heterophylliden, einer eigentümlichen paläozoischen Korallengruppe. *Paläontologische Zeitschrift*, v. 22, p. 213-306, pls. 9-16.

Schmidt, W.J., 1923, Über den Bau der Perlen mit besonderer Berücksichtigung ihrer kristallinischen Elementarteile. *Archiv für Mikroskopische Anatomie und Entwicklungsgeschichte*, v. 97, p. 251-282, pls. 15-16. {See also Schmidt, 1924, sometimes cited as 1923}

Silverman, H., Steffens, W.L., and Dietz, T.H., 1985, Calcium from extracellular concretions in the gills of freshwater unionid mussels is mobilized during reproduction. *Journal of Experimental Zoology*, v. 236, p. 137-142.

Simkiss, Kenneth, and Wilbur, K.M., 1989, *Biomineralization. Cell Biology and Mineral Deposition*. Academic Press, Inc., San Diego, New York, etc., 337 p.

Smith, A.B., 1980, Stereom microstructure of the echinoid test: *Special Papers in Palaeontology no. 25*, The Palaeontological Association, London, 81 p., 23 pls., 20 text-figures.

Smith, A.G., 1960, Amphineura, p. I41-I76 *in*: Knight, J.B., Cox, L.R., Keen, A.M., *et al.*, *Treatise on Invertebrate Paleontology, Part I, Mollusca 1*, Geological Society of America and University of Kansas Press, Lawrence, 351 p.

Sohn, I.G., and Kornicker, L.S., 1988, Ultrastructure of myodocopid shells (Ostracoda), p. 243-258 *in*: Hanai, Tetsuro; Ikeya, Noriyuki; and Ishizaki, Kunihiro, eds., *Evolutionary Biology of Ostracoda. Its Fundamentals and Applications*. Proceedings of the Ninth International Symposium on Ostracoda, held in Shizuoka, Japan, 29 July - 2 August, 1985. Elsevier, Kodansha, Tokyo, 1356 p.

Squires, R.L., 1976, Color pattern of *Naticopsis* (*Naticopsis*) *wortheniana*, Buckhorn Asphalt deposit, Oklahoma: *Journal of Paleontology*, v. 50, no. 2, p. 349-353.

Stehli, F.J., 1956, Shell mineralogy in Paleozoic invertebrates: *Science*, v. 123, p. 1031-1032.

Stricker, S.A., 1985, The ultrastructure and formation of the calcareous ossicles in the body wall of the sea cucumber *Leptosynapta clarki* (Echinodermata, Holothuroida): *Zoomorphology*, v. 105, p. 209-222.

Thiem, H., 1917, Beitrage zur Anatomie und Phylogenie der Docoglossen; II. Die Anatomie und Phylogenie der Monobranchen (Akmaiden und Scurriiden nach der Sammlung Plates): *Jenaische Zeitschrift für Naturwissenschaft*, v. 54, p. 405-630.

Thompson, E.H., 1969, *Morphology and taxonomy of* Cyclonema *Hall (Gastropoda), Upper Ordovician, Cincinnatian Province*: Masters Thesis, Miami University, Oxford, Ohio, p. 1-148.

Thompson, E.H., 1970, Morphology and taxonomy of *Cyclonema* Hall (Gastropoda), Upper Ordovician, Cincinnatian Province: *Bulletins of American Paleontology*, v. 58, no. 261, p. 219-284.

Tsujii, T., Sharp, D.G., and Wilbur, K.M., 1958, Studies on shell formation. VII. The submicroscopic structure of the shell of the oyster *Crassostrea virginica*. *Journal of Biophysical and Biochemical Cytology*, v. 4, p. 275-280, 4 pls.

Ulrich, E.O., and Scofield, W.H., 1897, The Lower Silurian Gastropoda of Minnesota: *Minnesota Geological Survey*, v. 3, pt. 2, p. 813-1081, pls. 61-82.

Von Brand, T., 1973, *Biochemistry of Parasites*. Second edition. Academic Press, New York, i-xii, 499 p.

Wängberg-Eriksson, 1979, K., 1979, Macluritacean gastropods from the Ordovician and Silurian of Sweden: *Sveriges Geologiska Undersökning*, ser. C, no. 758, Arsbok 72, no. 20, p. 1-33.

Ward, P.D., 1987, *The Natural History of* Nautilus. Allen and Unwin, Inc., Boston, London, etc., p. i-xiv, 267 p.

Waterhouse, J.B., 1963, Permian gastropods of New Zealand. Part 1 - Bellerophontacea and Euomphalacea: *New Zealand Journal of Geology and Geophysics*, v. 6, no. 1, p. 88-112, 37 figs.

Weller, J.M., 1930, A new species of *Euphemus*: *Journal of Paleontology*, v. 4, p. 14-21.

Wilde, Pat, 1987, Model of progressive ventilation of the late Precambrian - Early Paleozoic ocean. *American Journal of Science*, v. 287, p. 442-459.

Williams, Alwyn, 1971, Comments on the growth of the shell of articulate brachiopods, *in* Dutro, J.T., Jr., ed., *Paleozoic Perspectives: a Paleontological Tribute to G. Arthur Cooper. Smithsonian Contributions to Paleobiology*, v. 3, p. 47-67.

Wilmot, N.V., and Fallick, A.E., 1989, Original mineralogy of trilobite exoskeletons: *Palaeontology*, v. 32, no. 2, p. 297-304.

Worms, D., and Weiner, Stephen, 1986, Mollusk shell organic matrix: fourier transform infrared study of the acidic macromolecules. *Journal of Experimental Zoology*, v. 237, p. 11-20.

Yochelson, E.L., 1967, Quo vadis, *Bellerophon*? p. 141-161 *in*: Teichert, Curt, and Yochelson, E.L., eds., *Essays in Paleontology and Stratigraphy*, University of Kansas Press, Lawrence.

Yochelson, E.L., 1975, Discussion of Early Cambrian "mollusks": *Journal of the Geological Society of London*, v. 131, no. 6, p. 661-662.

Yochelson, E.L., and Nuelle, L.M., 1985, *Strepsodiscus* (Gastropoda) in the Late Cambrian of Missouri: *Journal of Paleontology*, v. 59, no. 3, p. 733-740.

Yü, Wen, 1984, On merismoconchids: *Acta Paleontologica Sinica*, v. 23, no. 4, p. 432-446, 3 pls.

Ziegler, B., and Rietschel, S., 1970, Phylogenetic relationships of fossil calcisponges. *Symposia of the Zoological Society of London*, no. 25, p. 23-40.

Part 12

Taxonomic Index

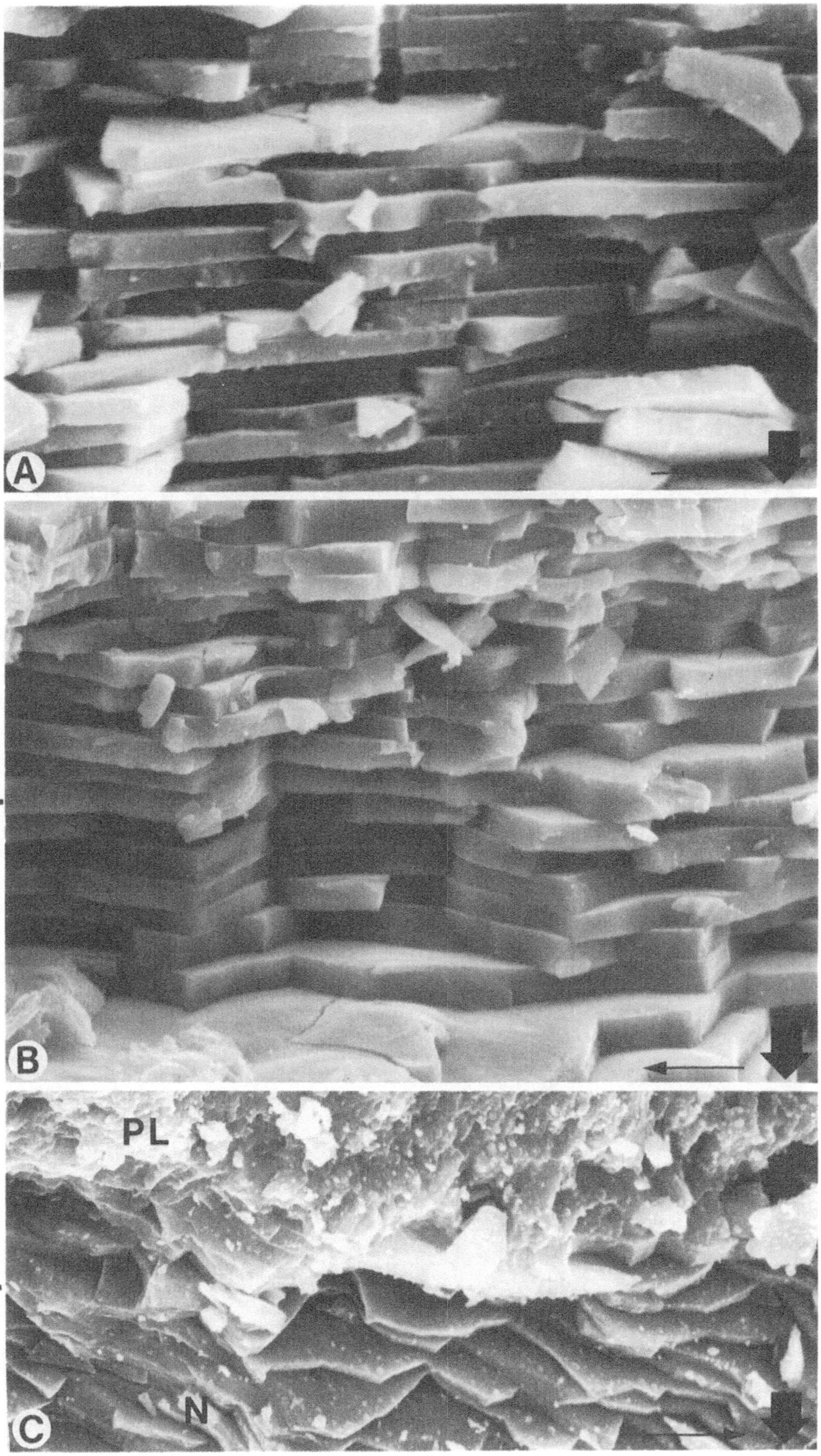

1. R 1

21. R 5

85. R 5

PLATE 2

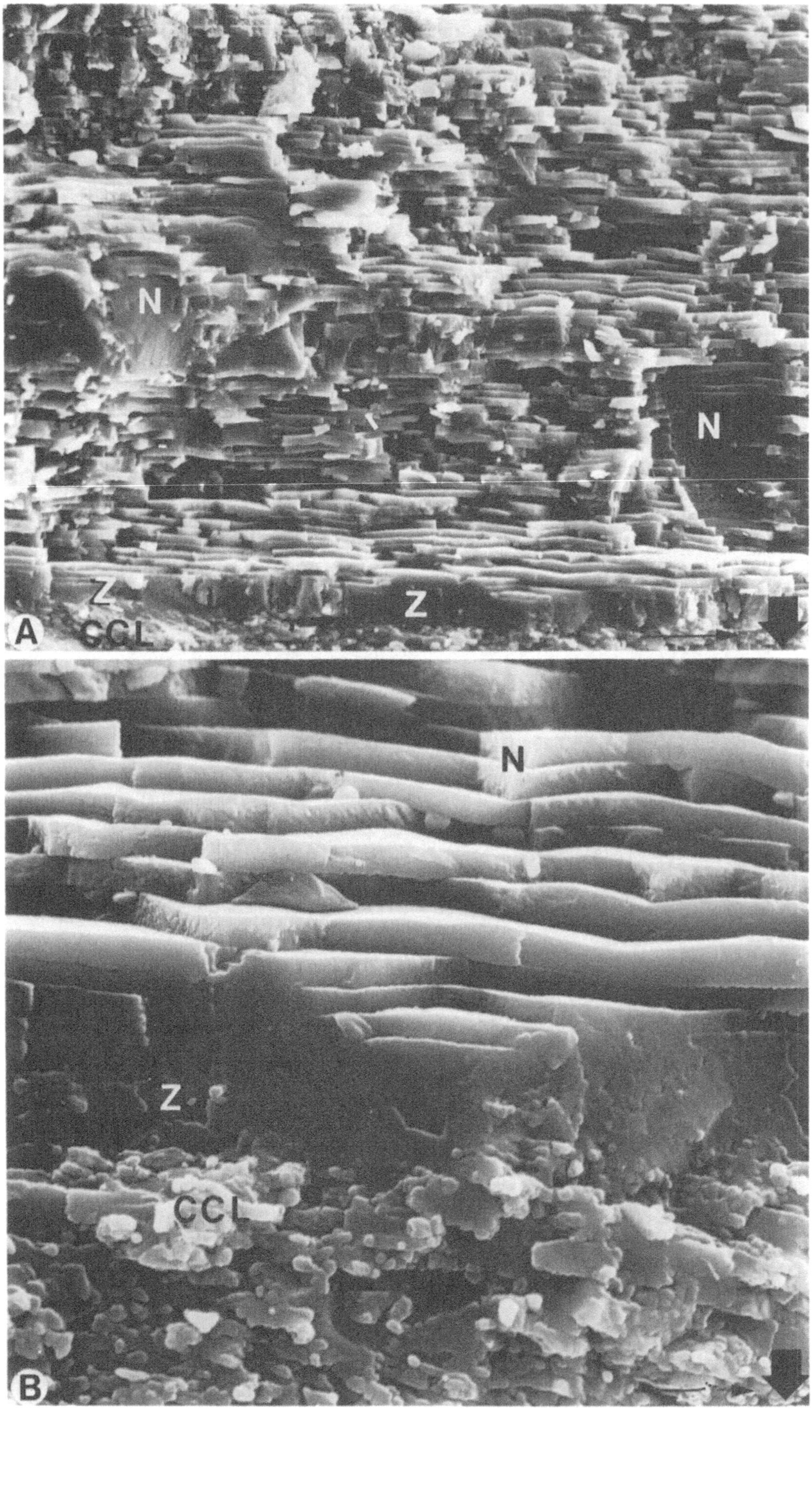

100. R 5

100. R 1

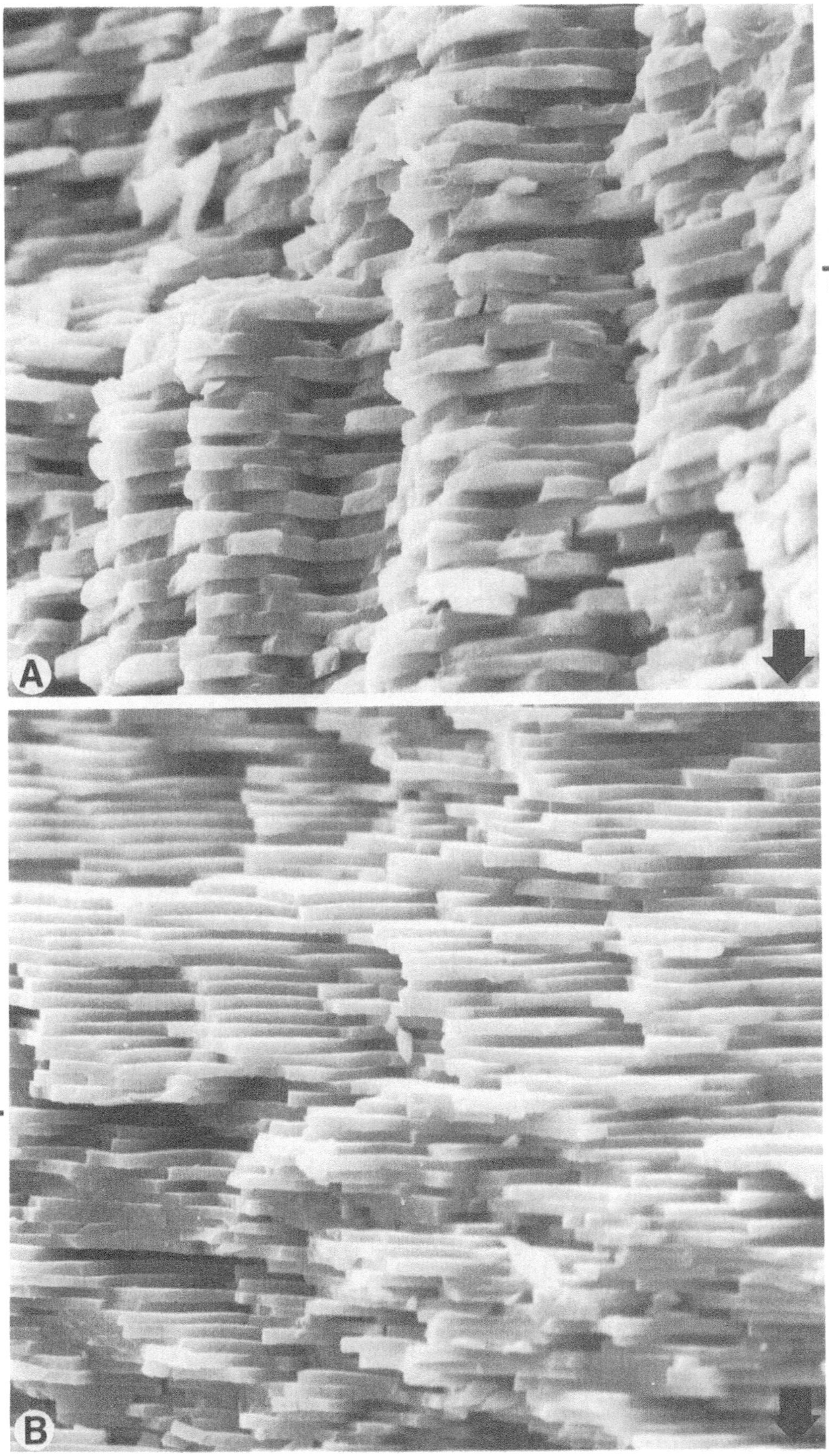

20. T 2

20. T 2

11.

O

1

11.

O

1

17.

T

1

17.

H

1

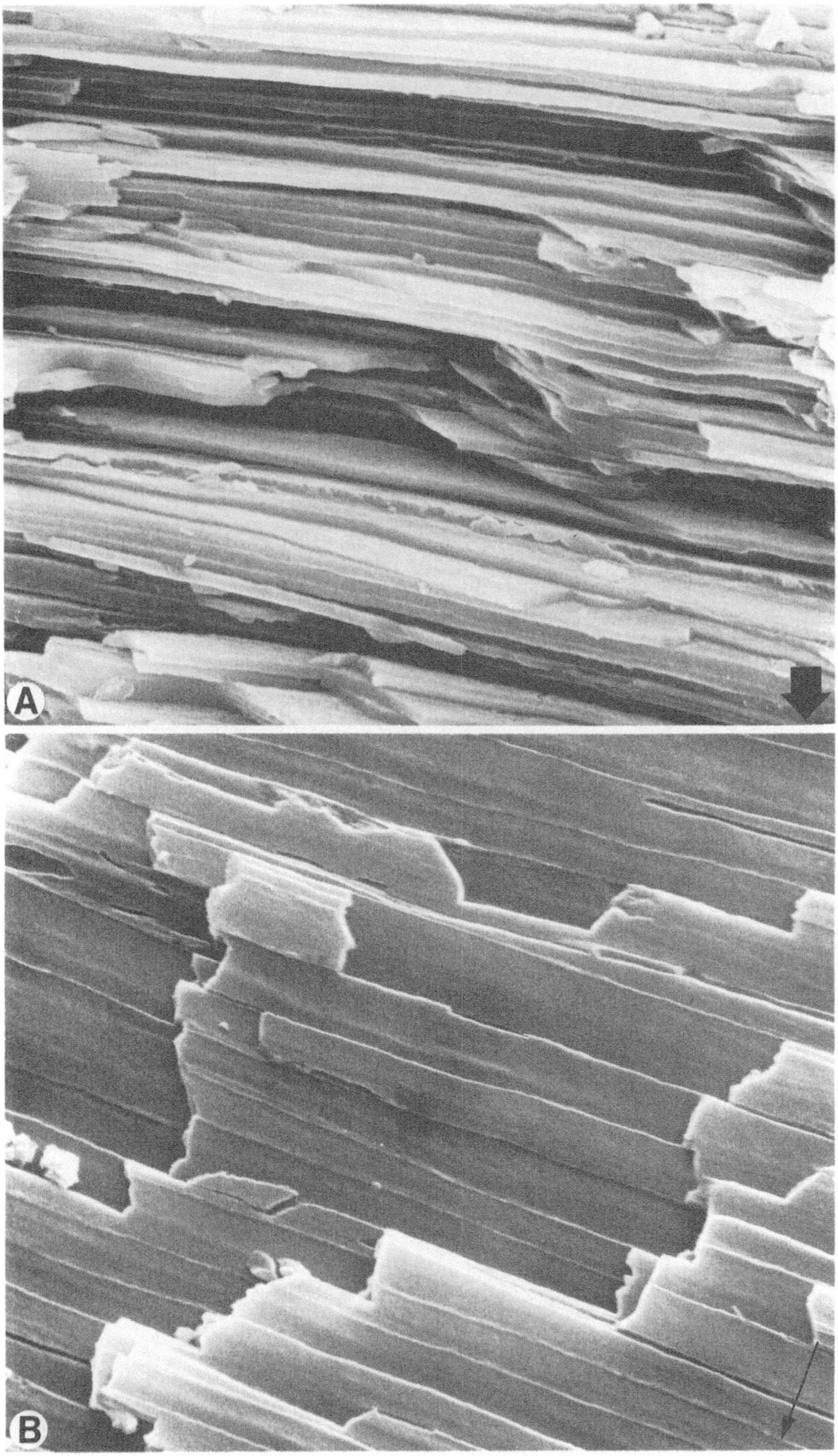

16.

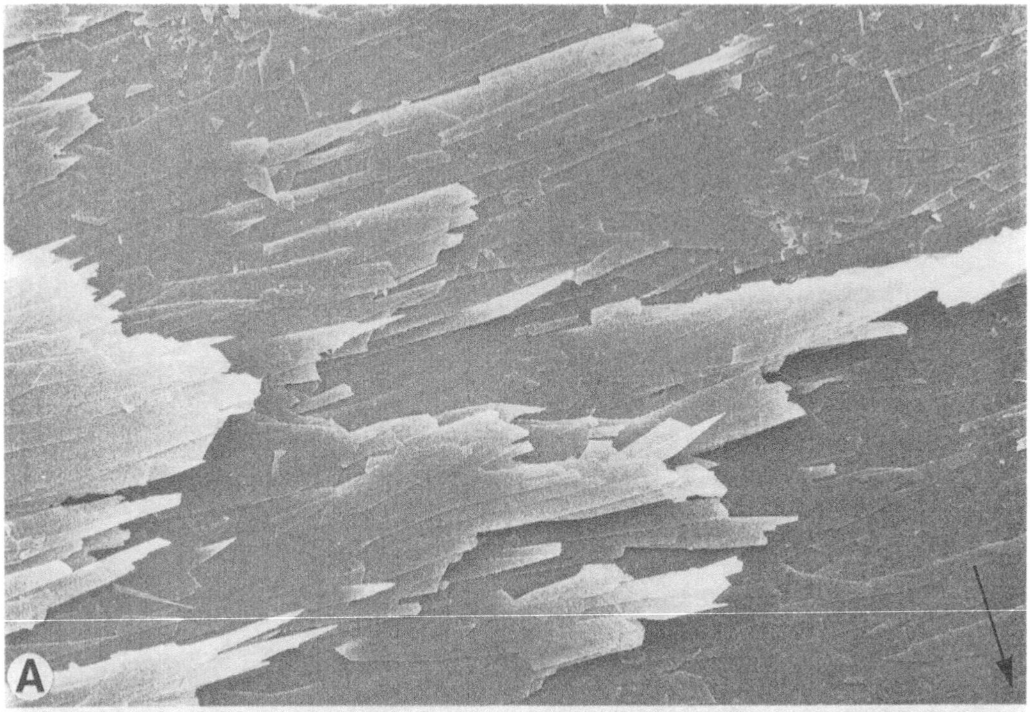

H

20

16.

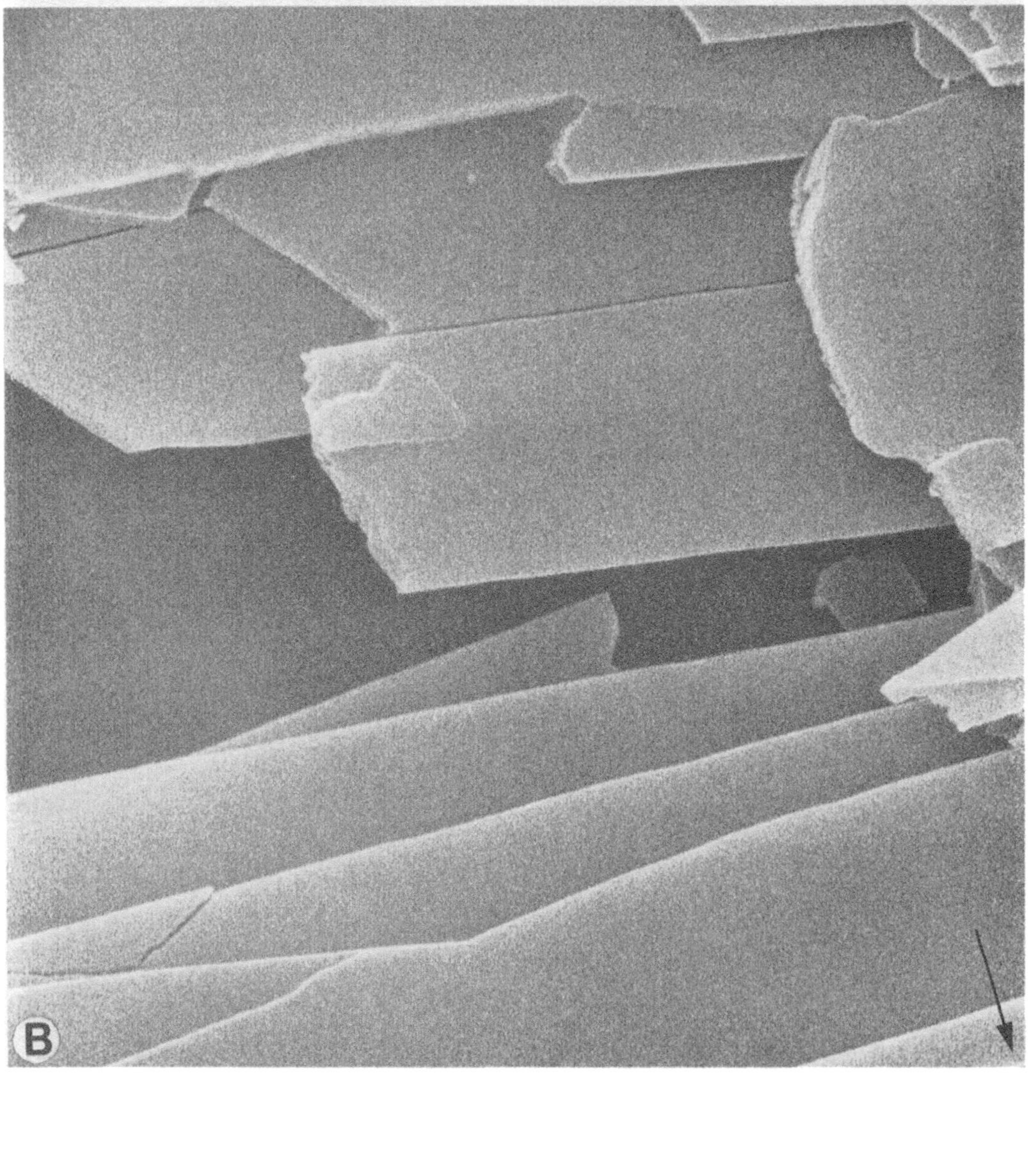

H

1

13.

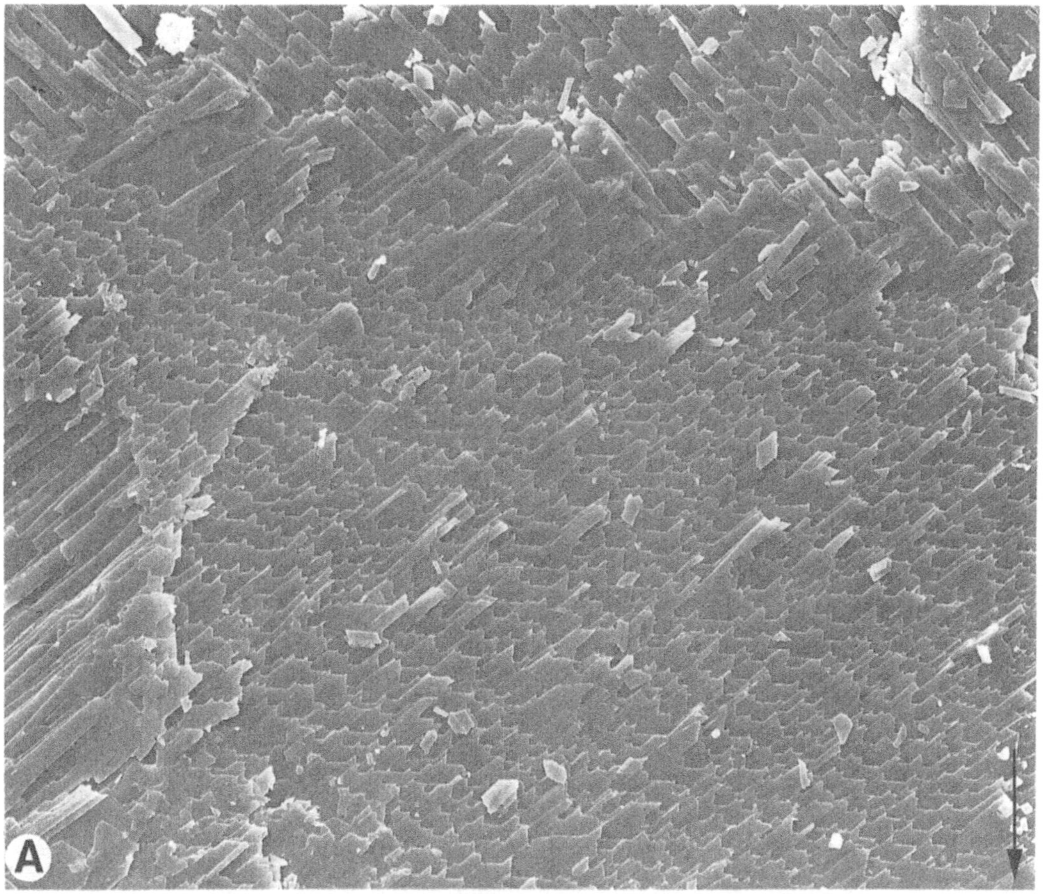

H

20

13.

H

2

PLATE 8

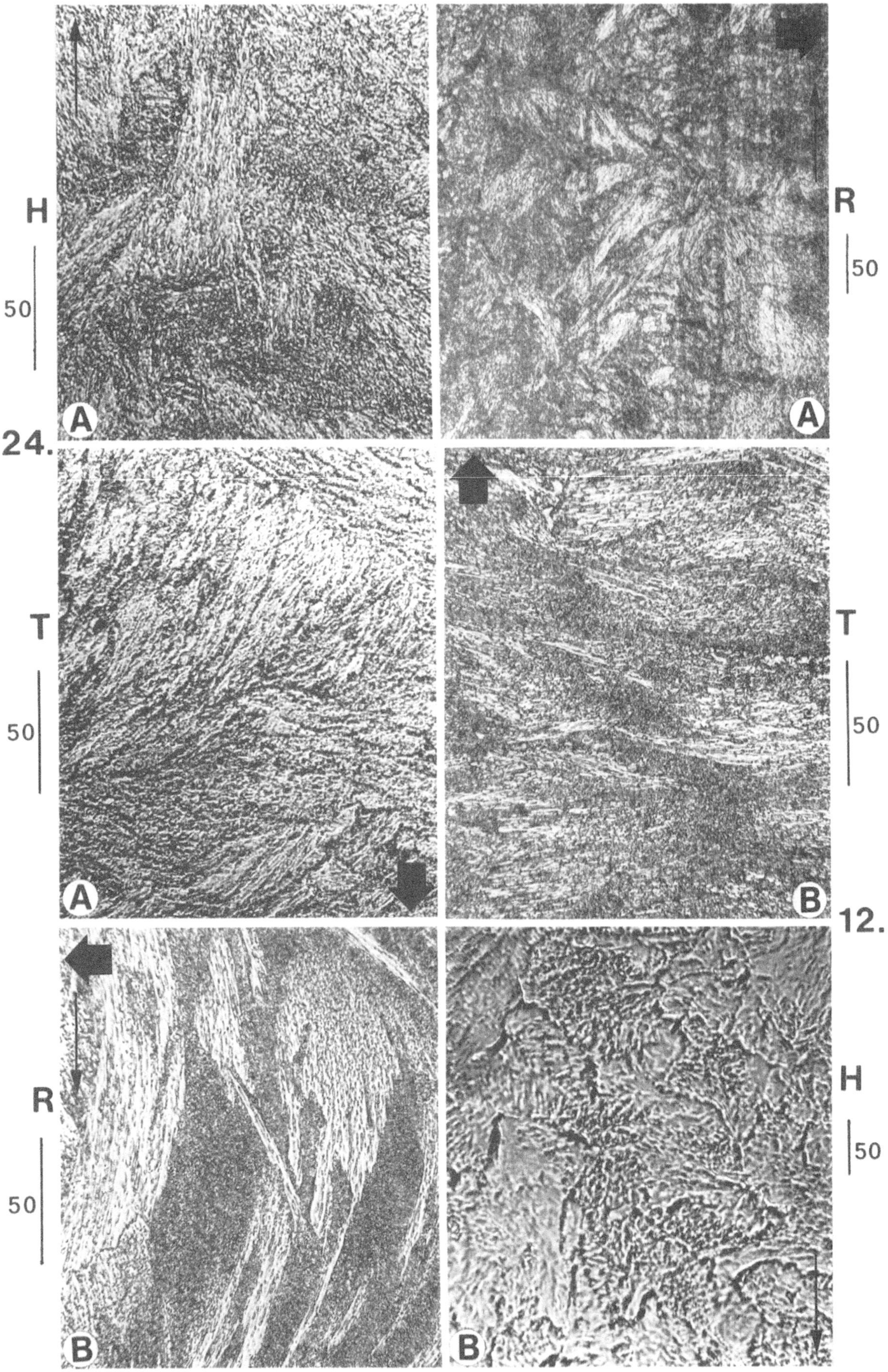

32.

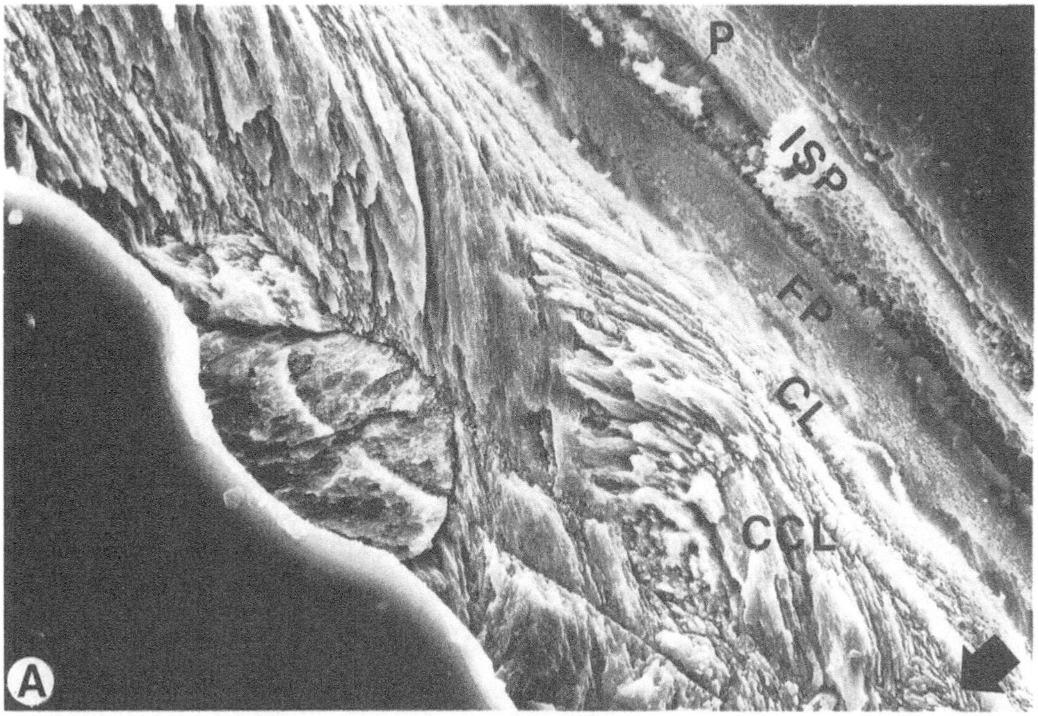

T
5

32.

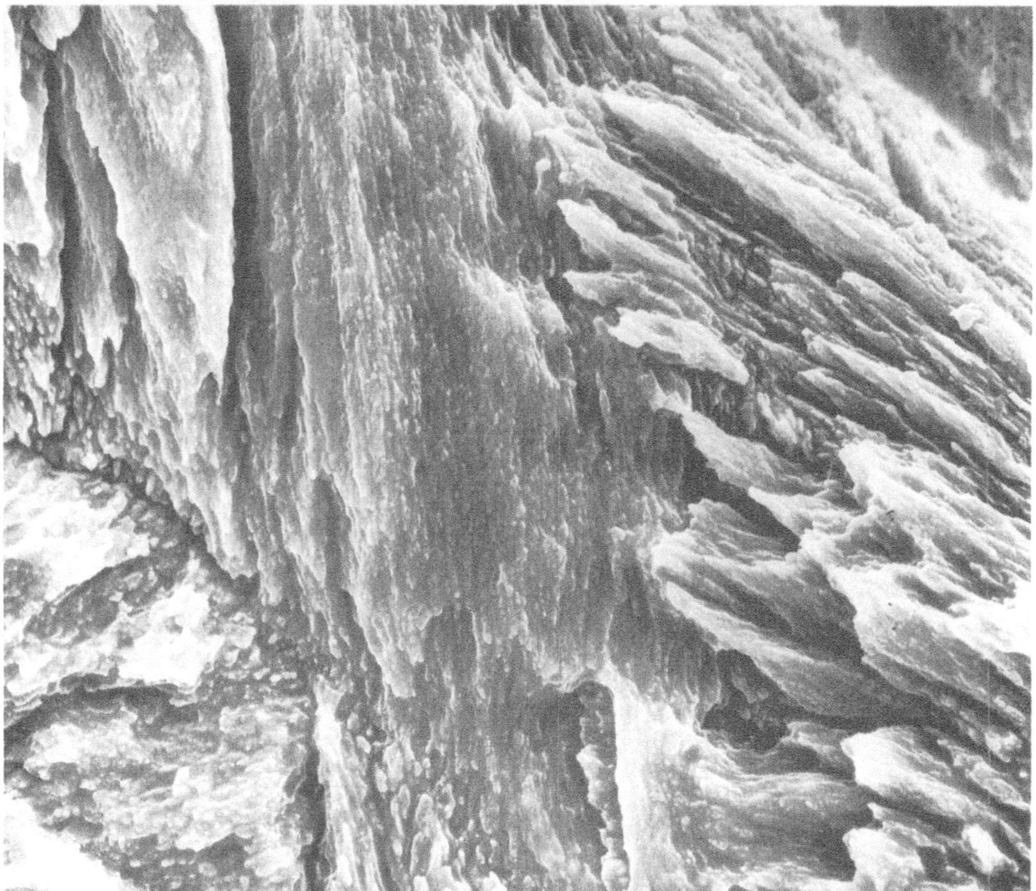

T
5

85.

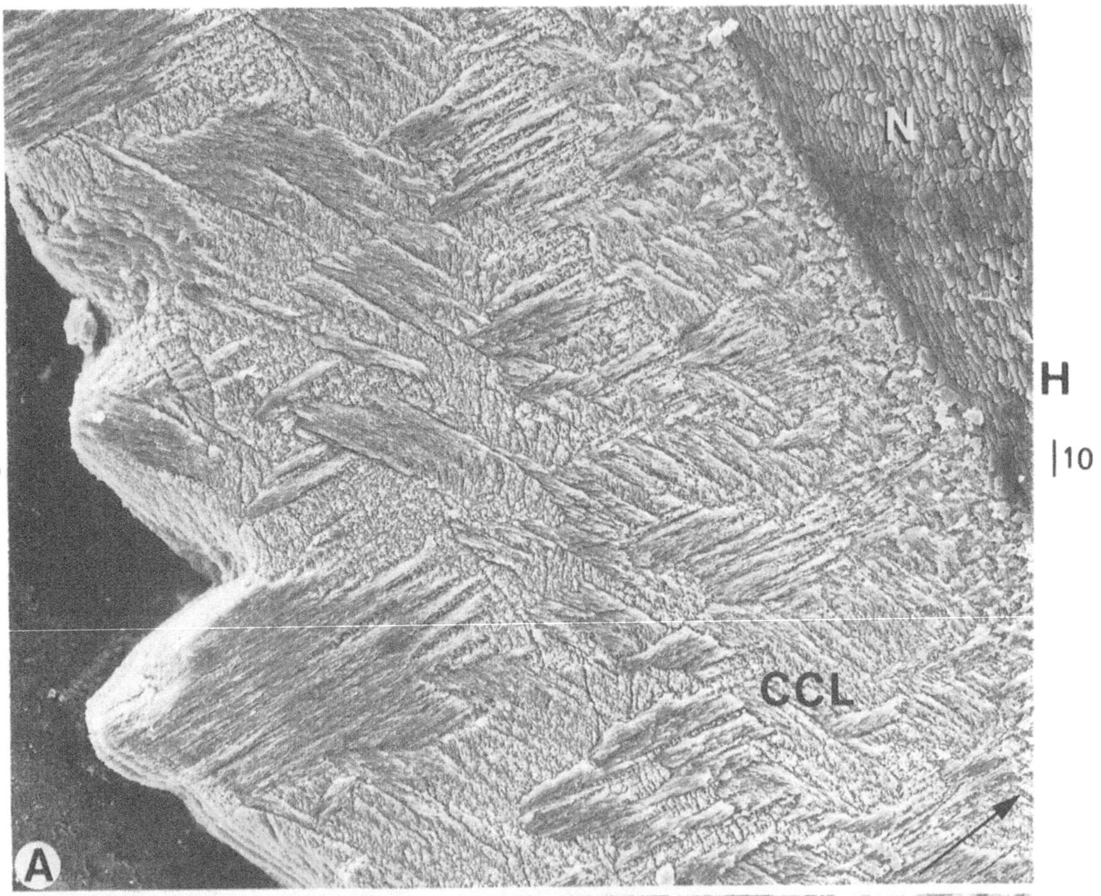

85.

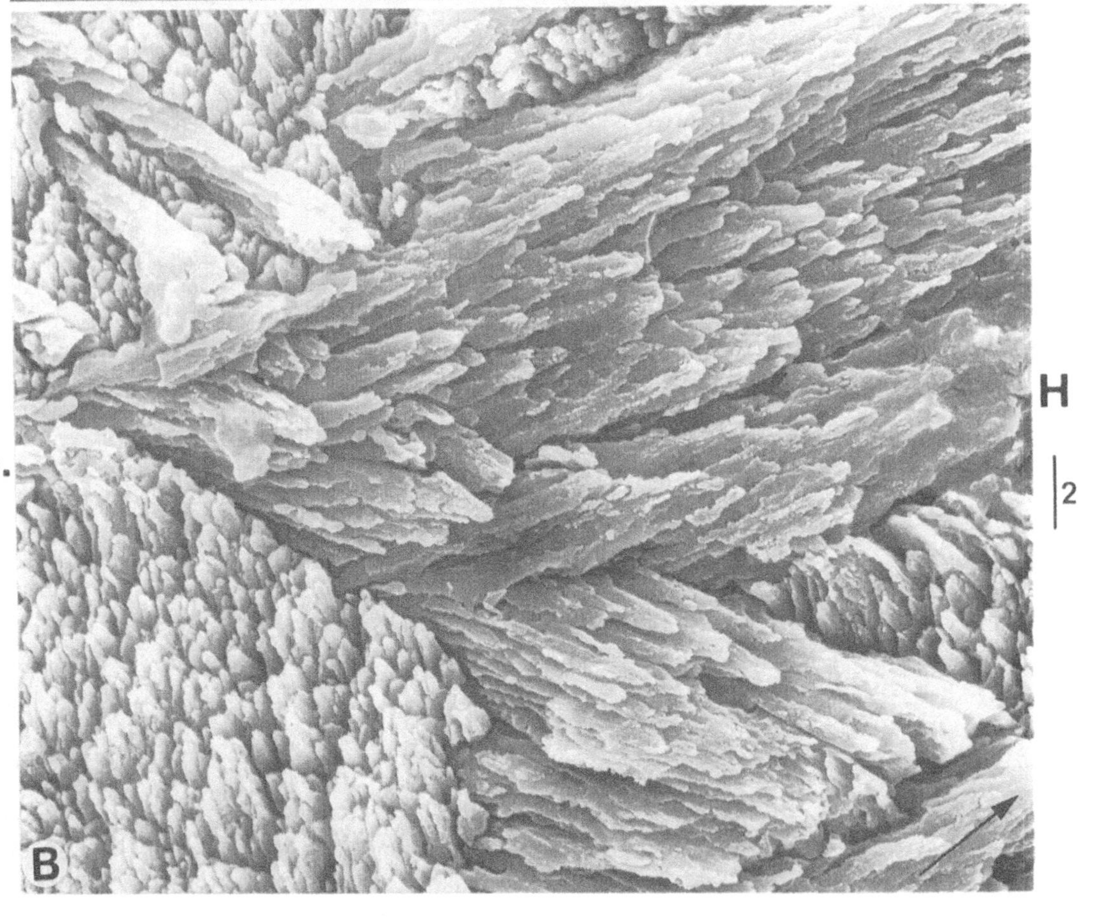

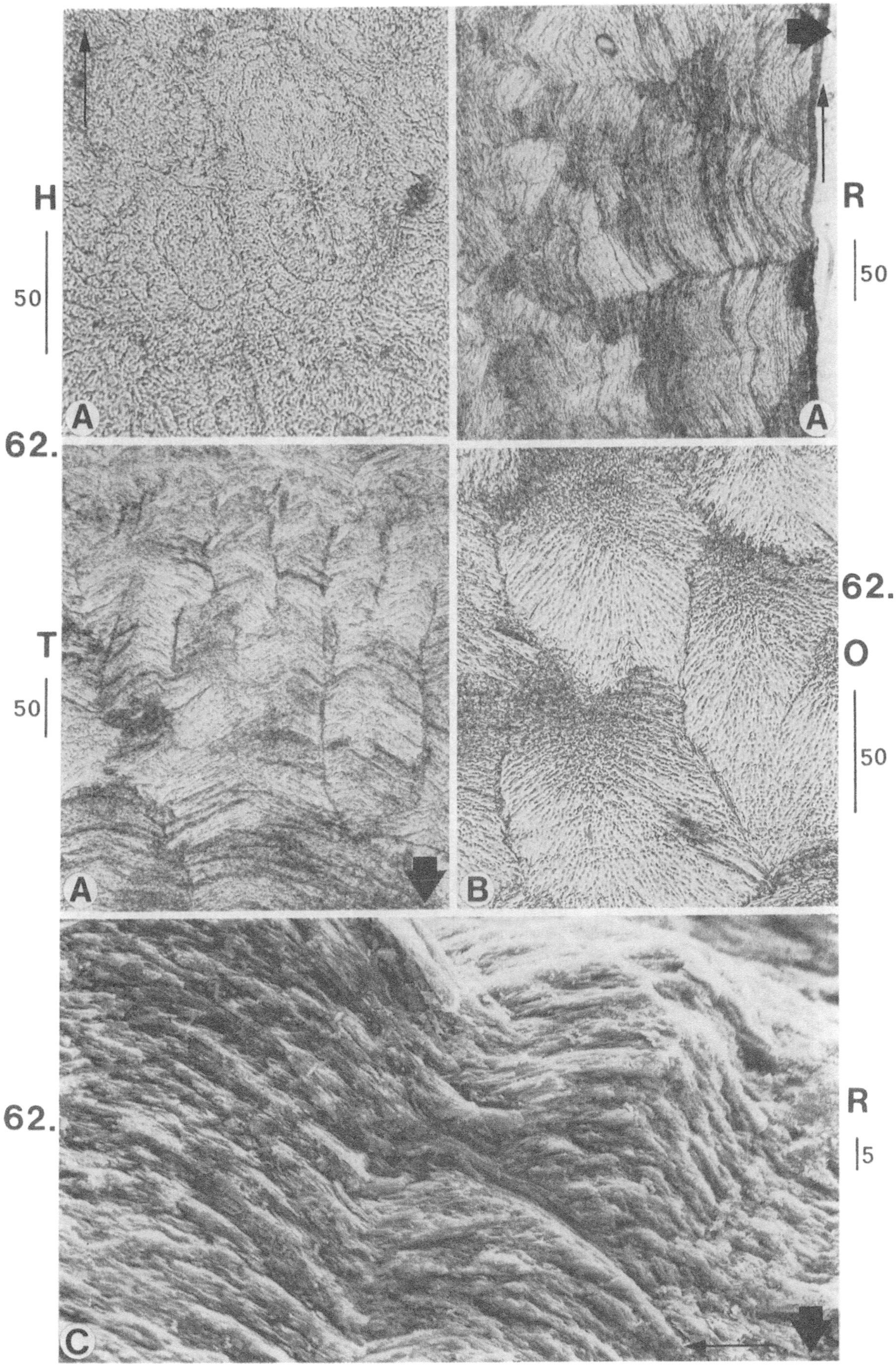
H
50
A
R
50
A
62.
T
50
A
62.
O
50
B
62.
R
5
C

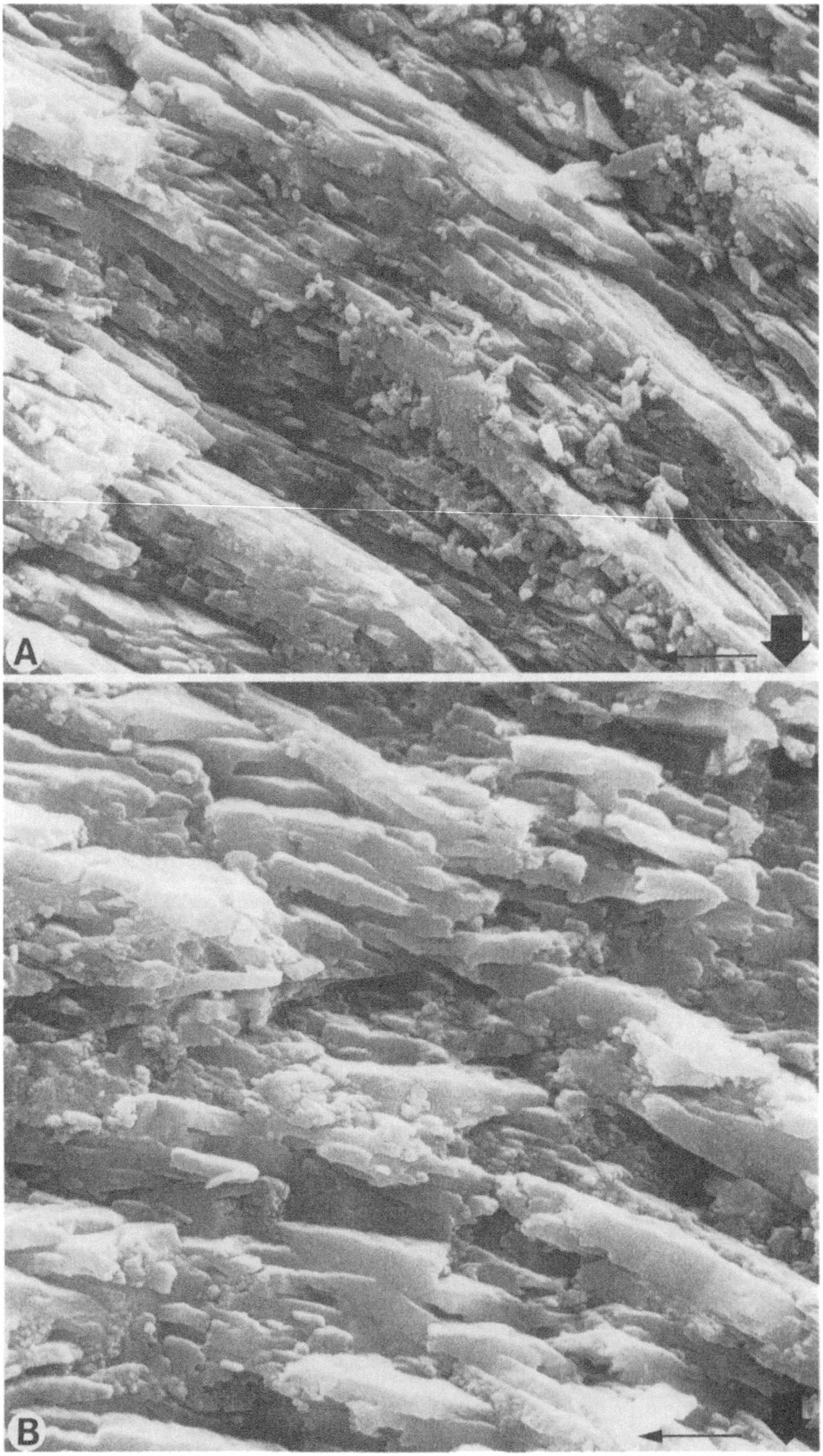

62.

R

1

62.

R

1

45.

R

1

45.

R

1

92.

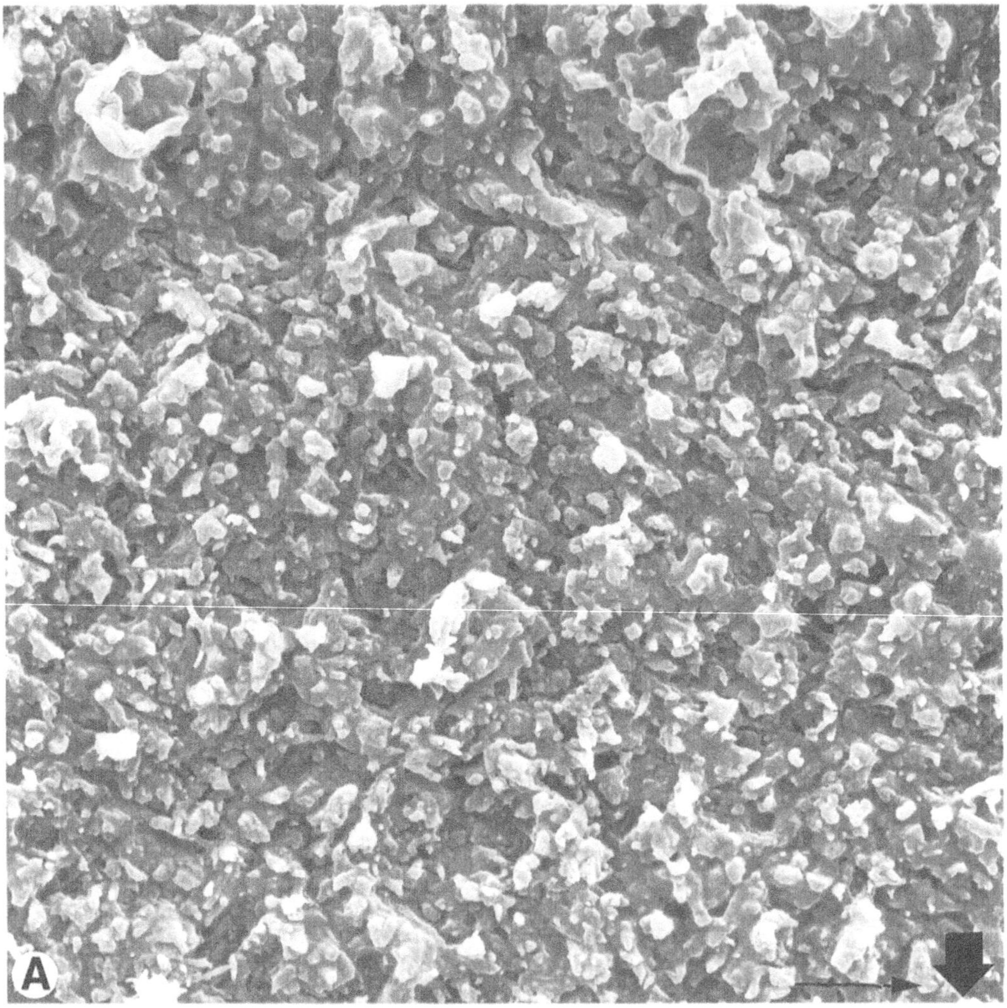

R

1

91.

R

1

92.

R

1

20.

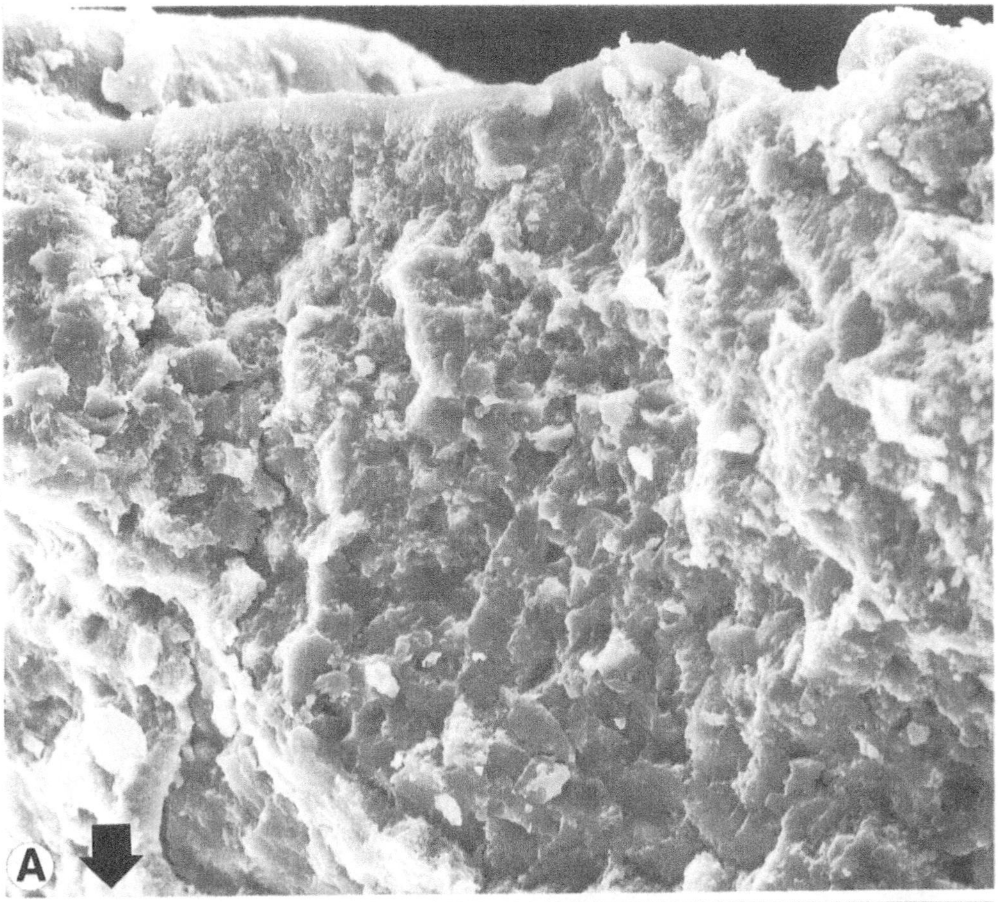

T

5

47.

H

1

75.

91.

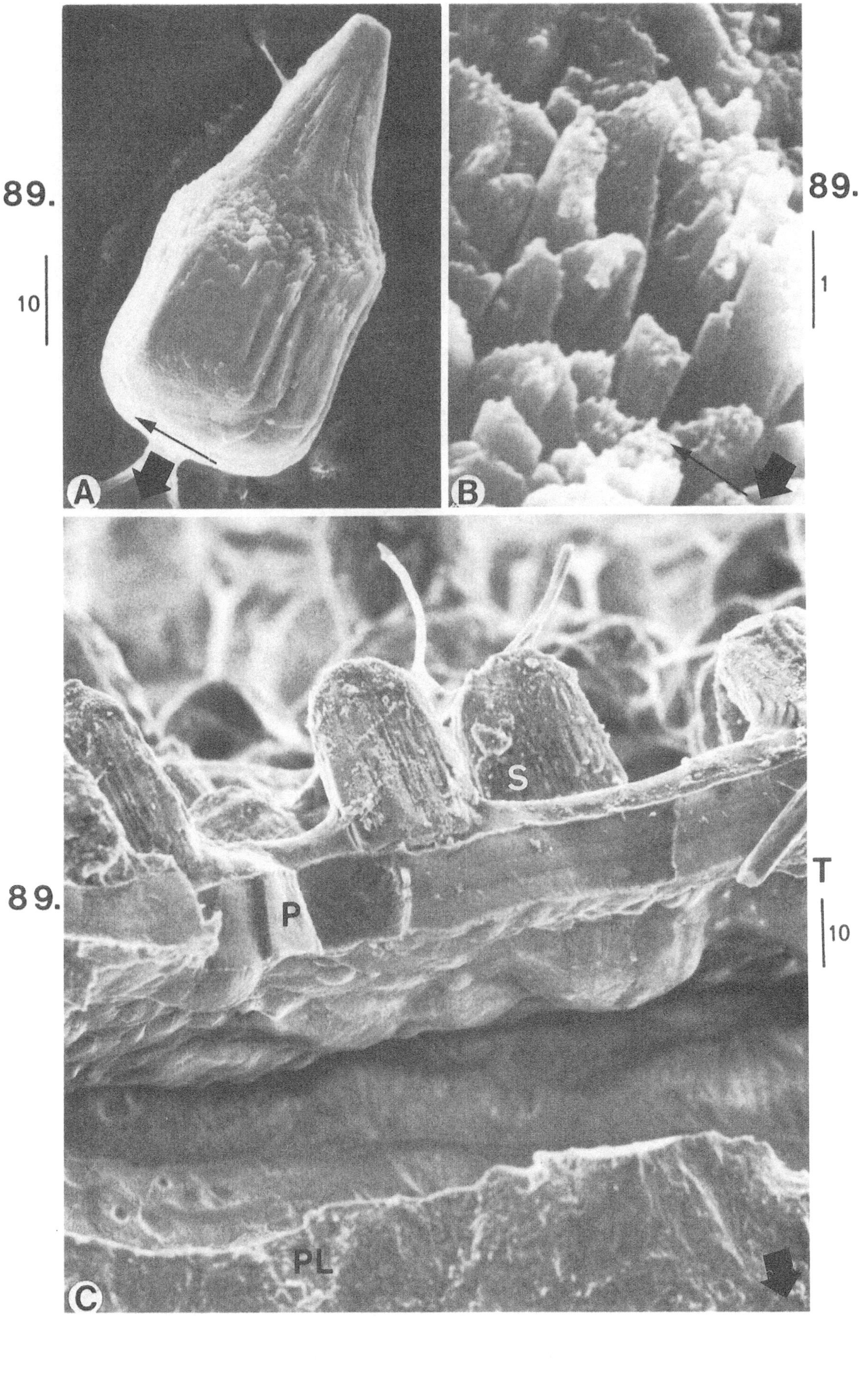
89.
10
A
89.
1
B
89.
S
P
T
10
PL
C

89.

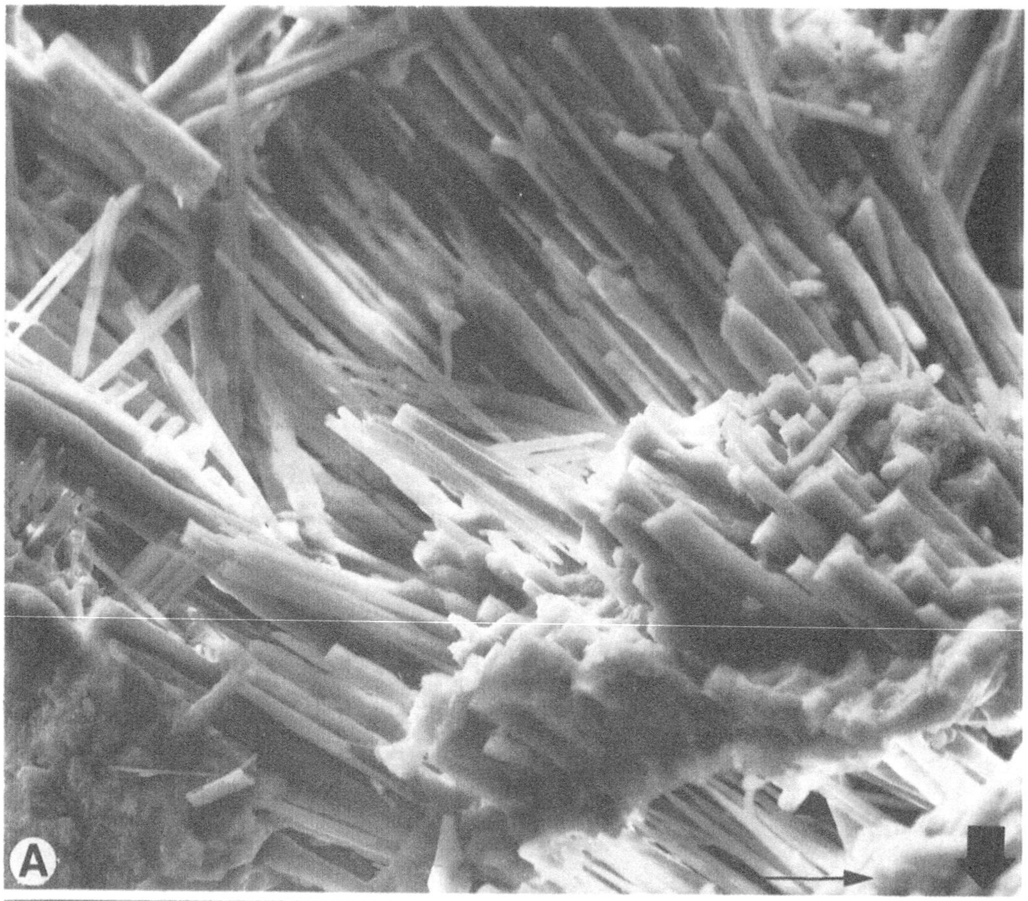

89.

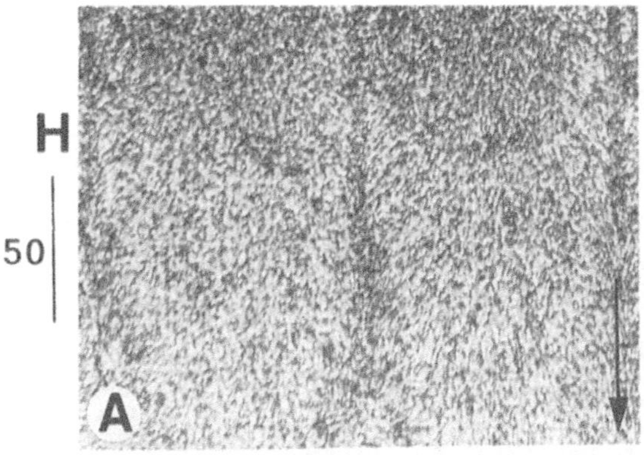

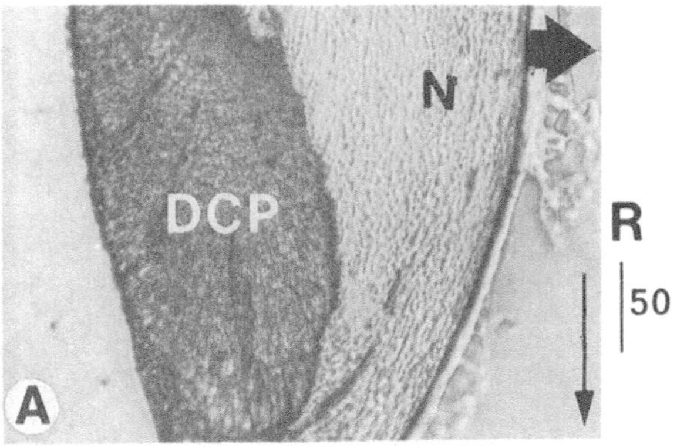

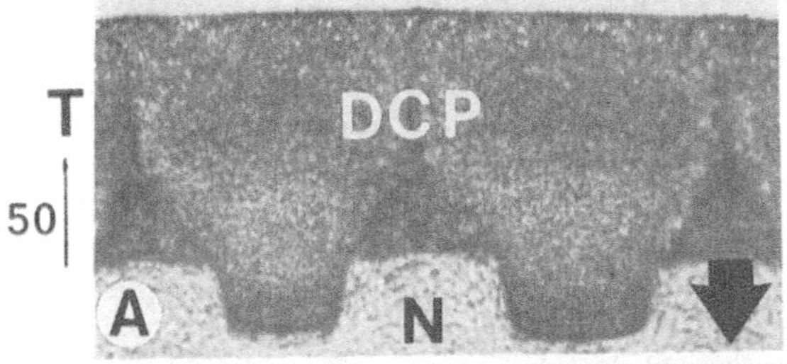

1.
Nuculidae
Nucula

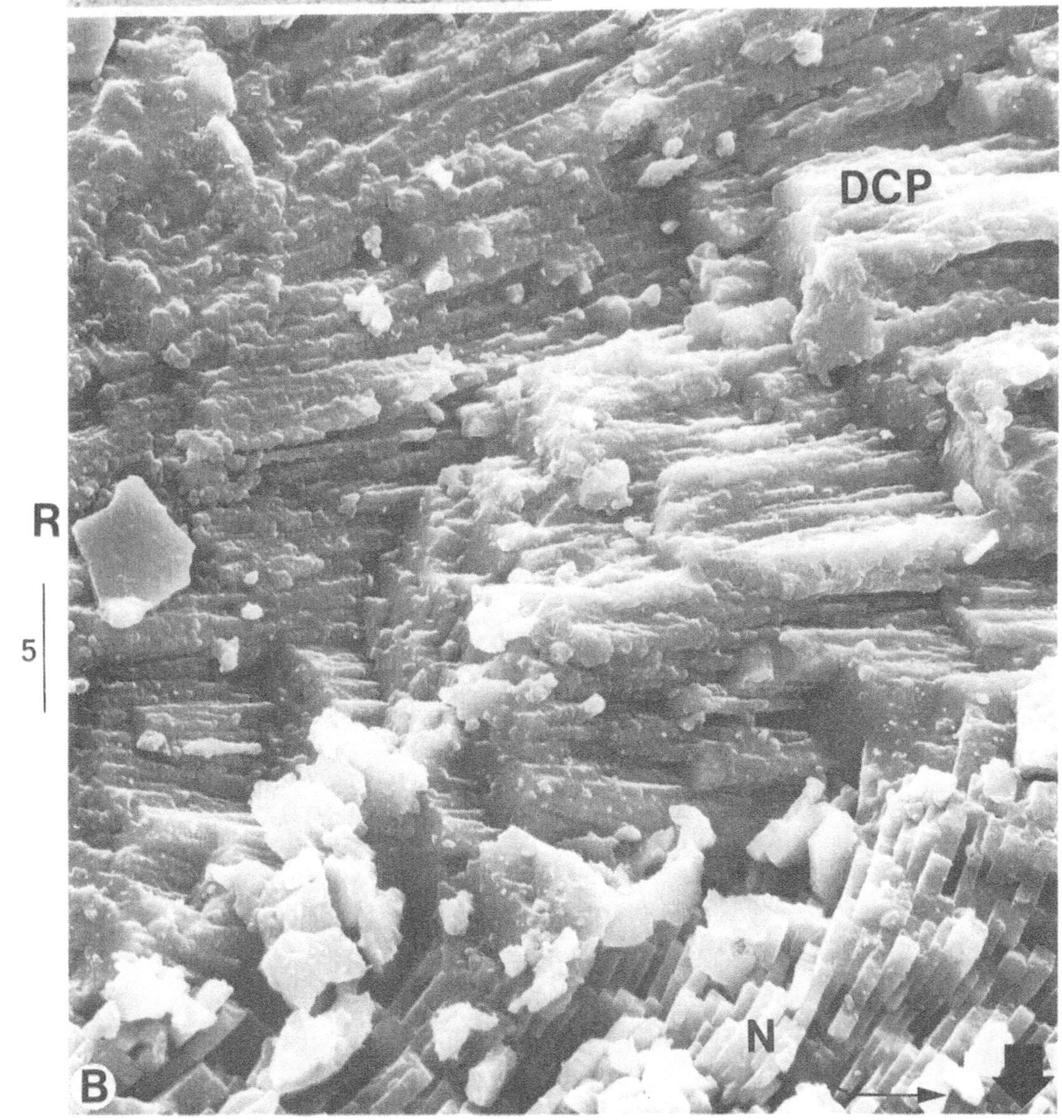

PLATE 20

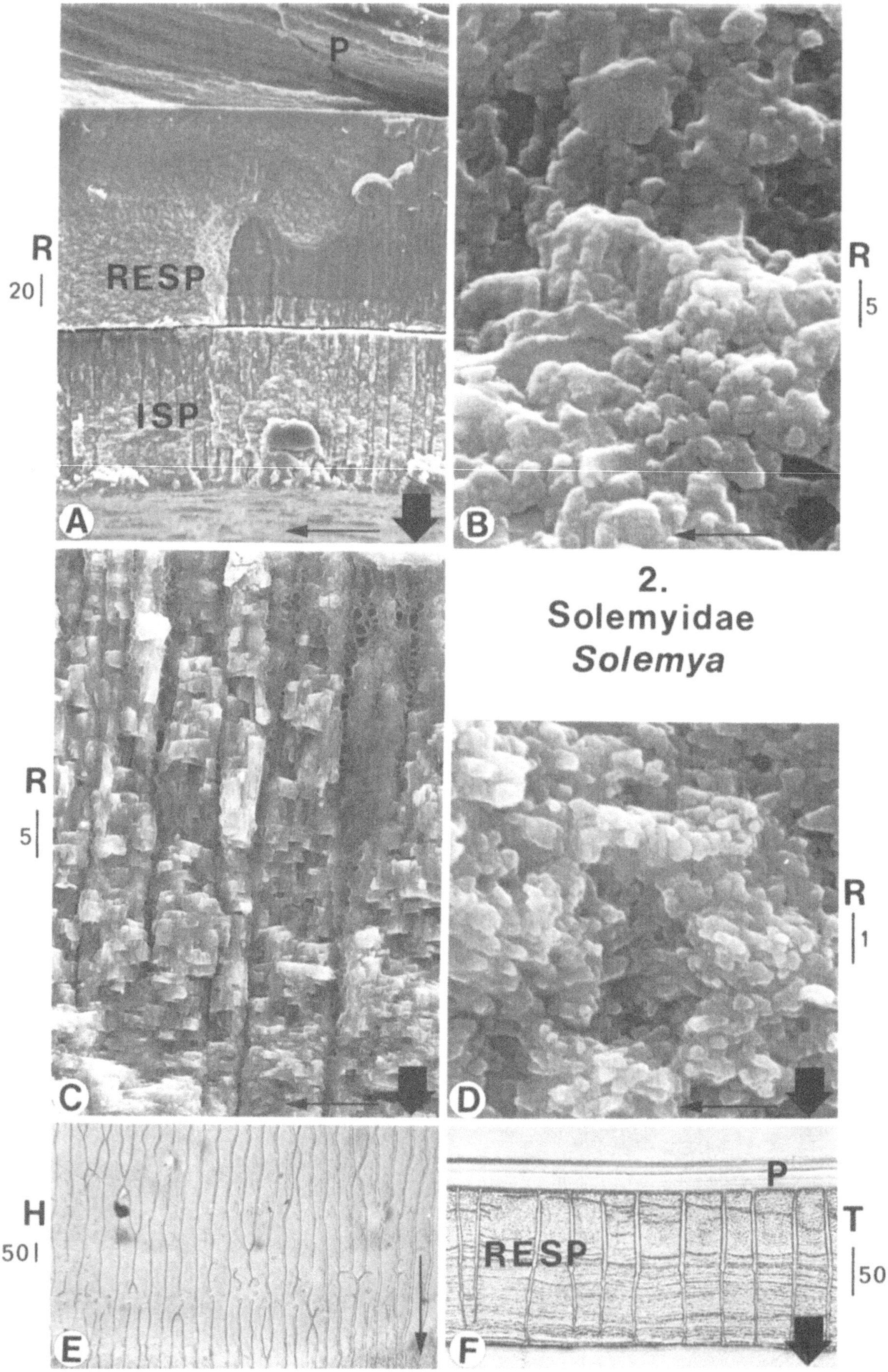

2.
Solemyidae
Solemya

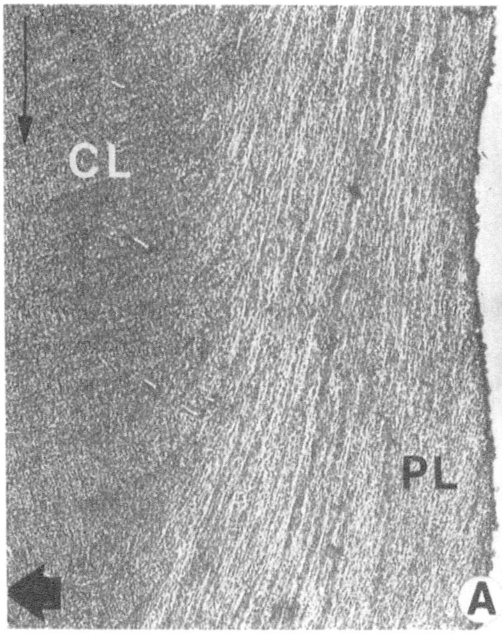

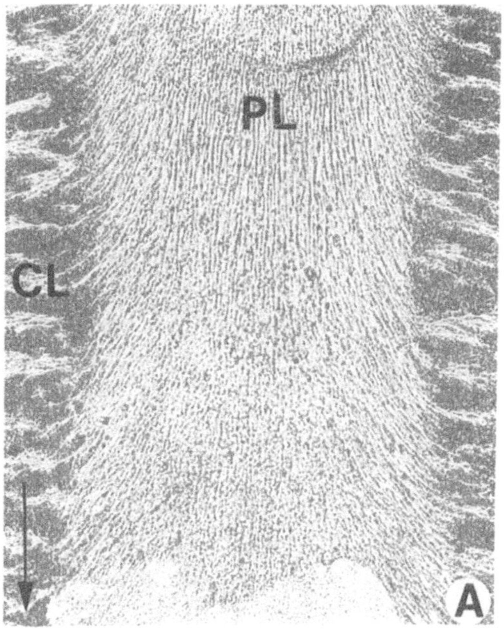

3.
Arcidae
Barbatia

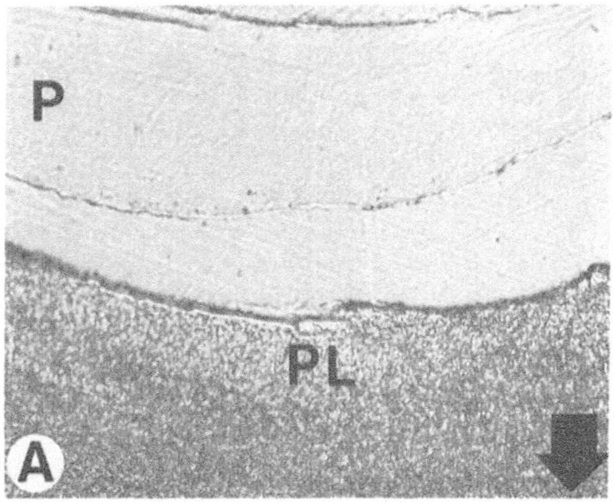

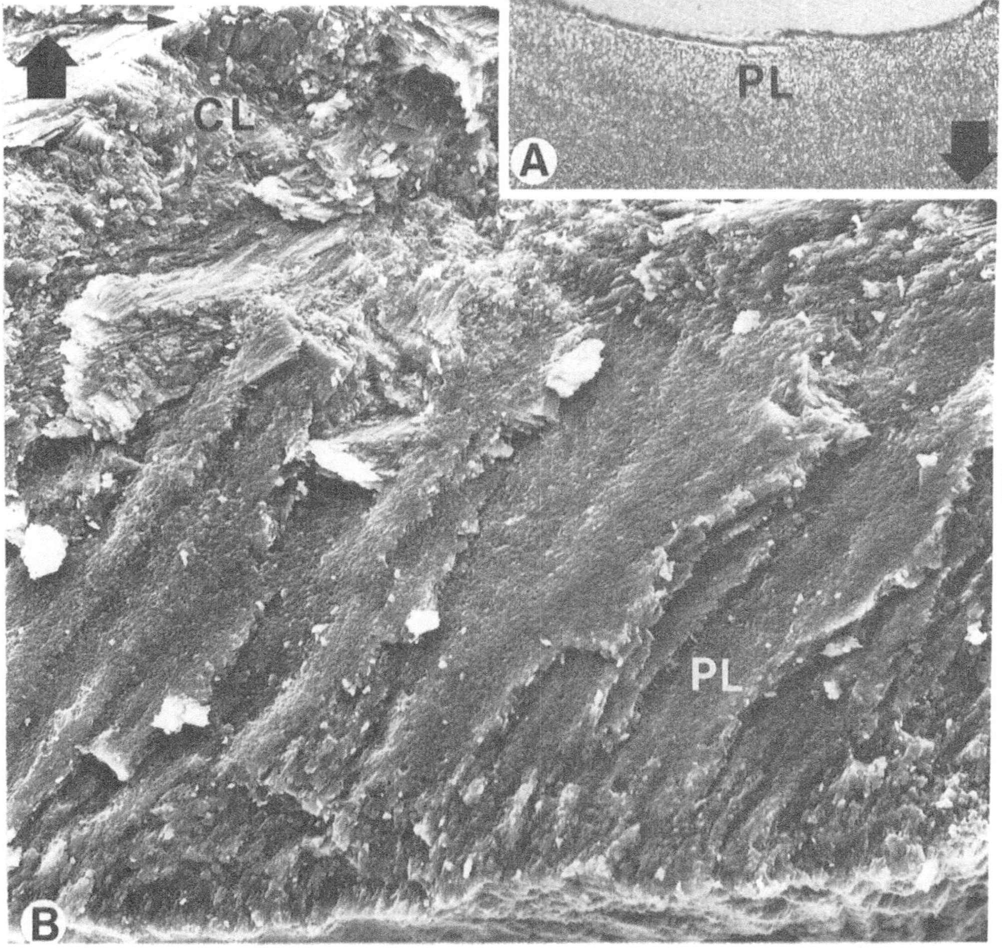

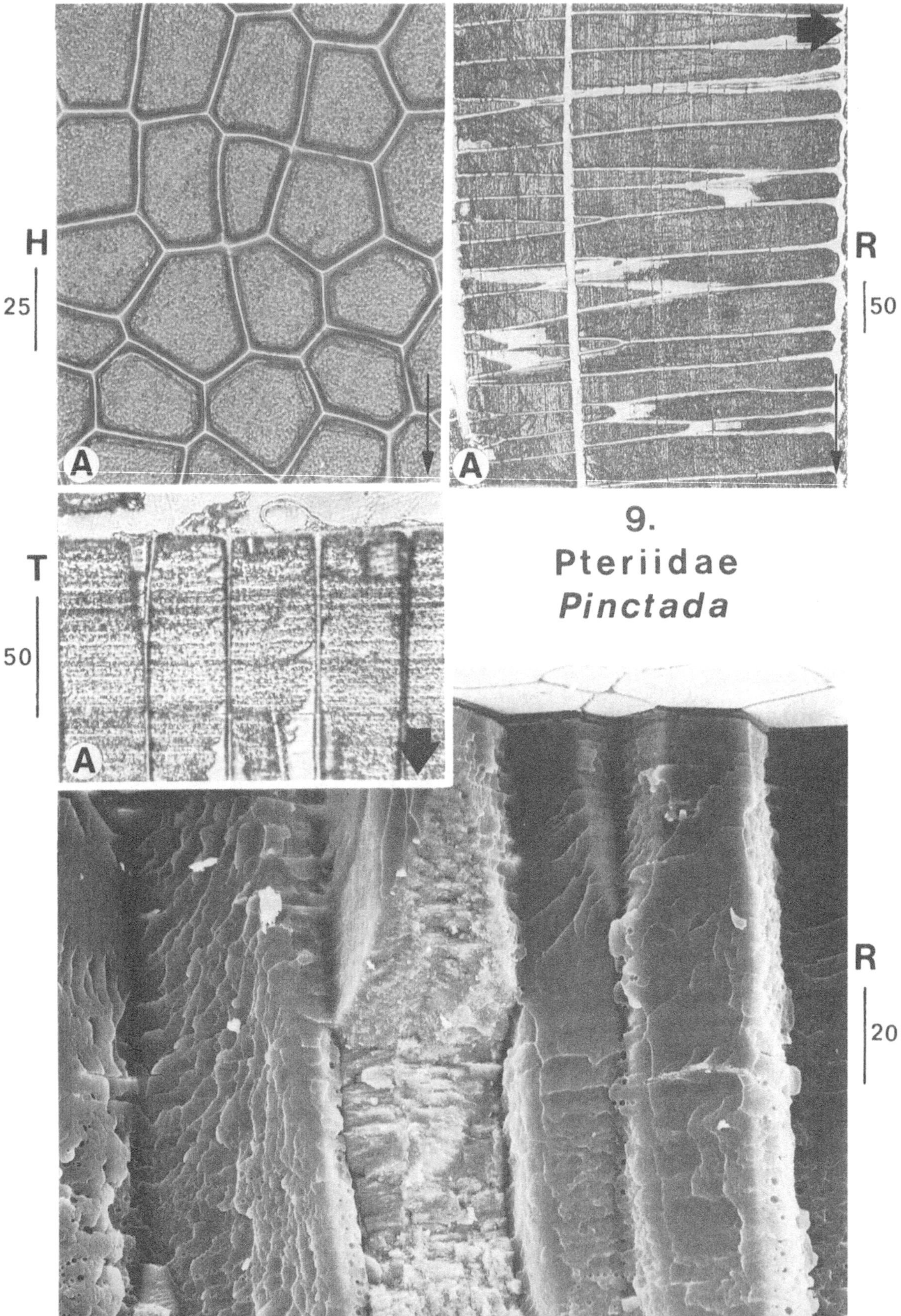

9.
Pteriidae
Pinctada

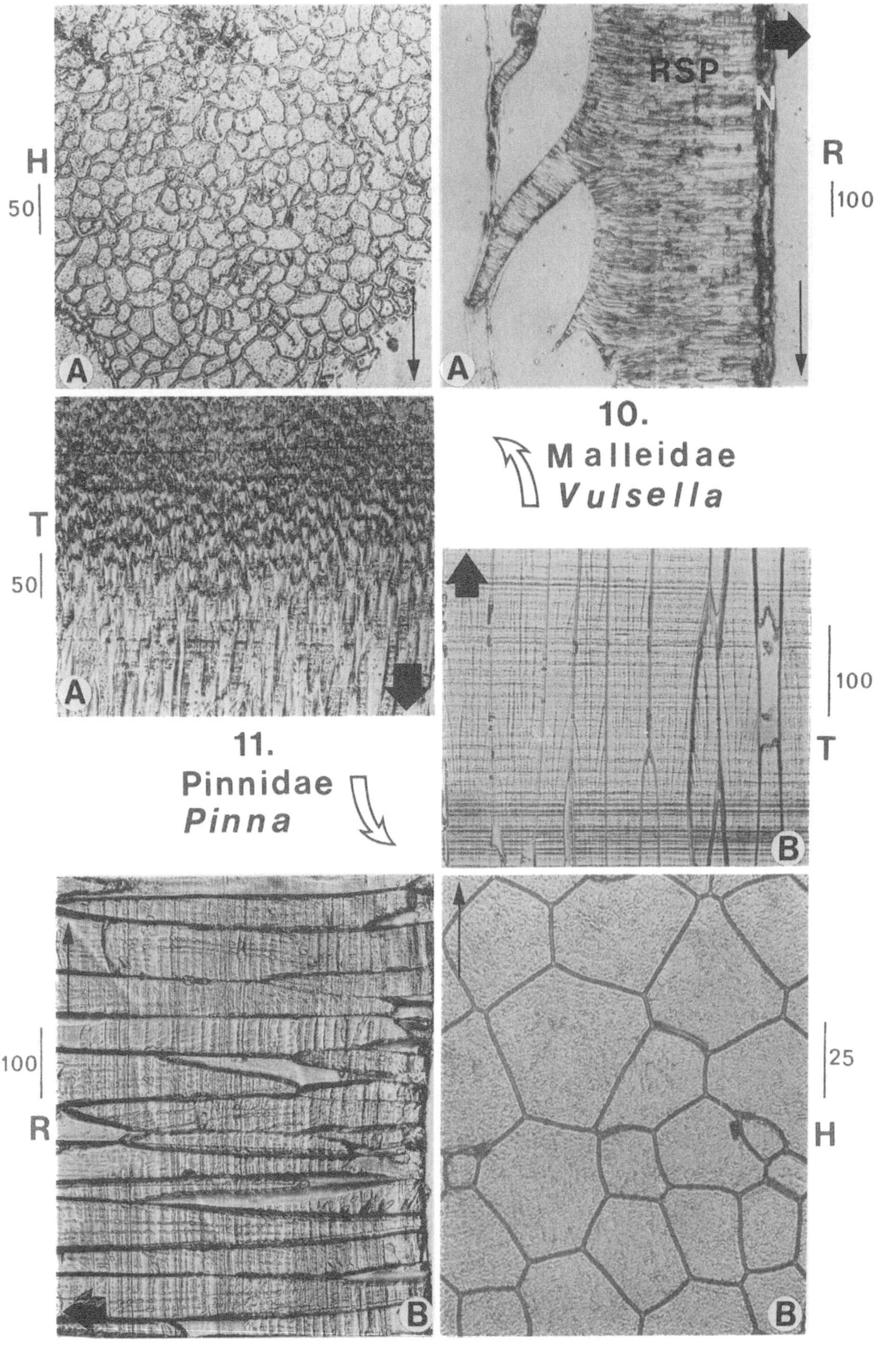

10. Malleidae *Vulsella*

11. Pinnidae *Pinna*

11.
Pinnidae
Pinna

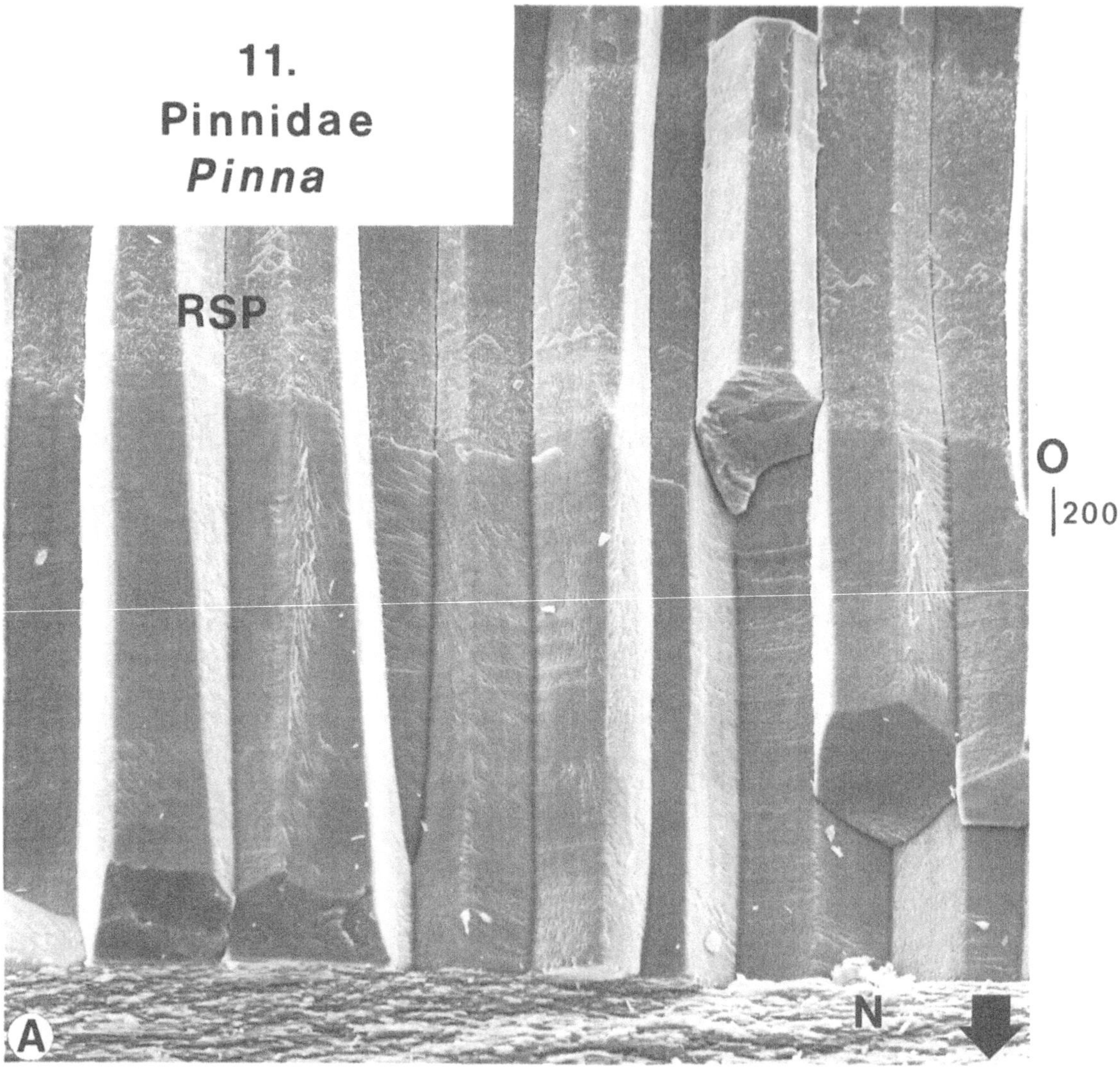

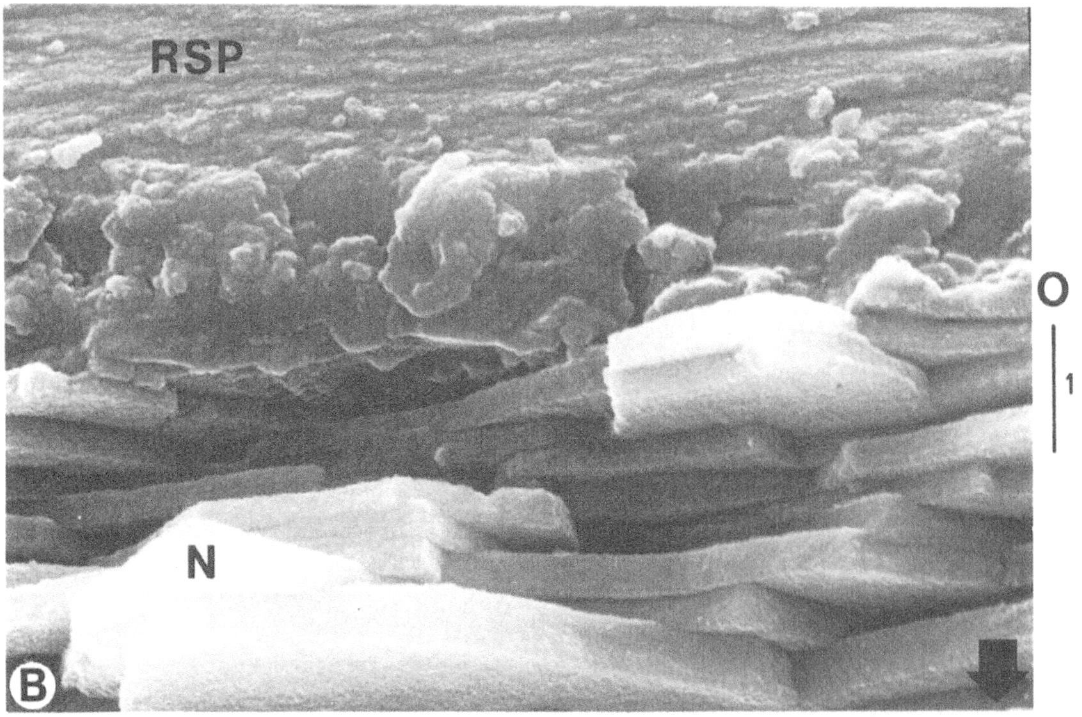

H 50

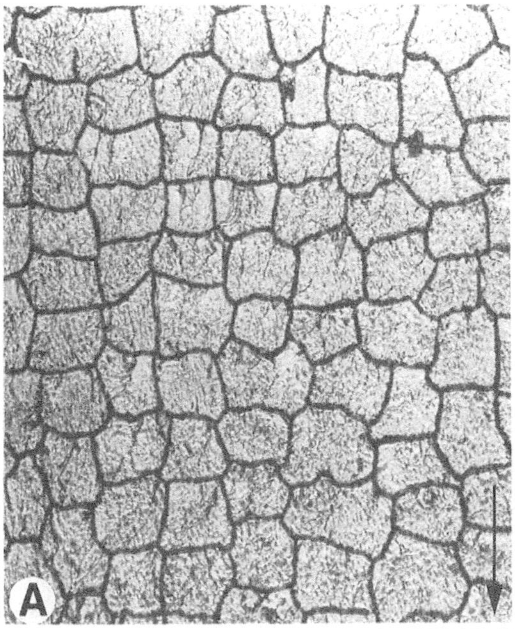

R 50

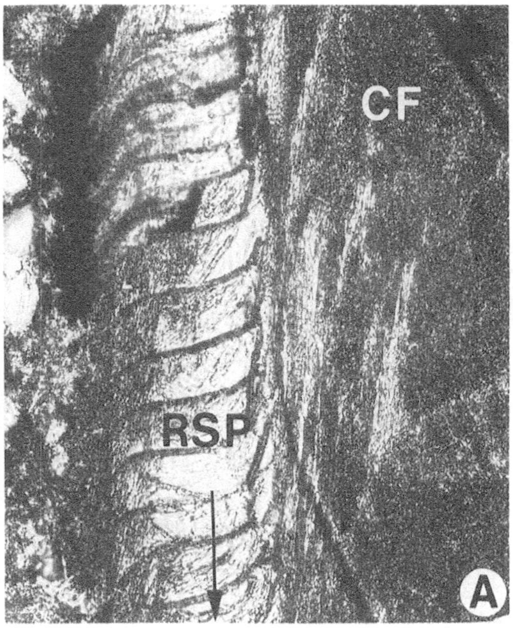

12.
Oxytomidae
Oxytoma

T 50

R 50

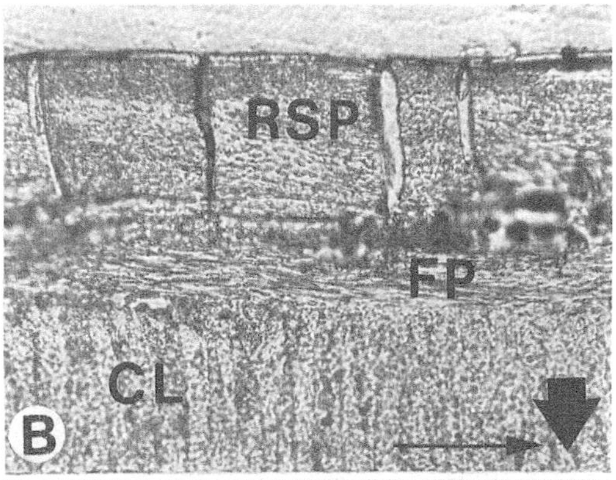

13.
Propeamussiidae
Propeamussium

R 5

R 50

16.
Anomiidae
Anomia

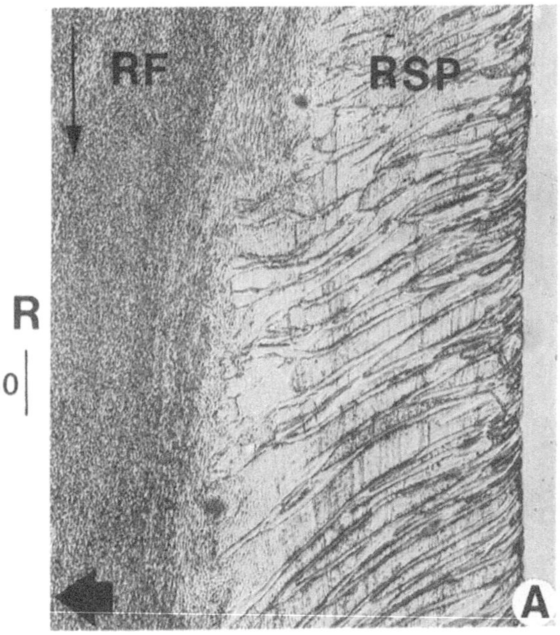

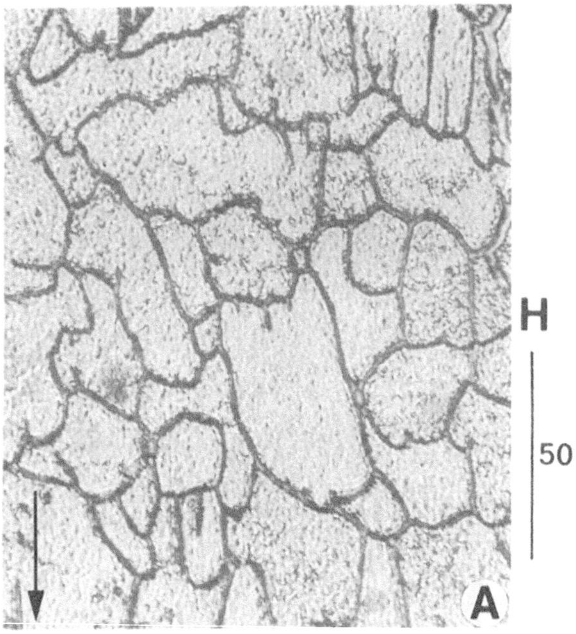

17.
Ostreidae
Crassostrea

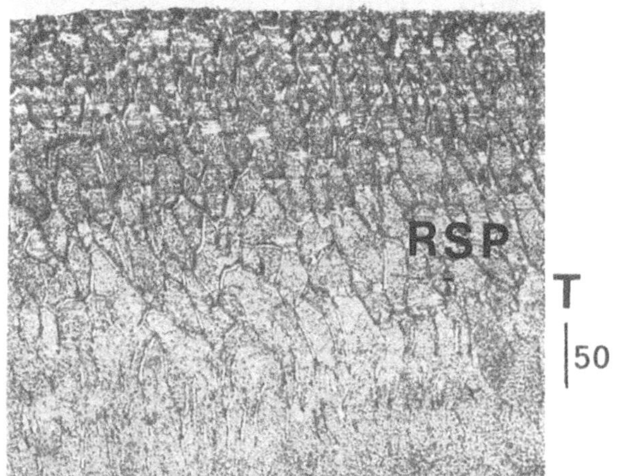

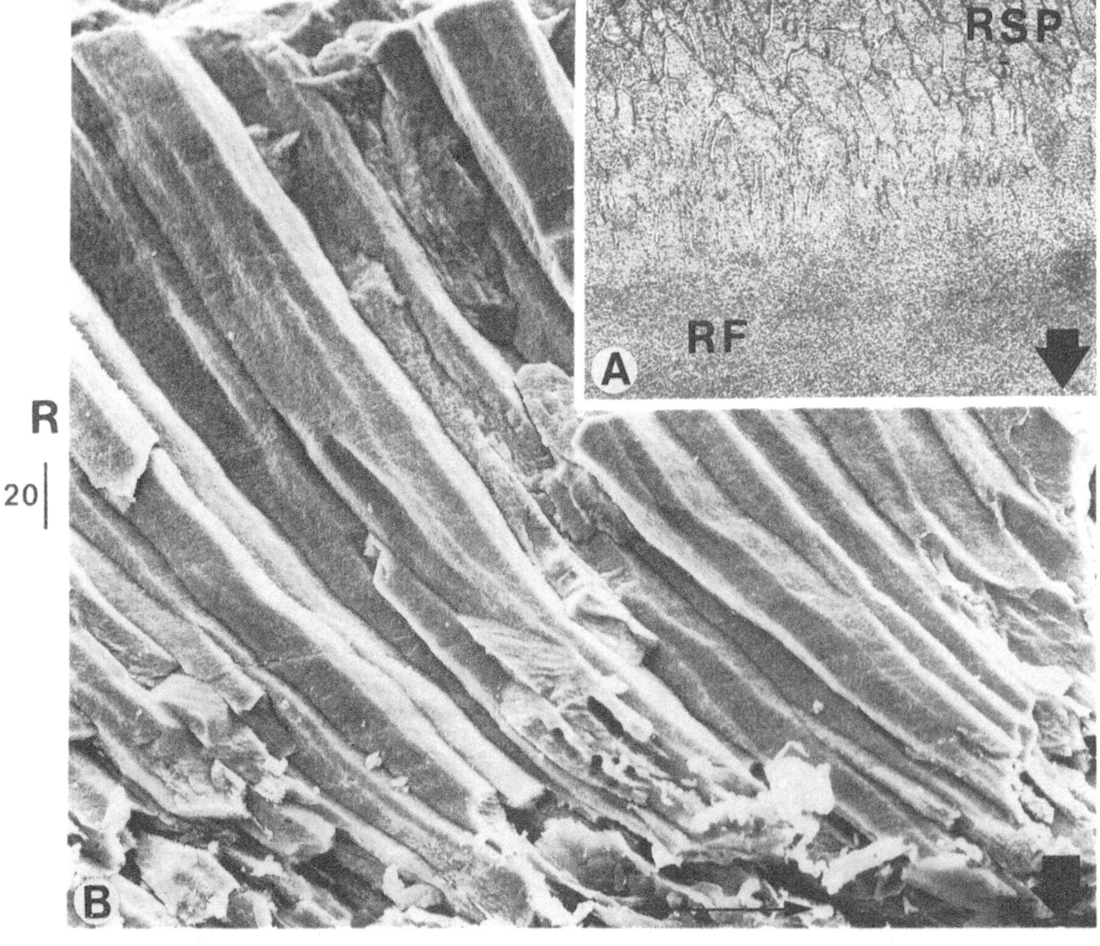

18.
Limidae
Ctenoides

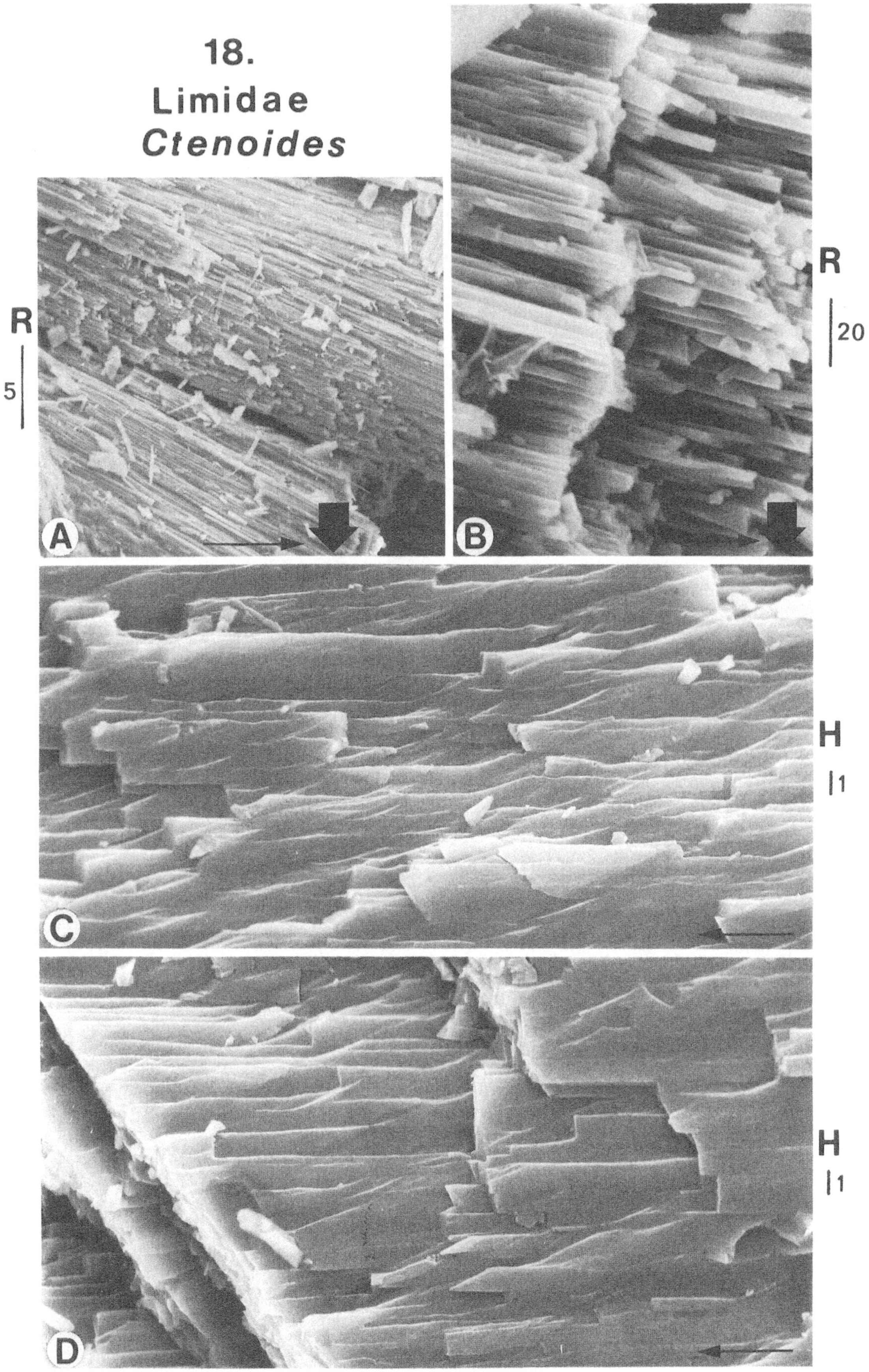

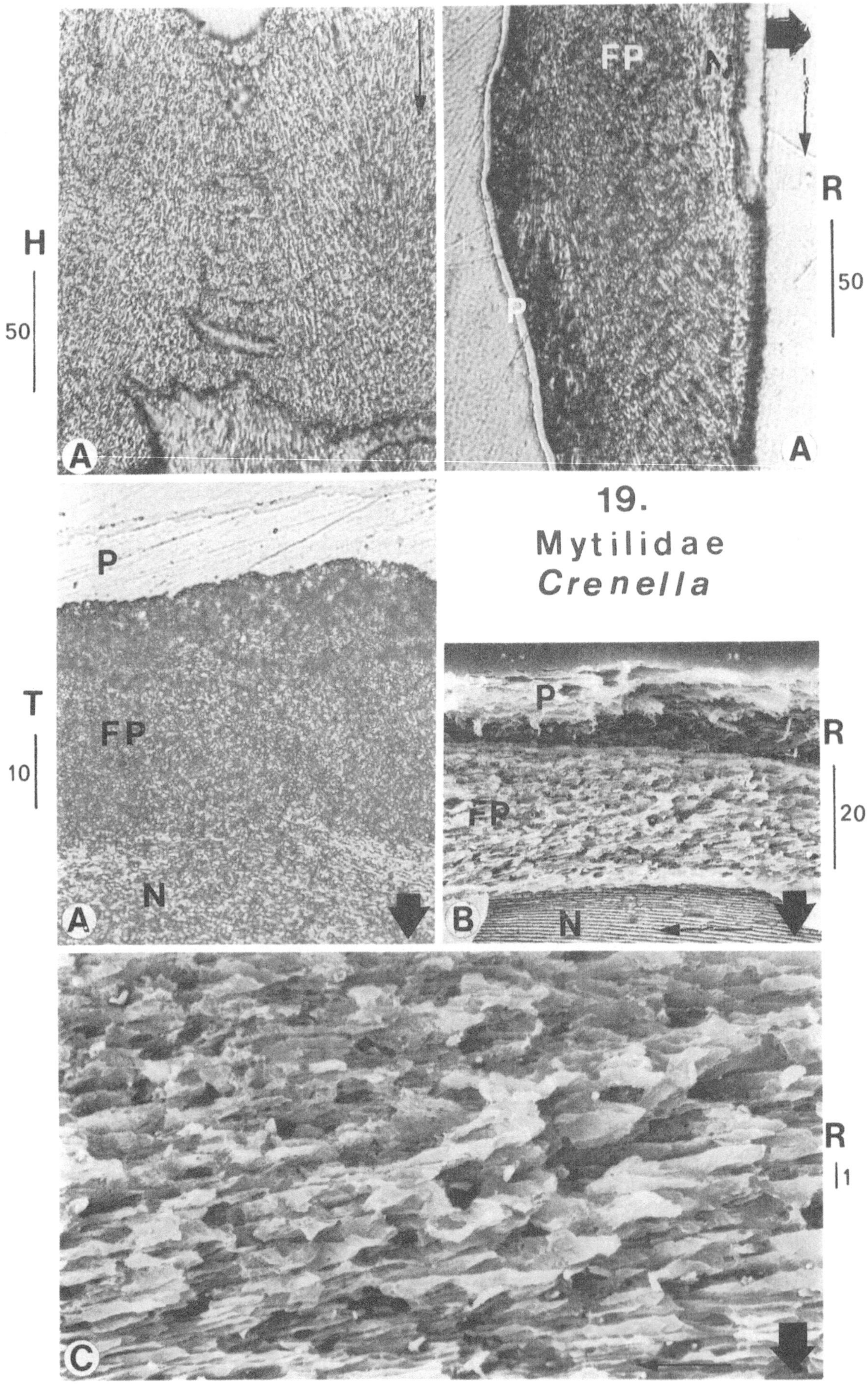

19.
Mytilidae
Crenella

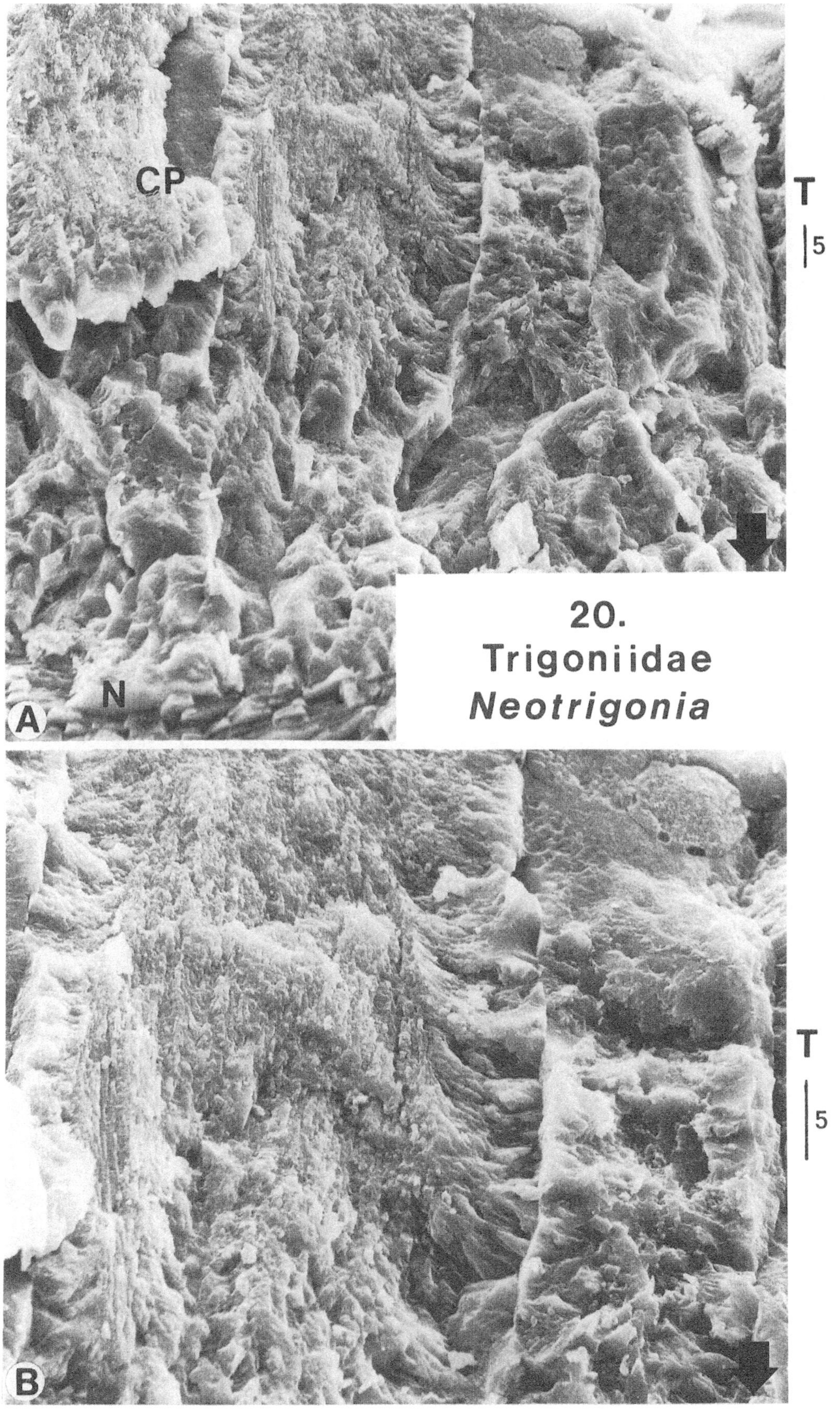

20.
Trigoniidae
Neotrigonia

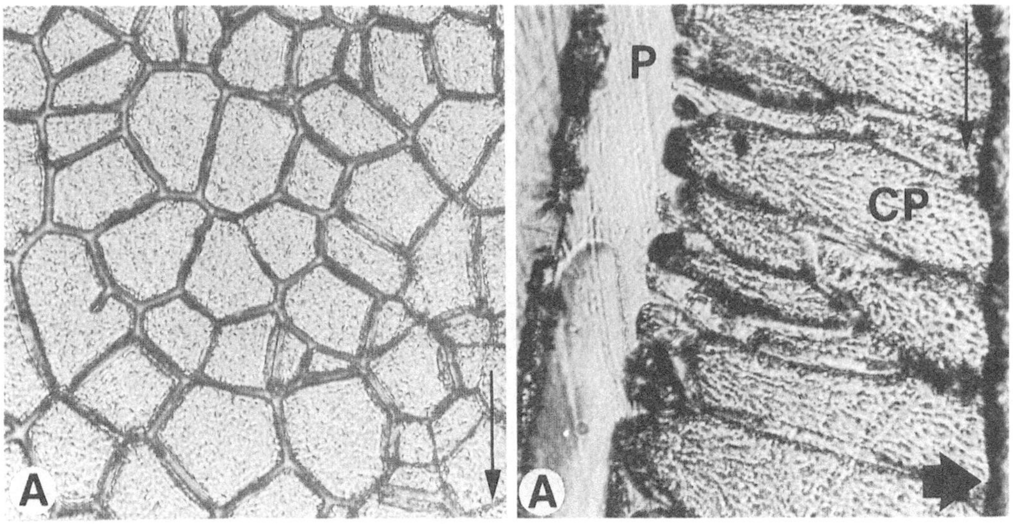

21. Unionidae *Obliquaria*

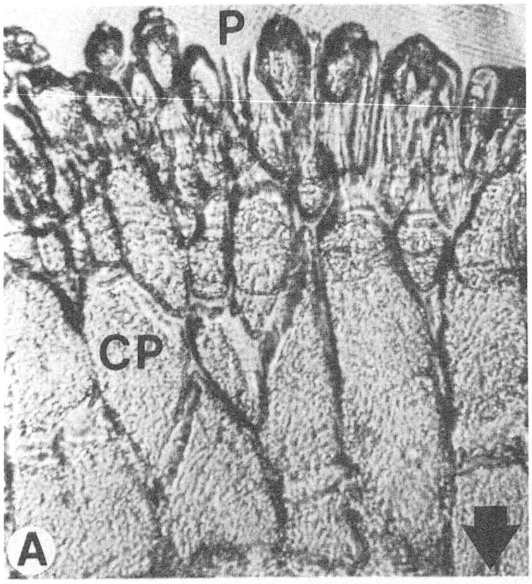

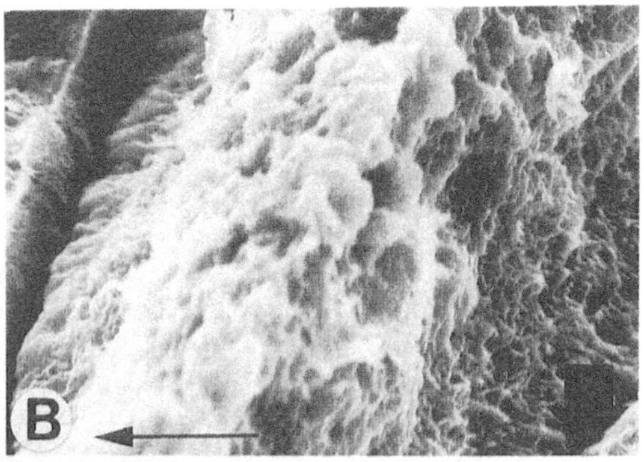

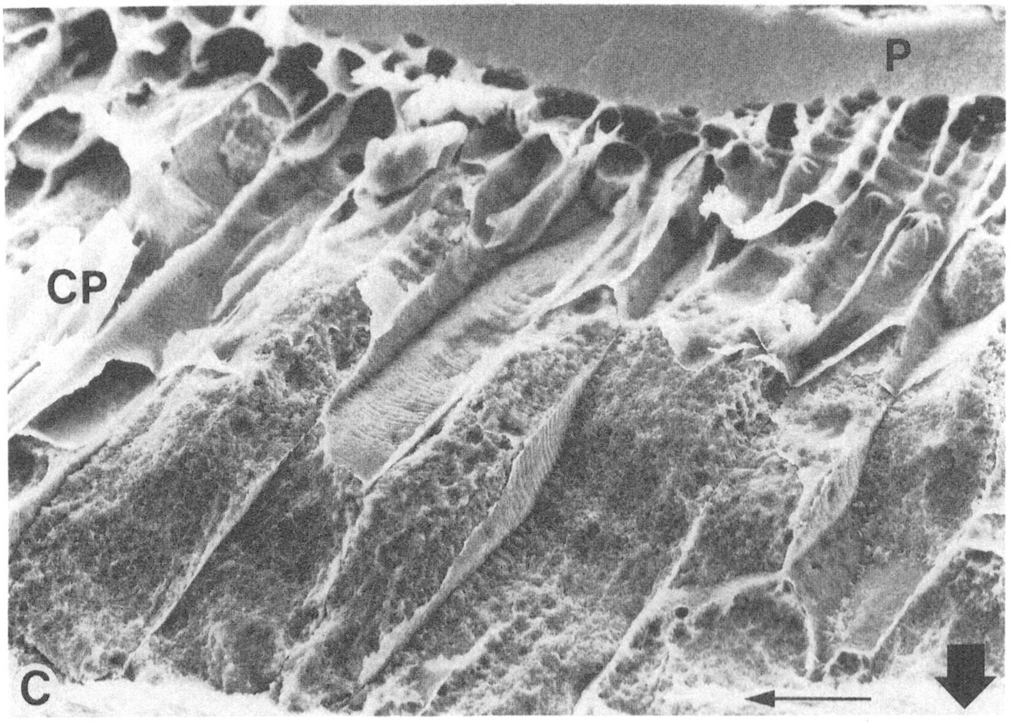

H
50

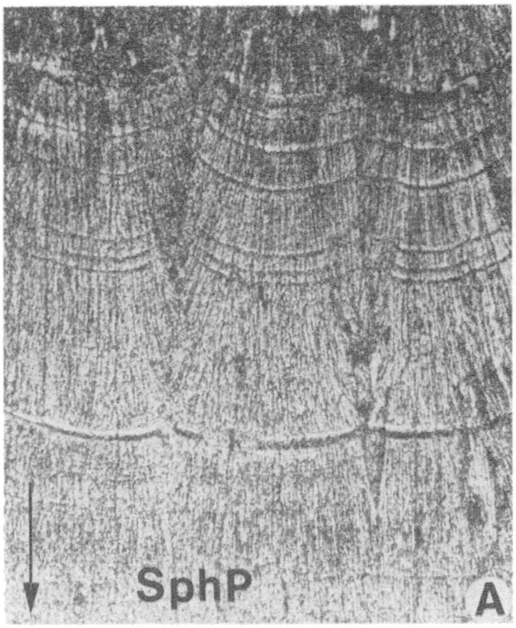

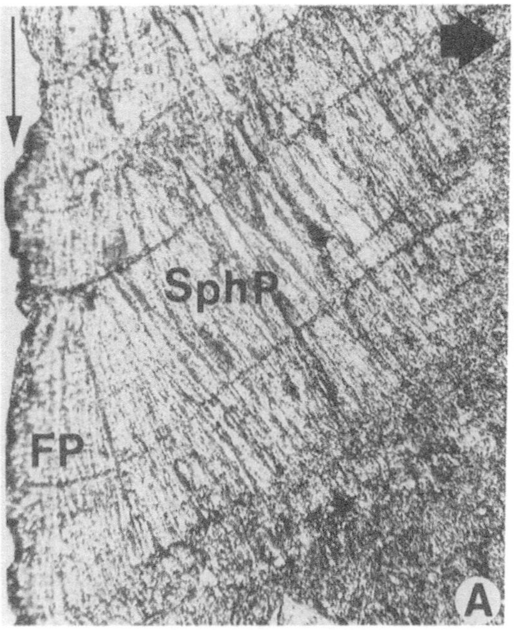

R
50

T
50

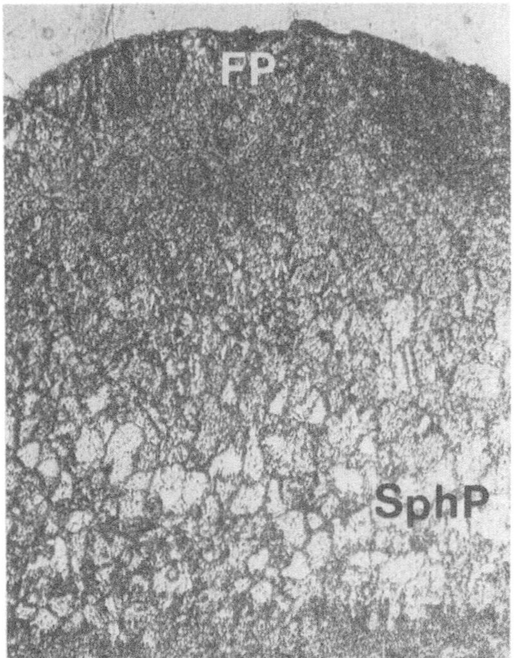

R
50

25.
Lucinidae
Divalucina

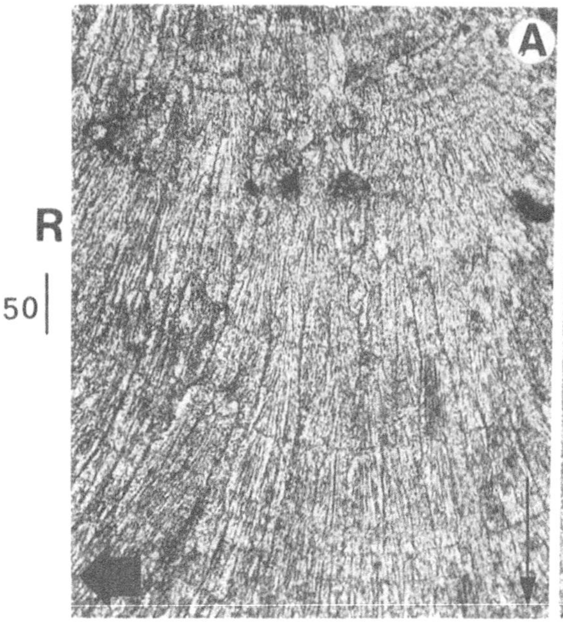

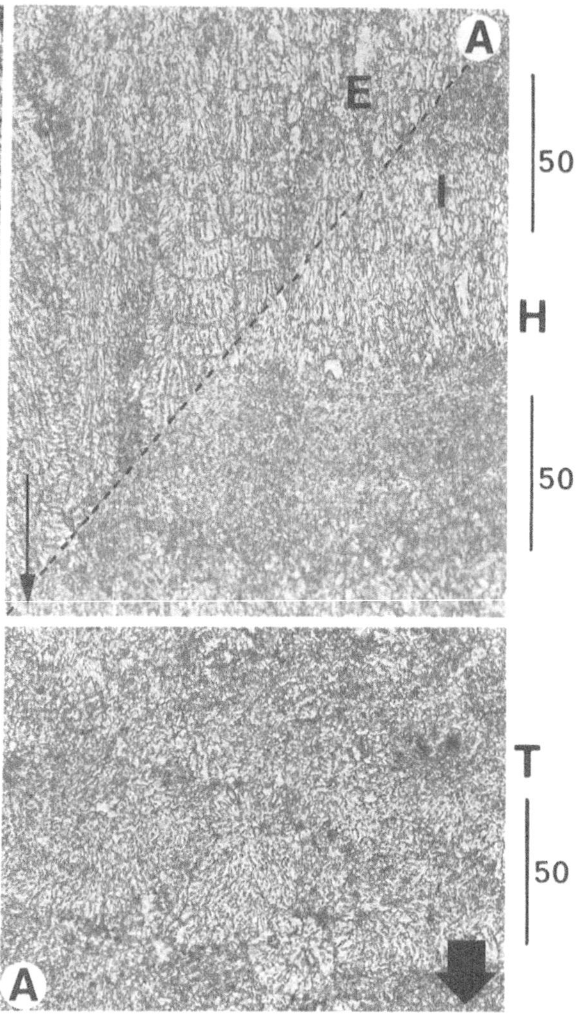

27.
Fimbriidae
Fimbria

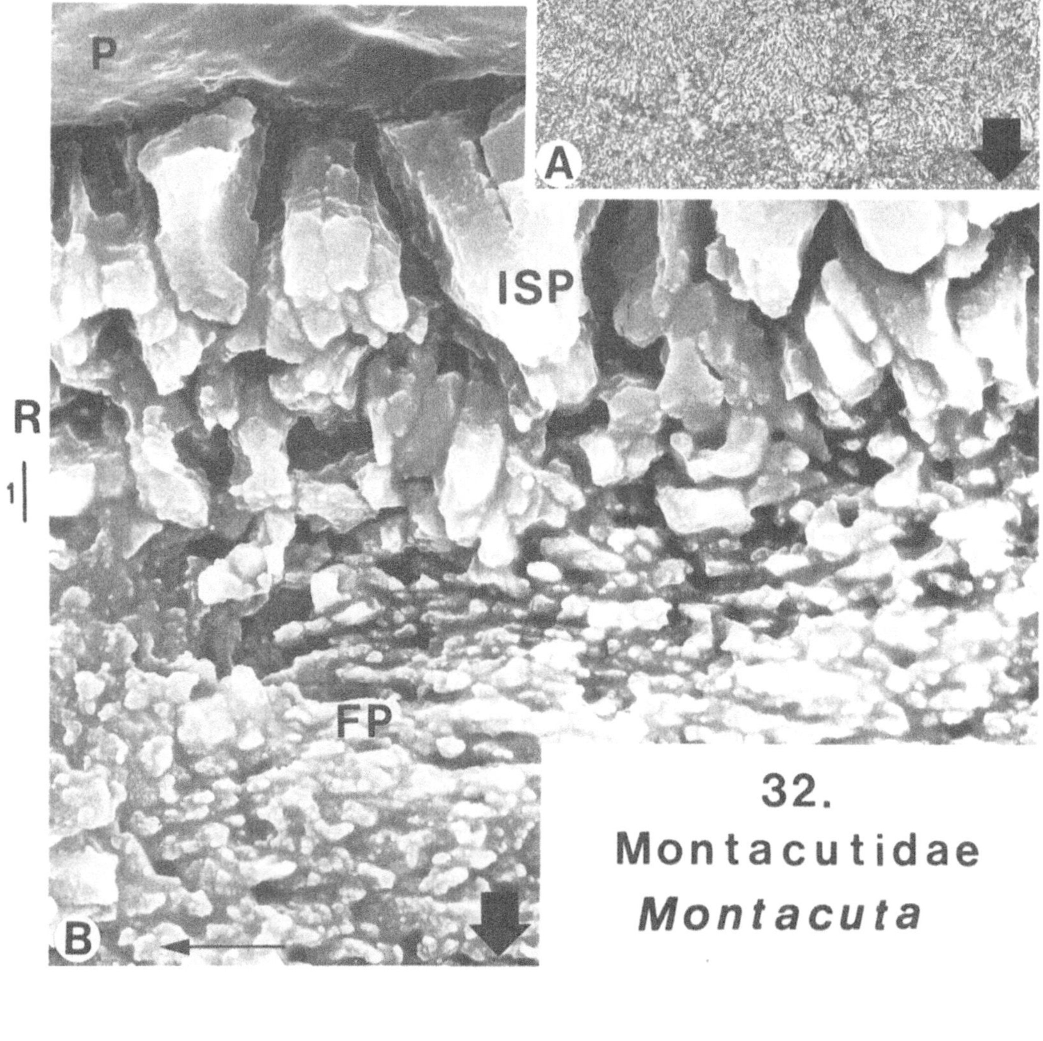

32.
Montacutidae
Montacuta

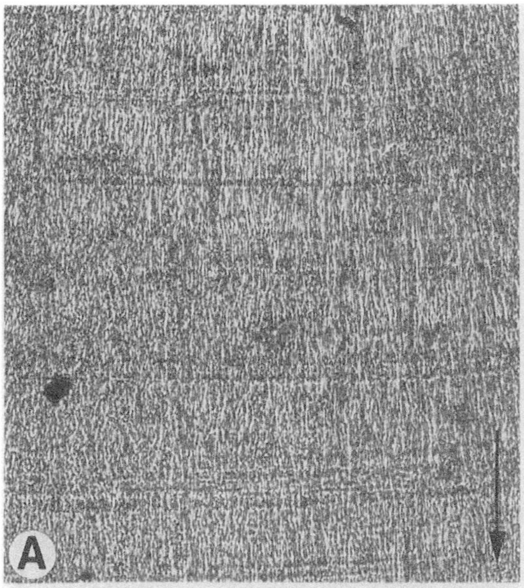

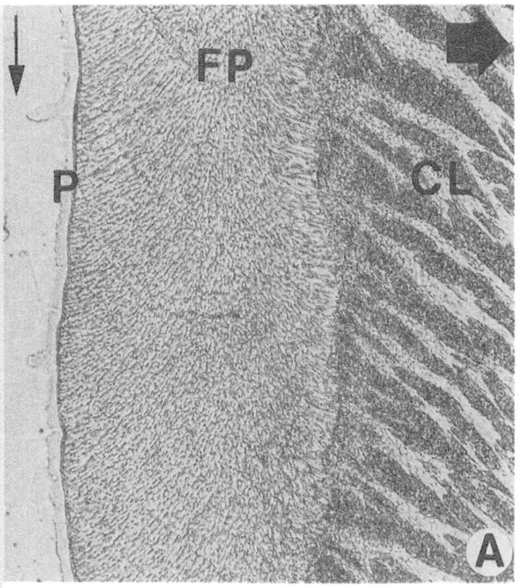

41. Cardiidae *Fragum*

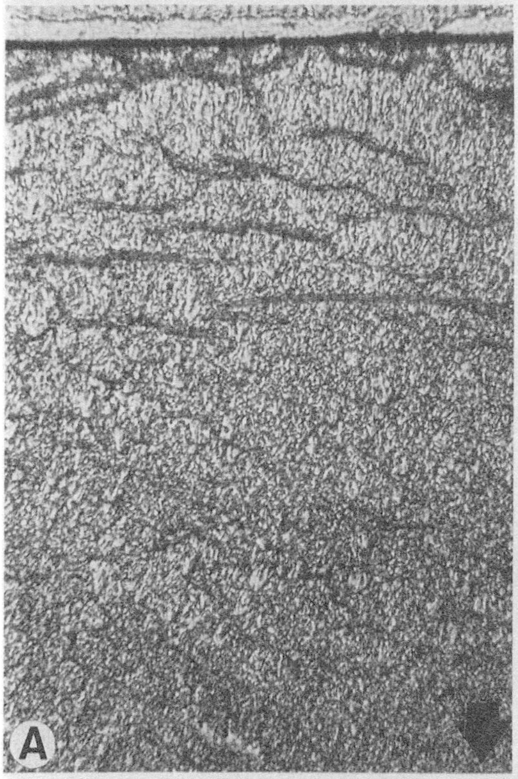

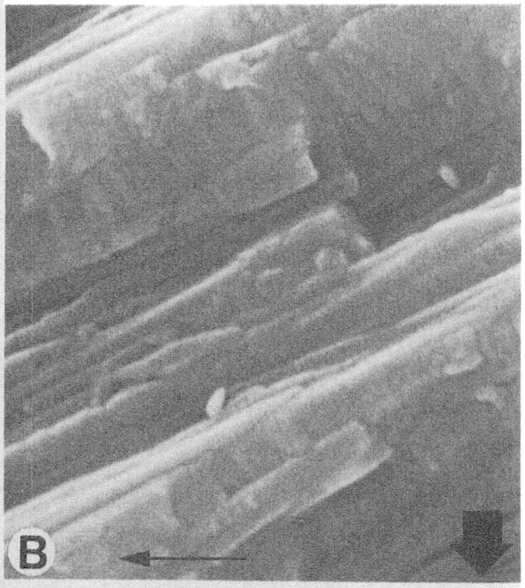

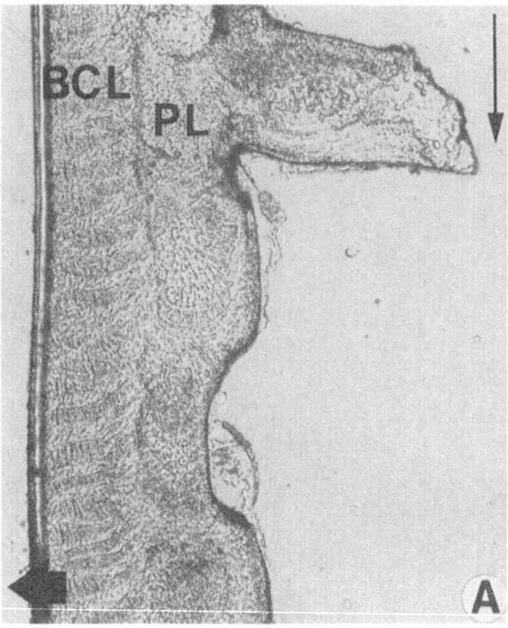

R
50

H
50

42.
Cardiidae
Nemocardium

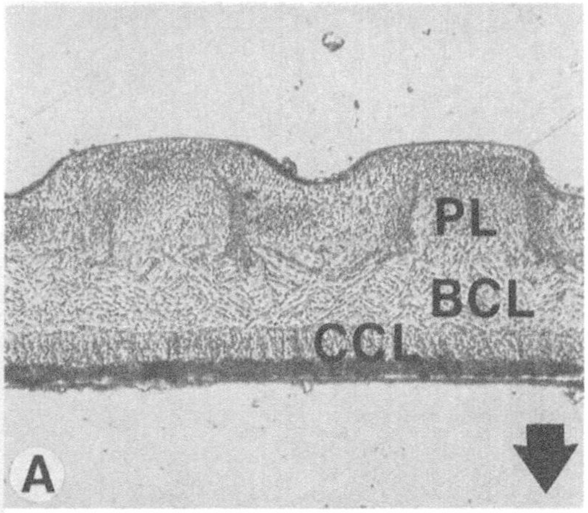

T
50

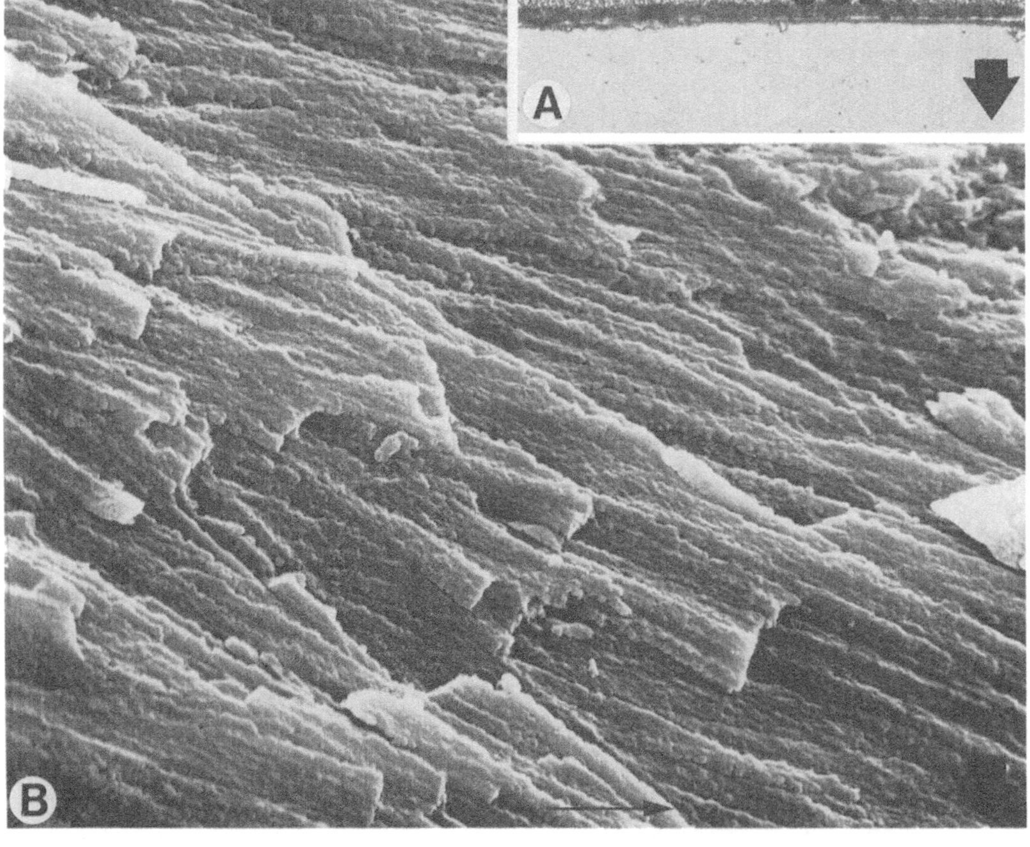

R
1

45.
Mactridae
Lutraria

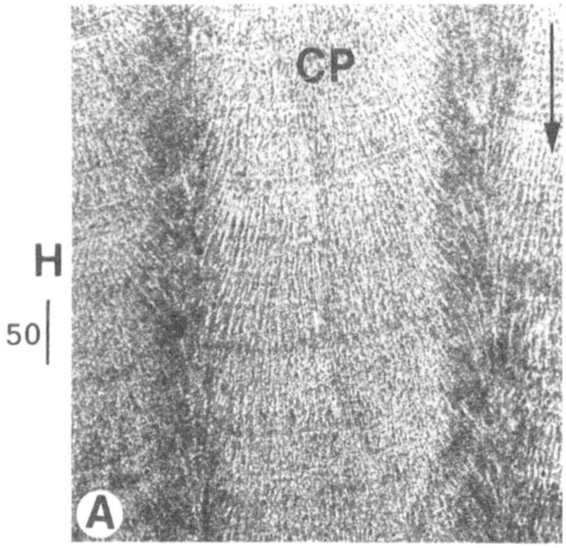

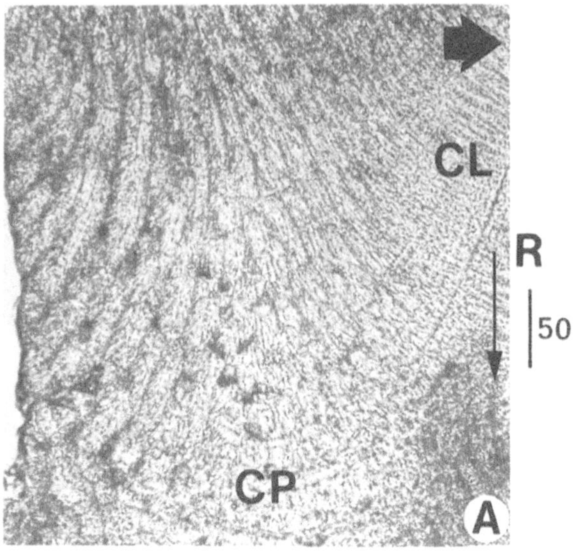

53.
Donacidae
Donax

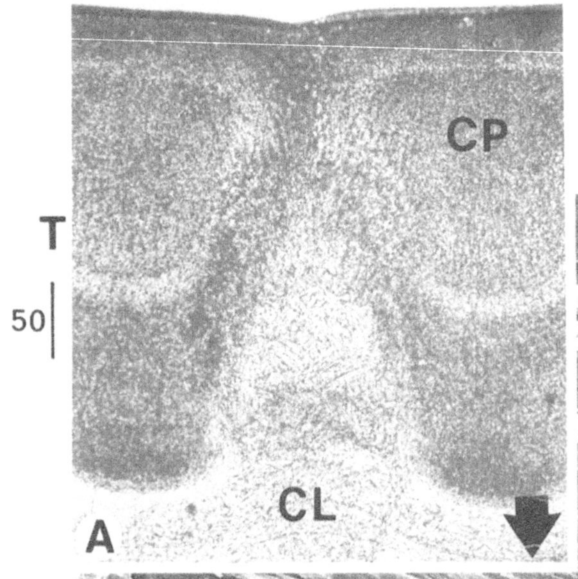

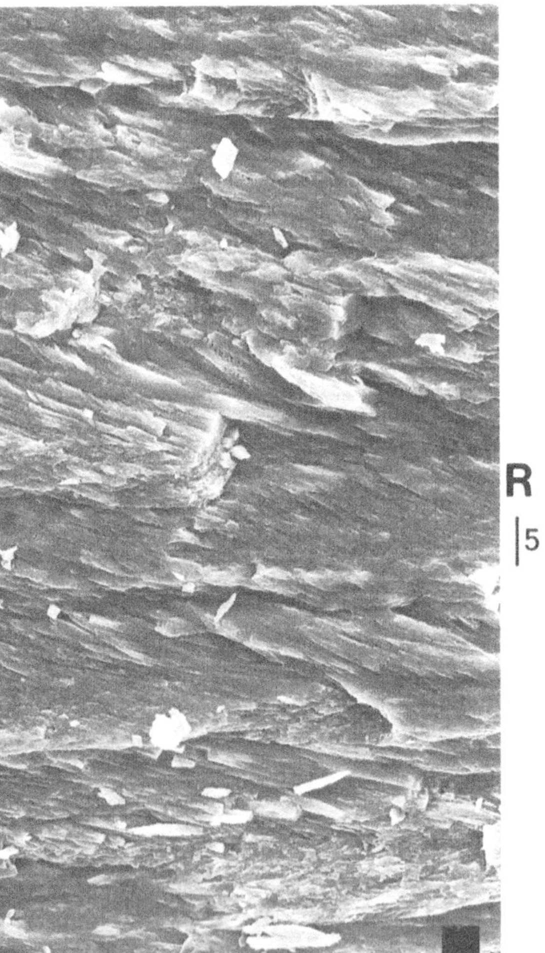
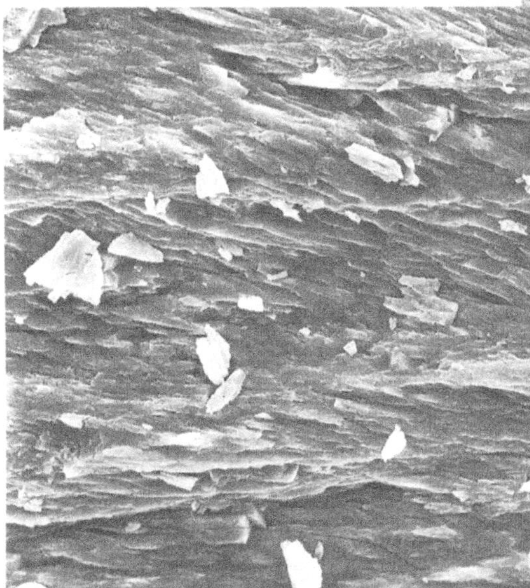

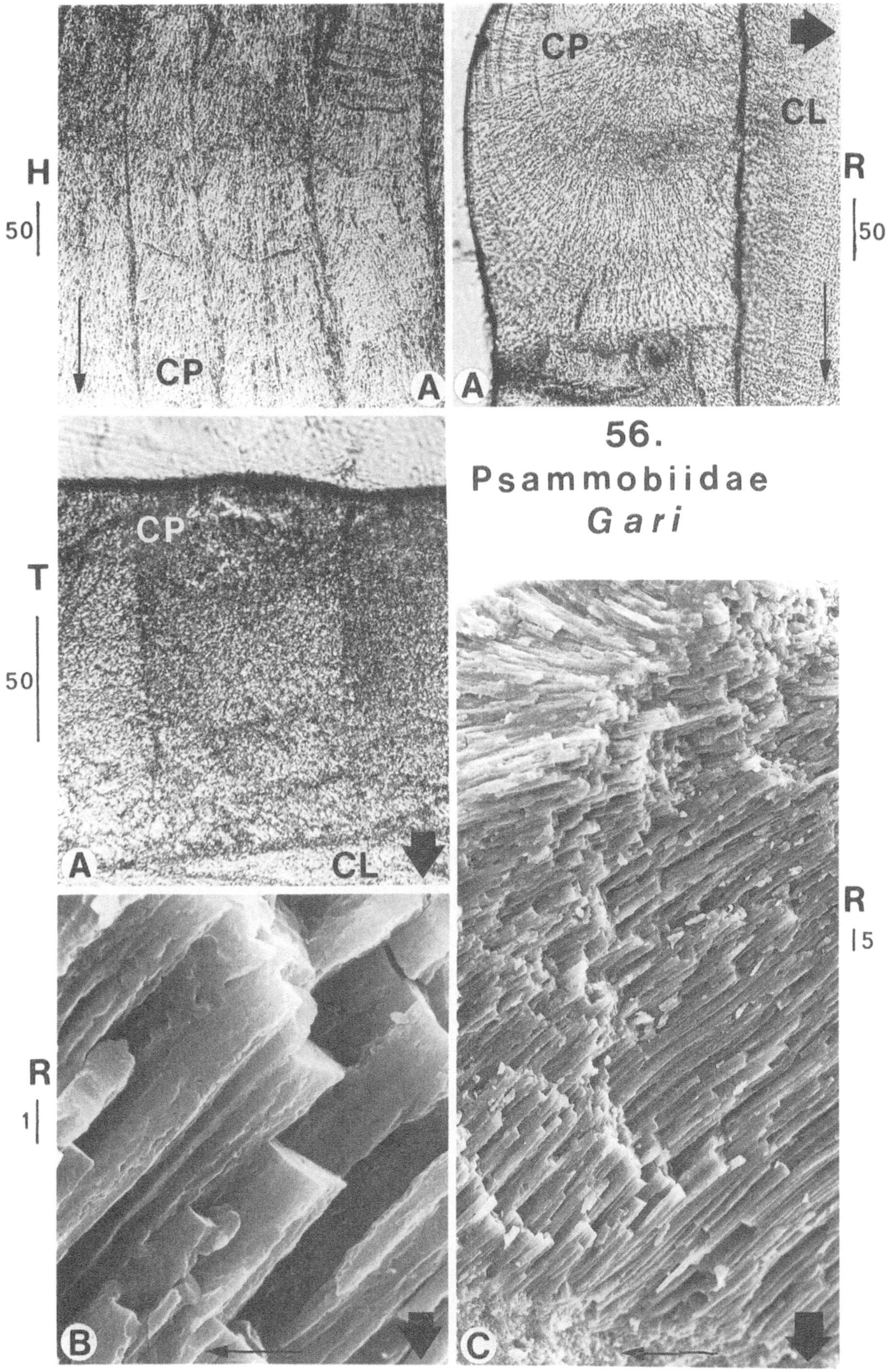

56.
Psammobiidae
Gari

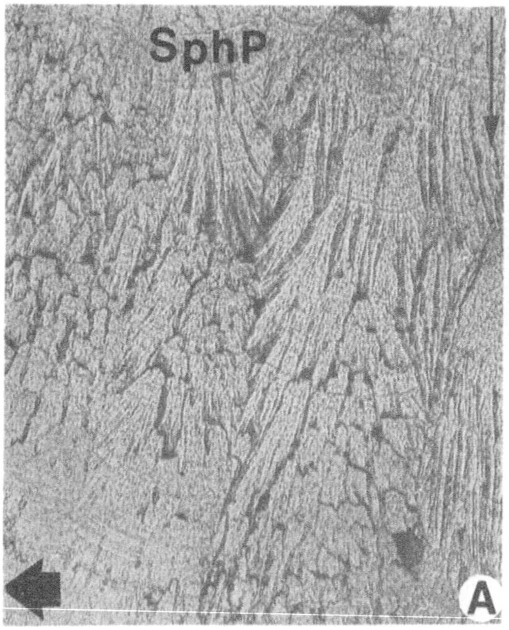

R

50

H

50

57.
Semelidae
Semele

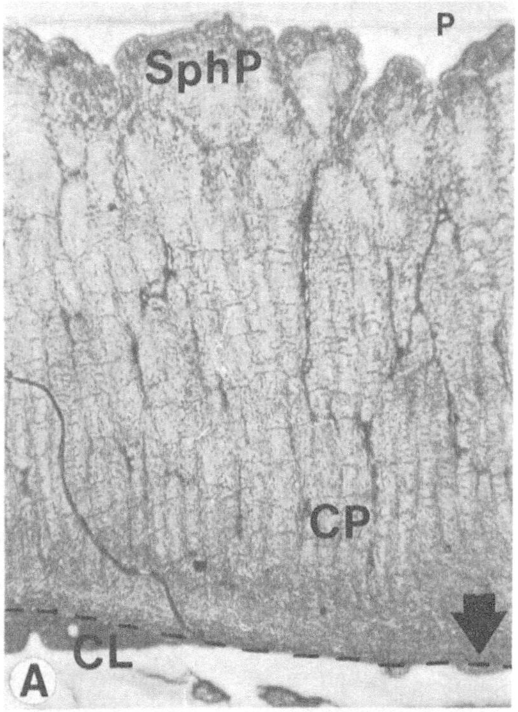

T

100

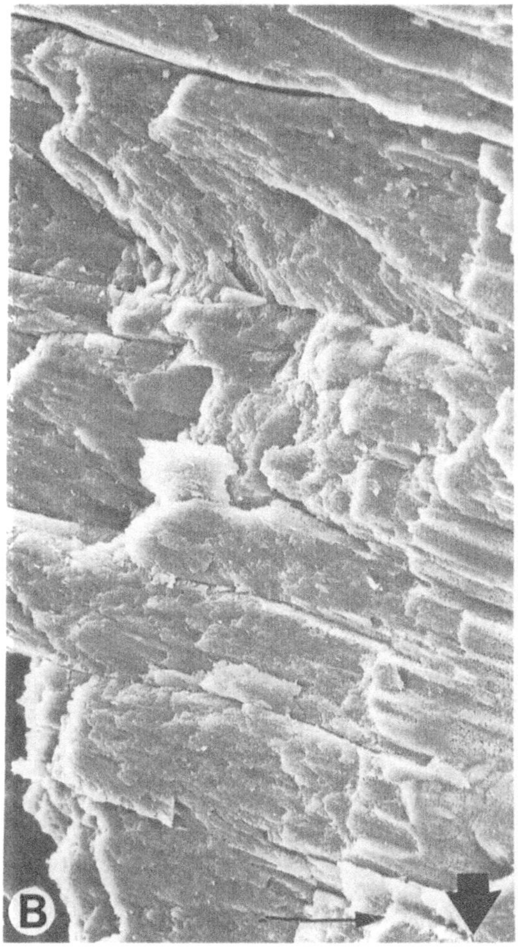

R

5

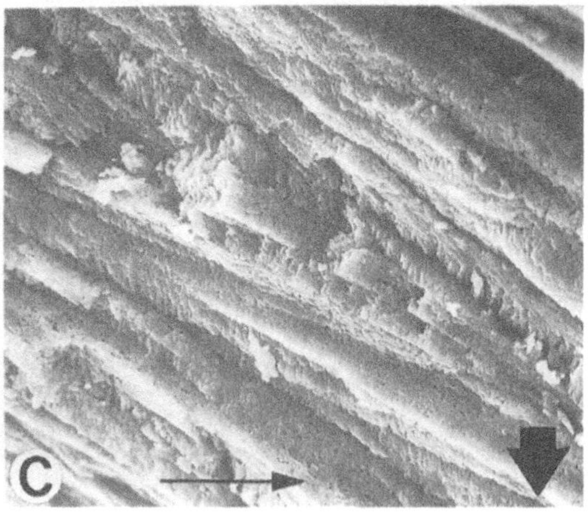

R

5

H

50

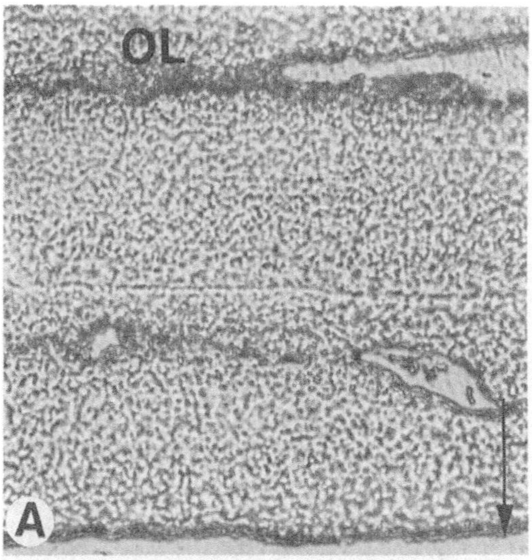

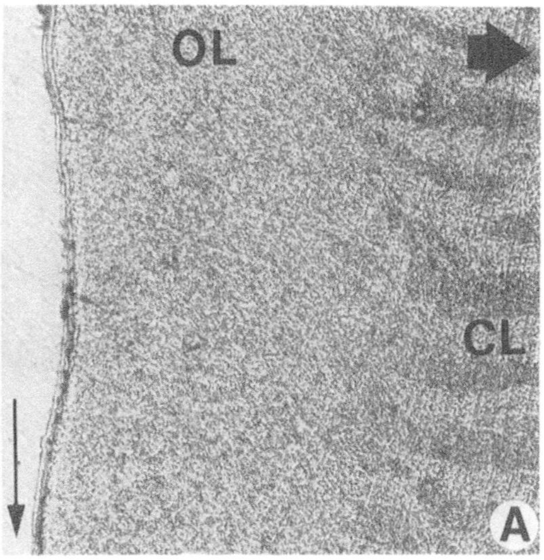

R

50

64.
Glossidae
Glossus

T

50

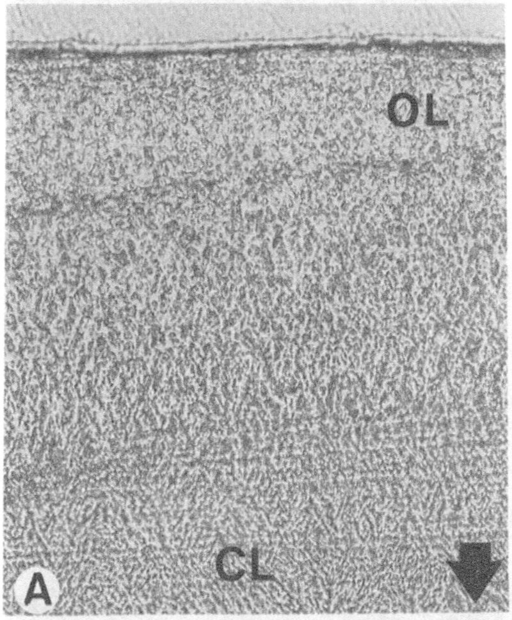

O

2

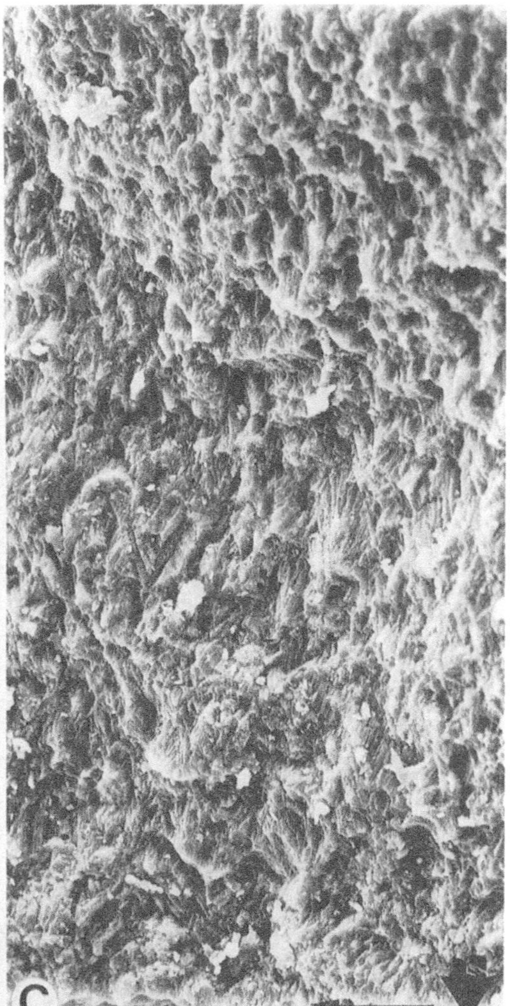

O

10

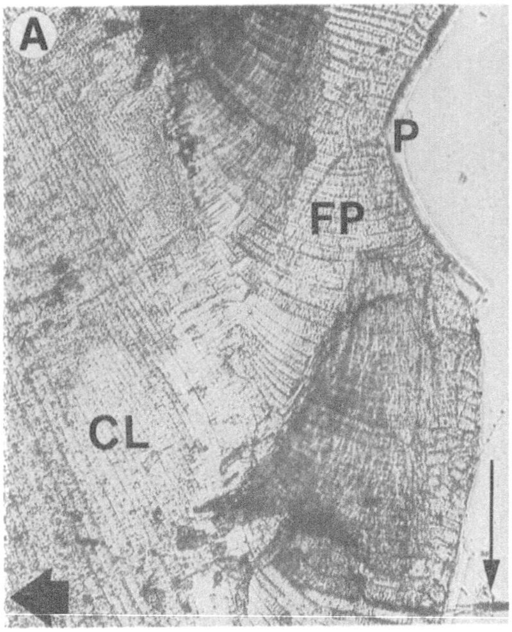

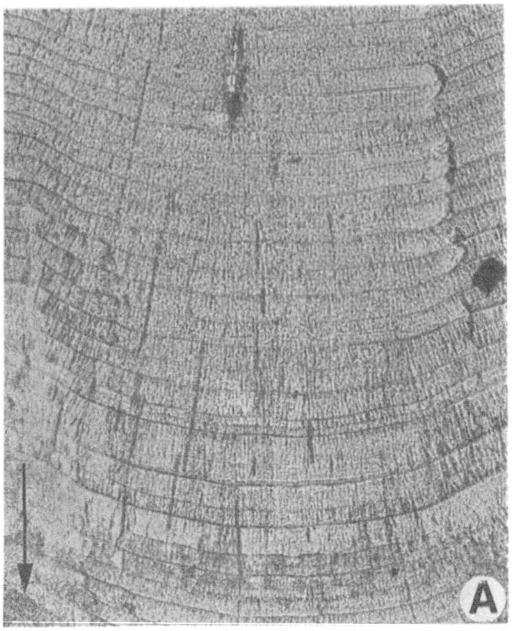

67.
Veneridae
Cyclina

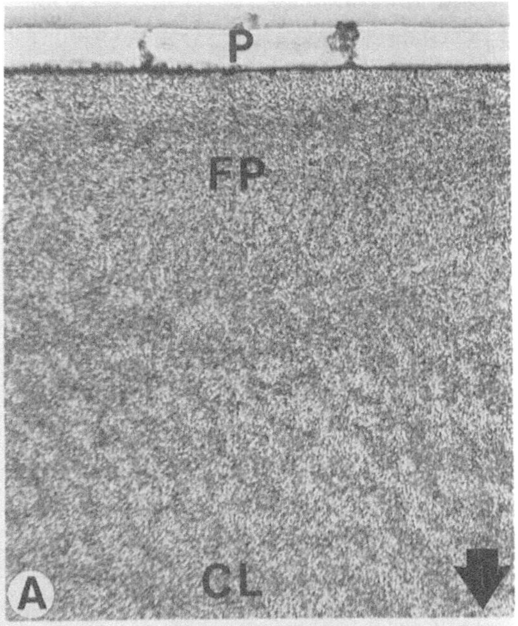

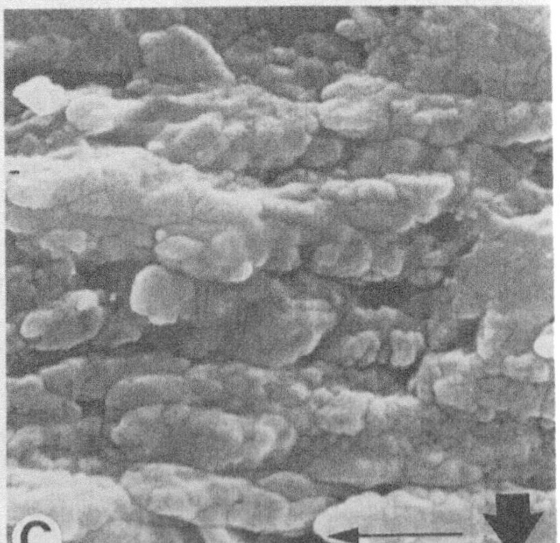

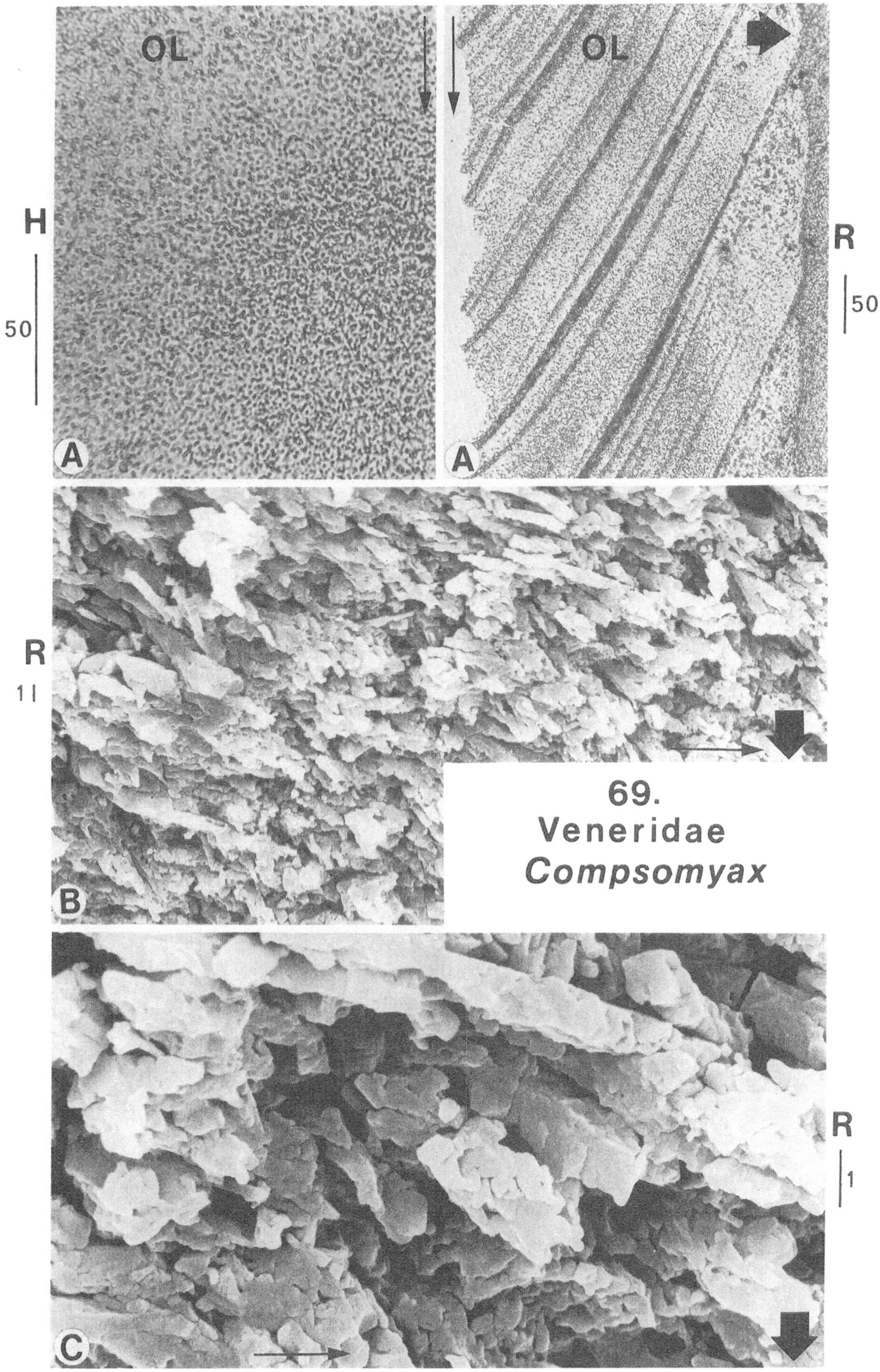

69.
Veneridae
Compsomyax

PLATE 42

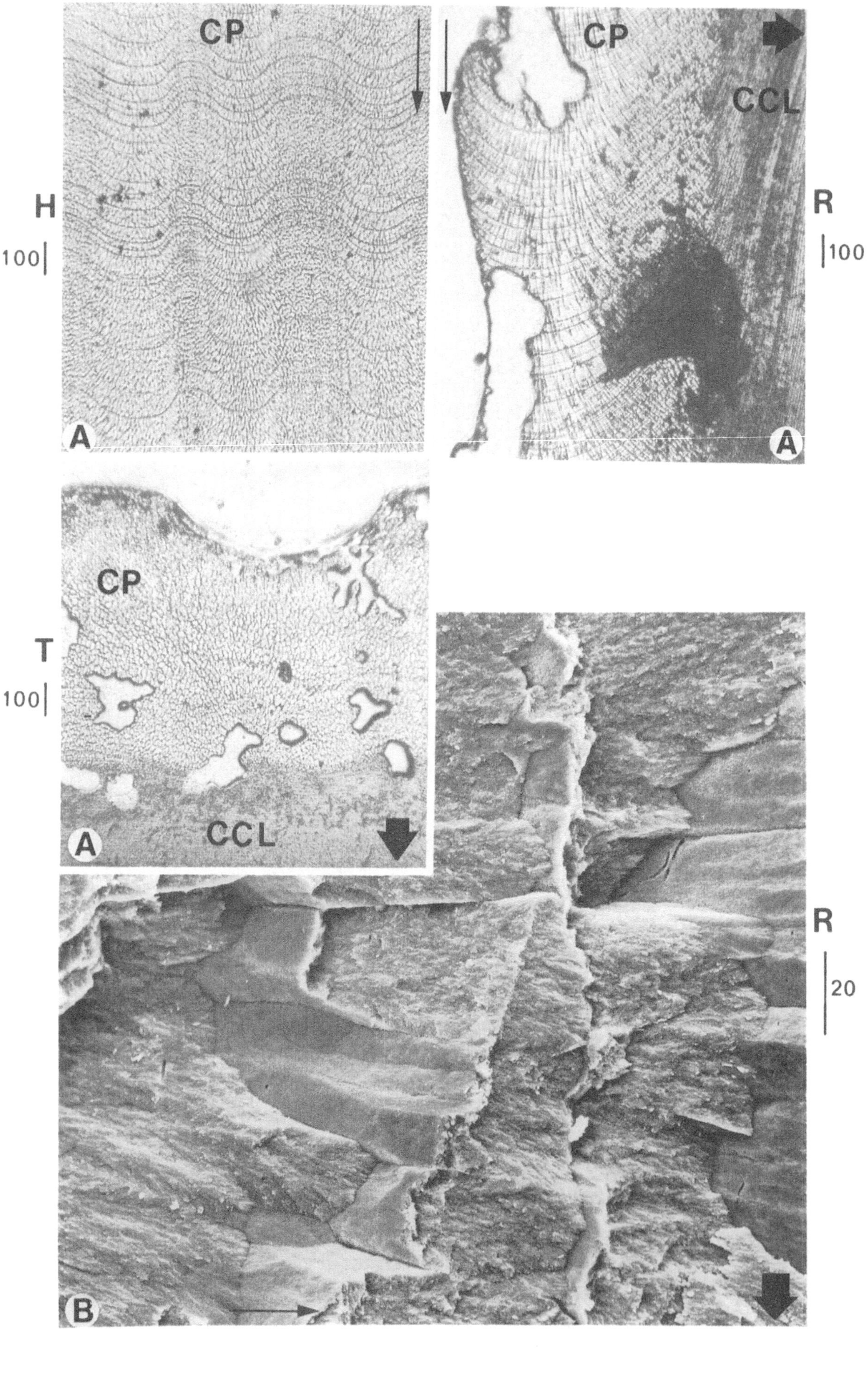

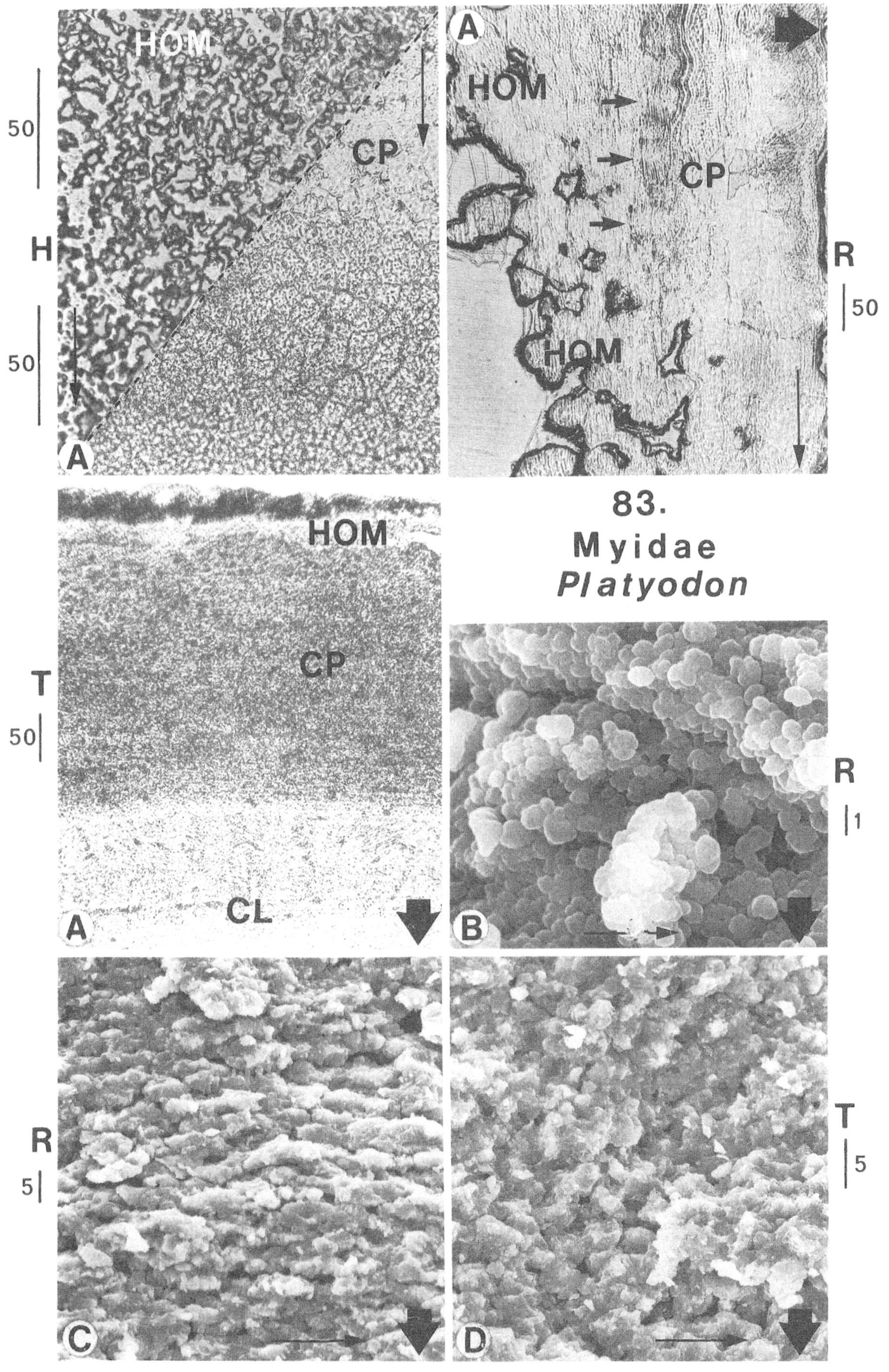

83.
Myidae
Platyodon

85. Poromyidae *Poromya*

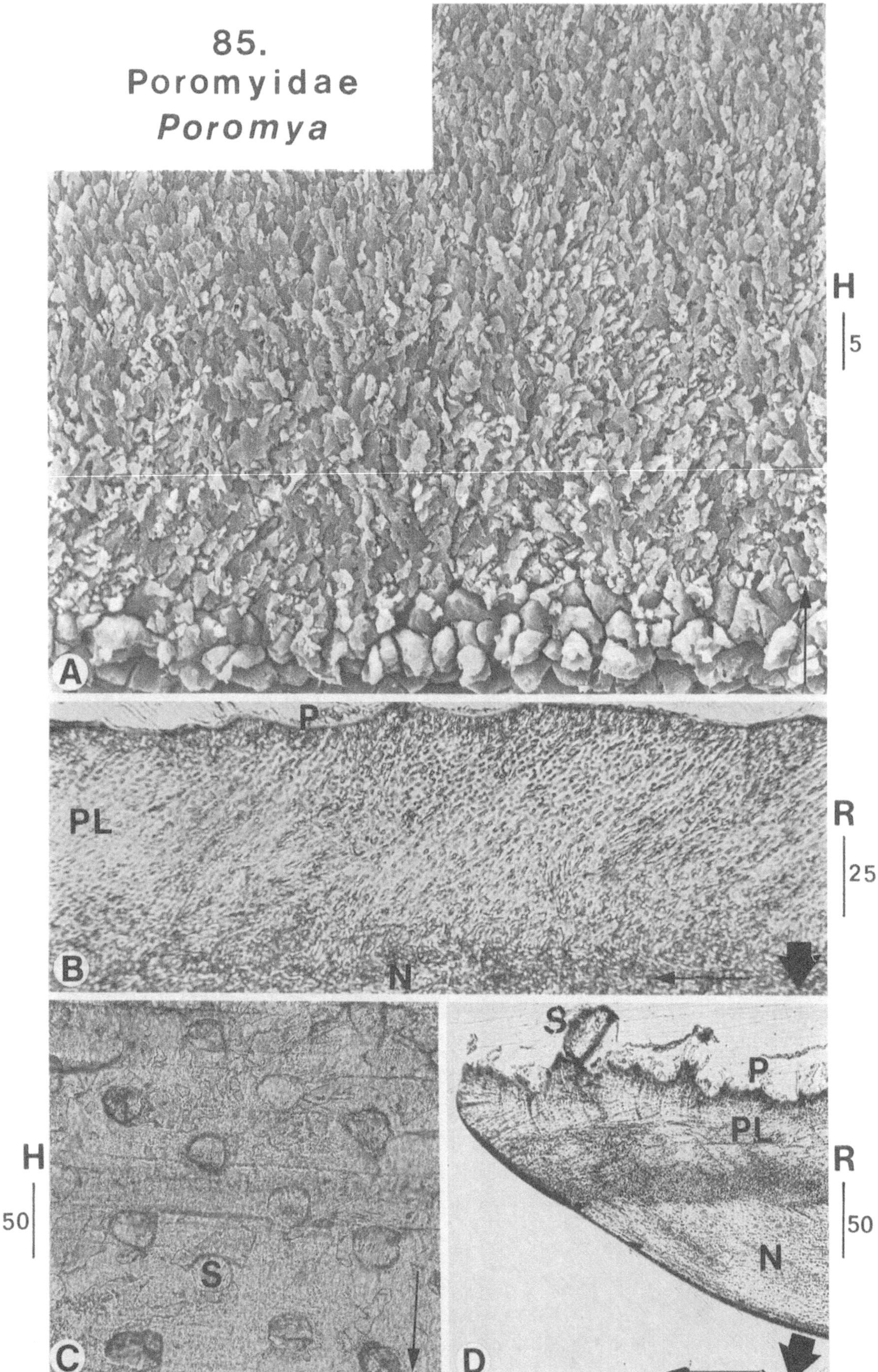

H
50

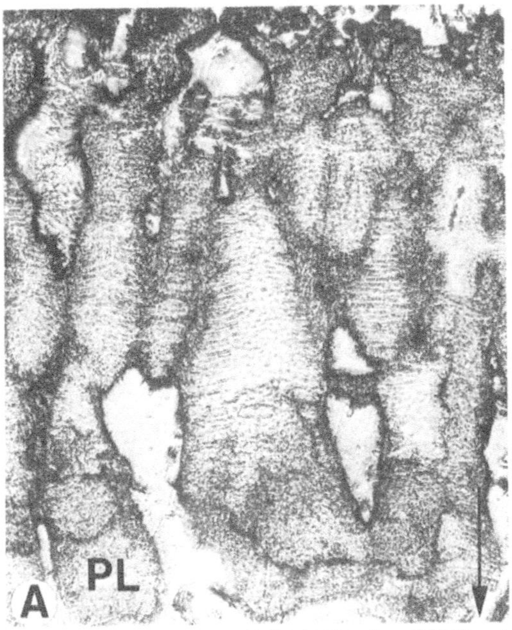

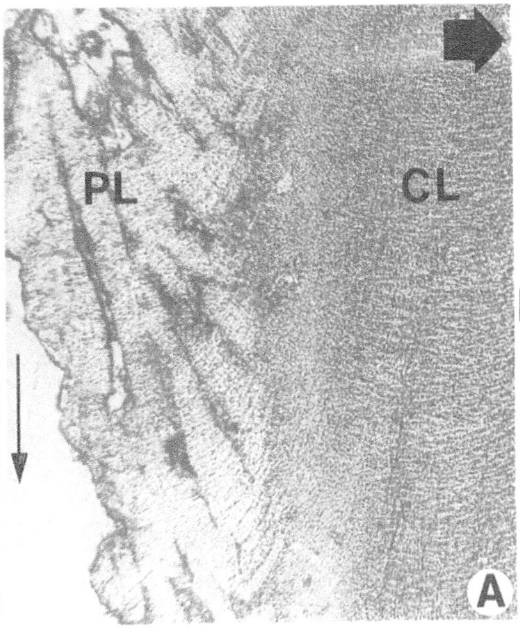

R
50

89.
Gastrochaenidae
Spengleria

T
50

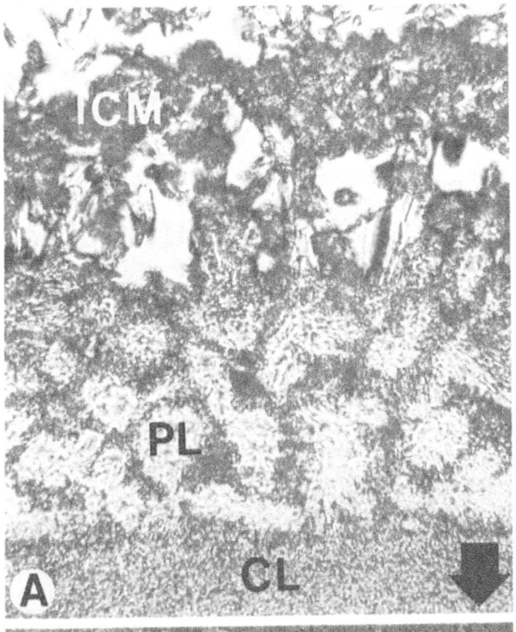

R
5

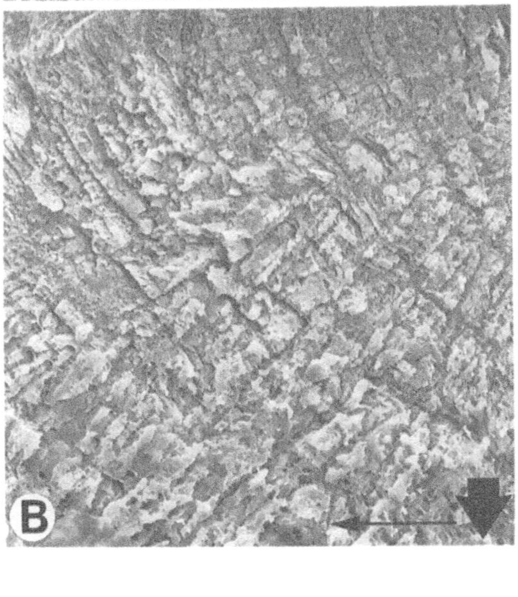

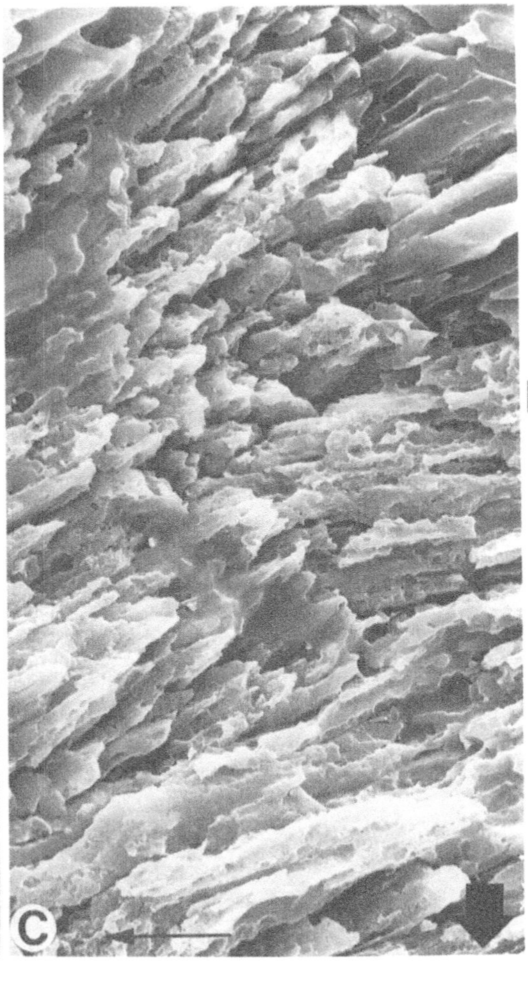

R
5

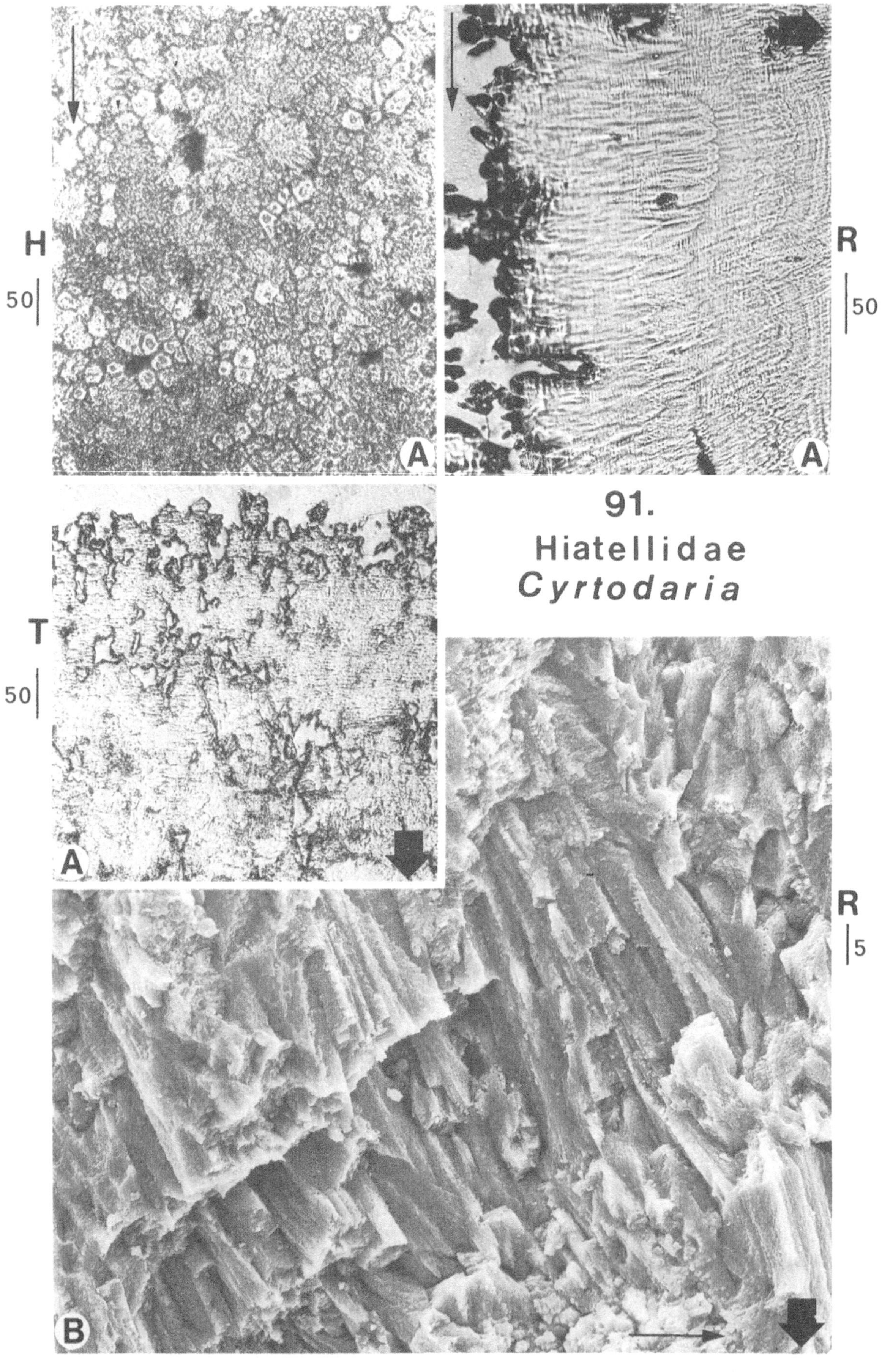

91.
Hiatellidae
Cyrtodaria

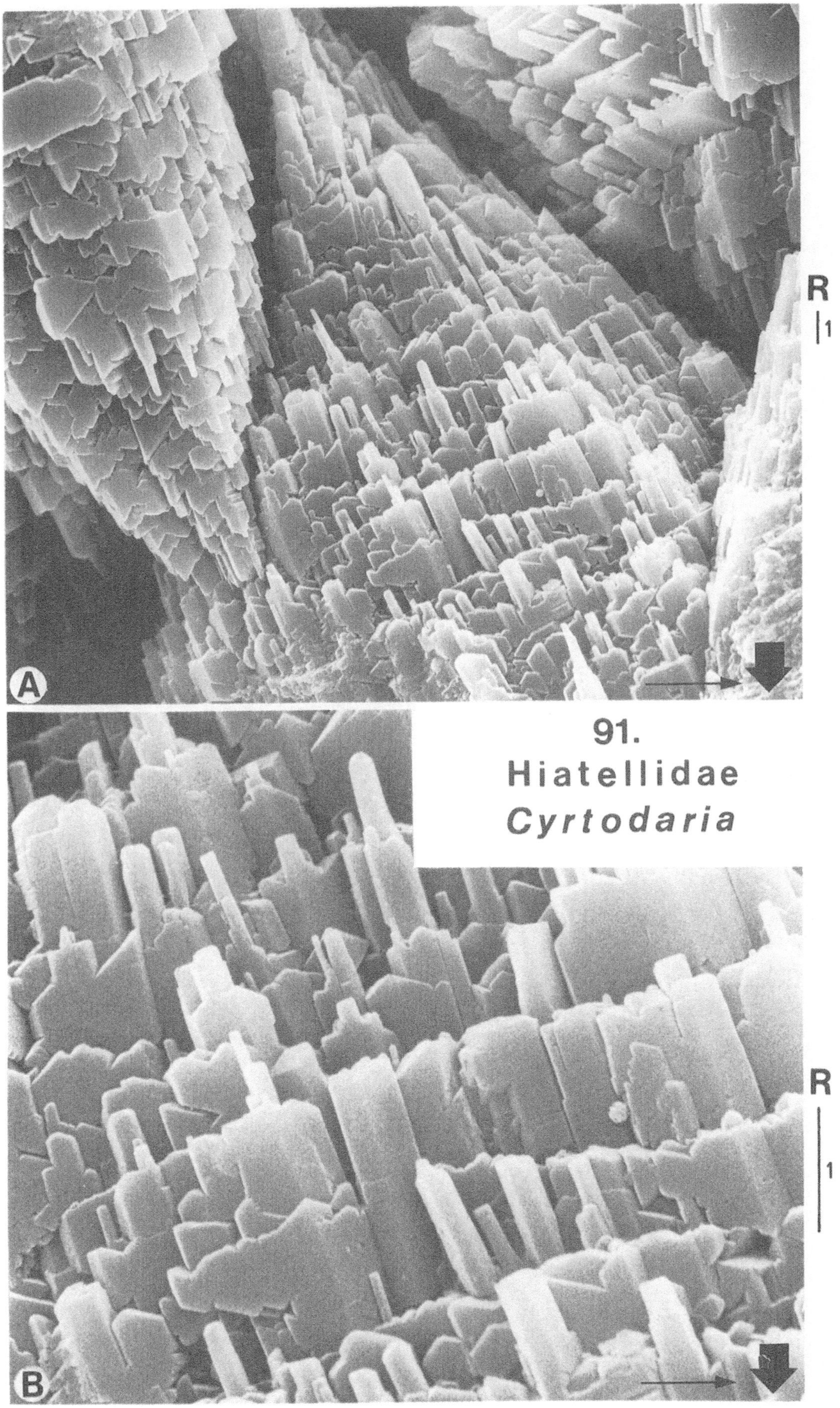

91.
Hiatellidae
Cyrtodaria

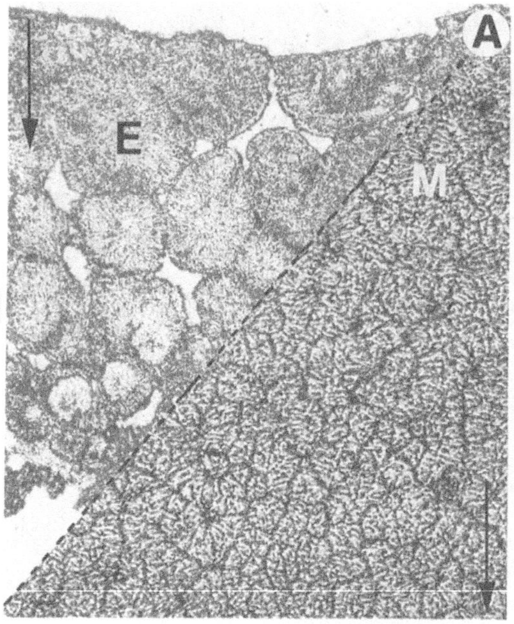

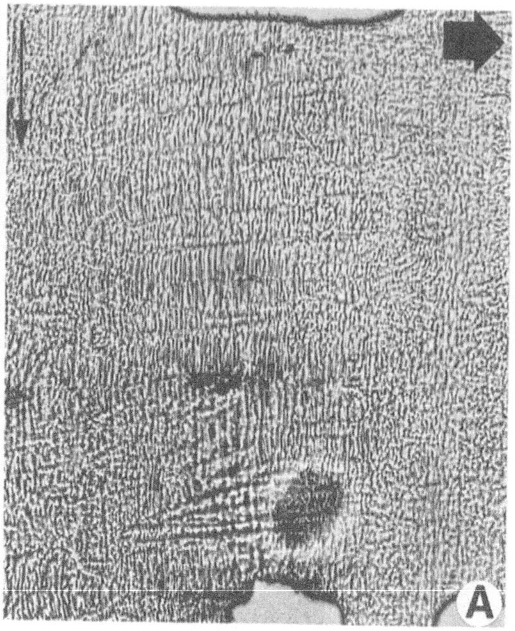

92.
Hiatellidae
Panopea

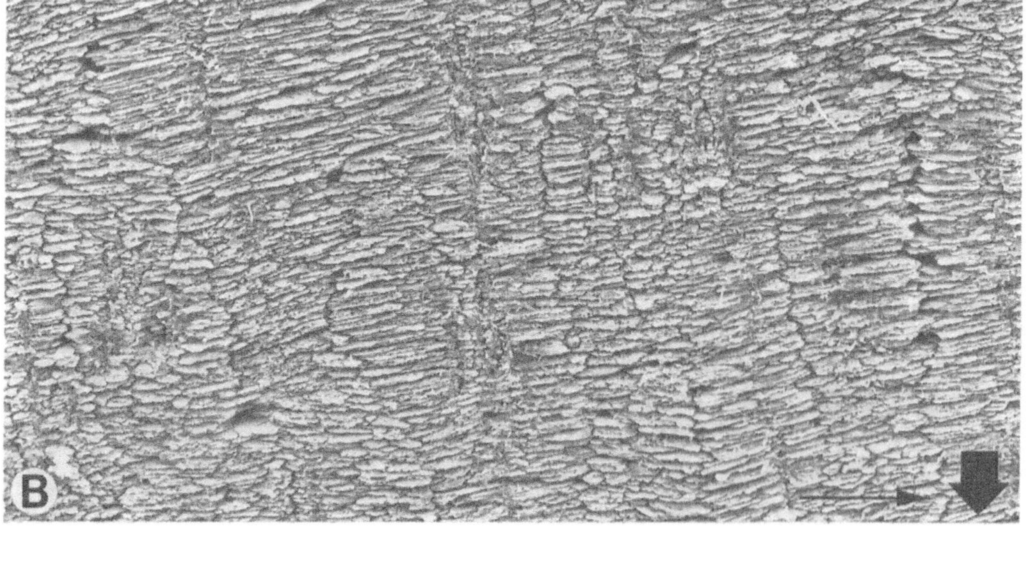

H 50

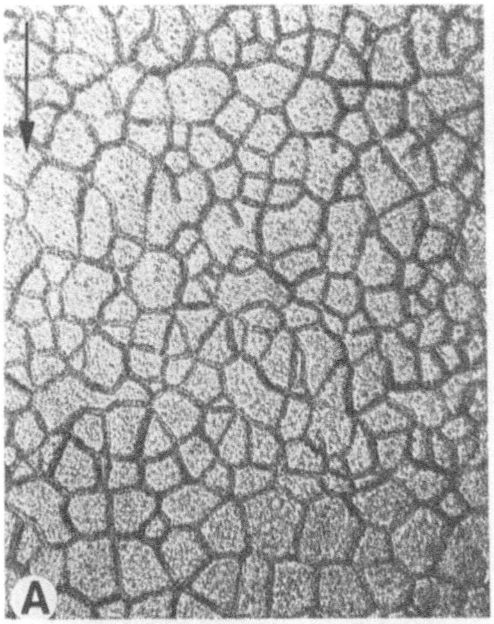

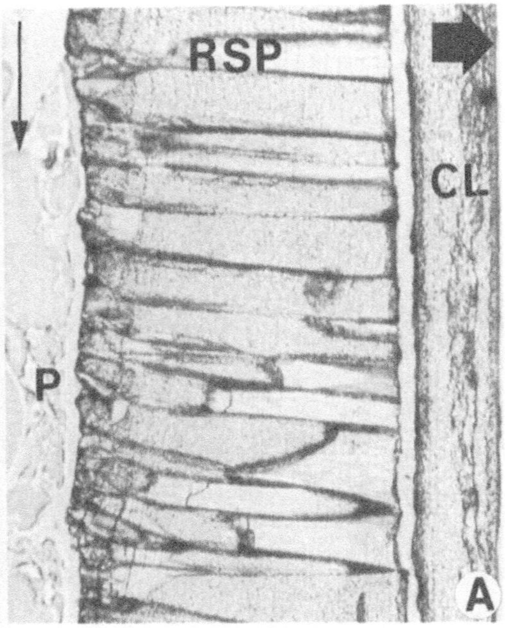

R 50

98.
Pholadidae
Parapholas

T 50

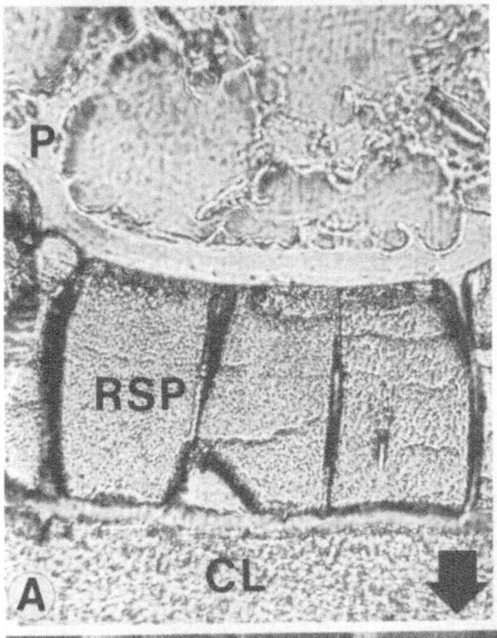

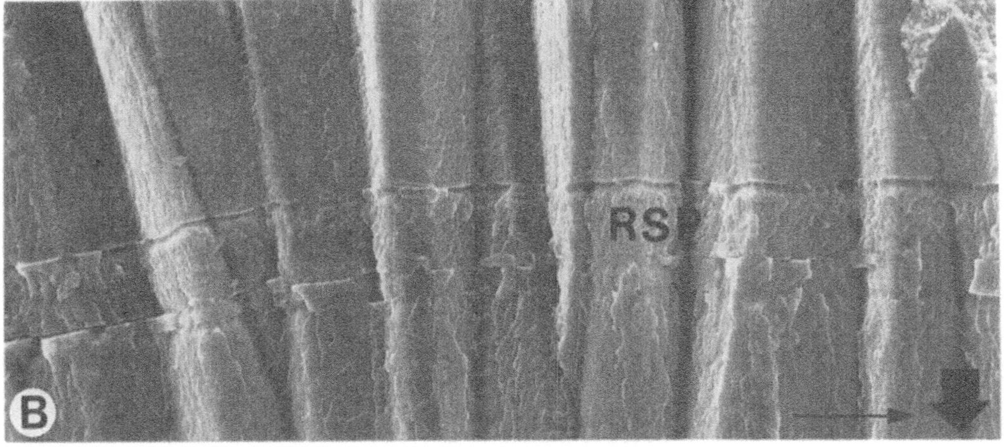

R 20

R
20

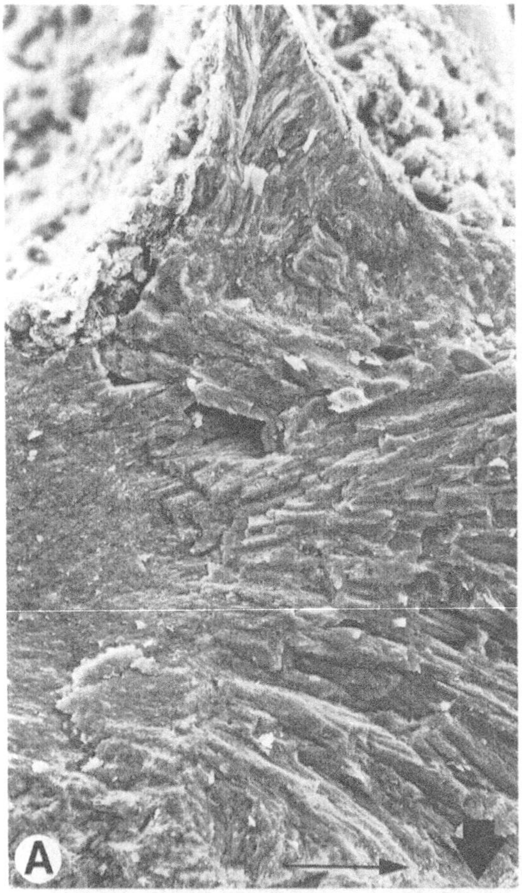

R
5

H
100

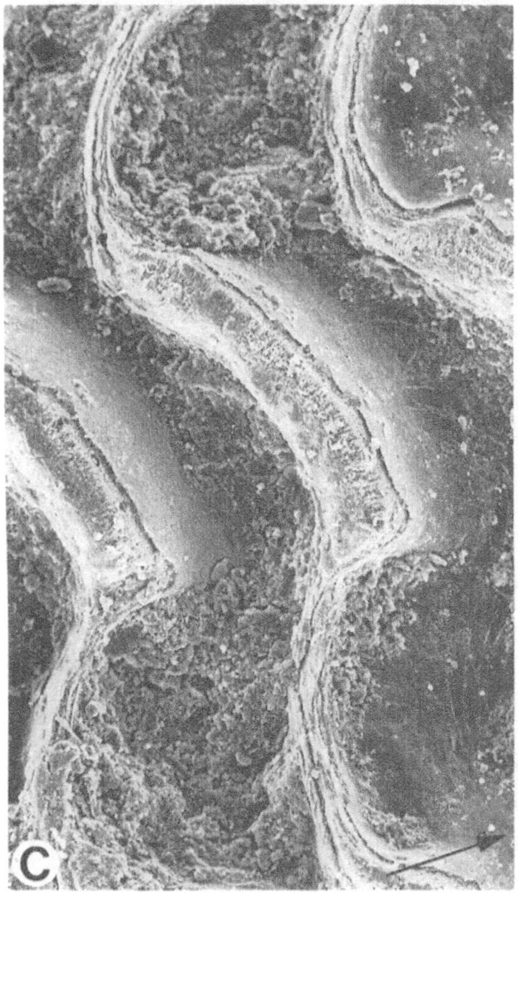

98.
Pholadidae
Parapholas

H
5

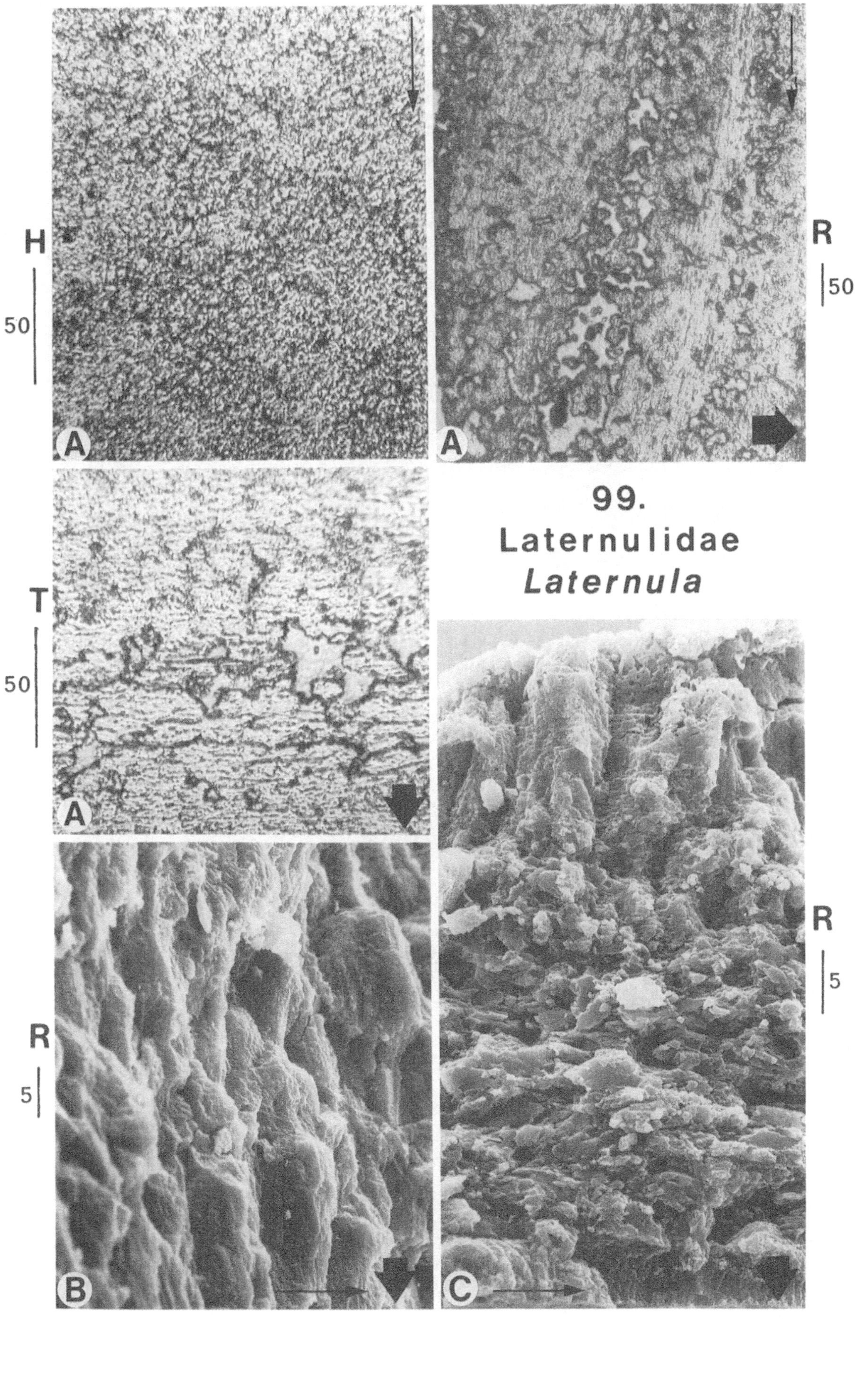

99.
Laternulidae
Laternula

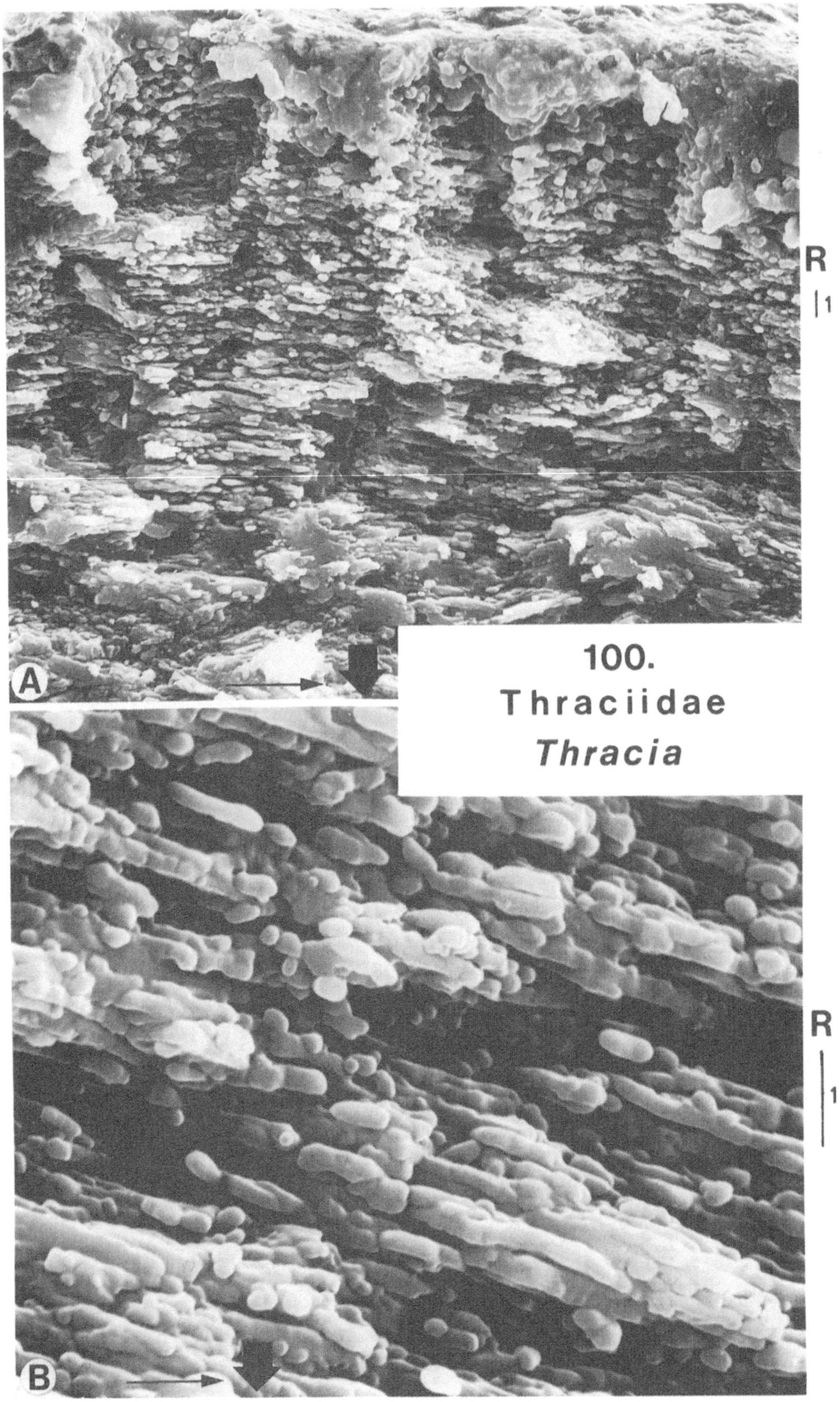

100.
Thraciidae
Thracia

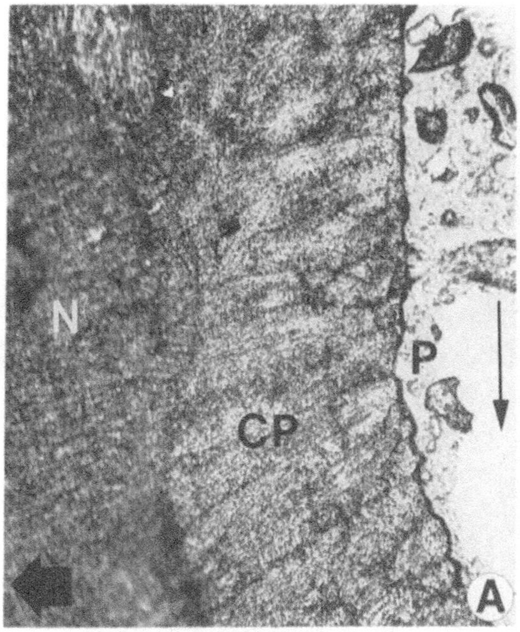

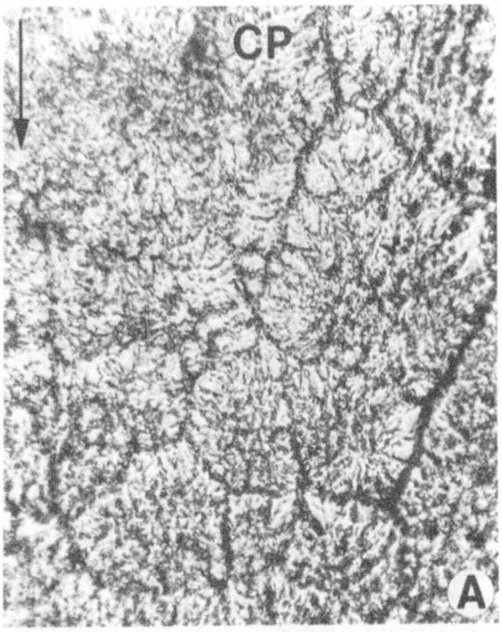

101.
Verticordiidae
Euciroa

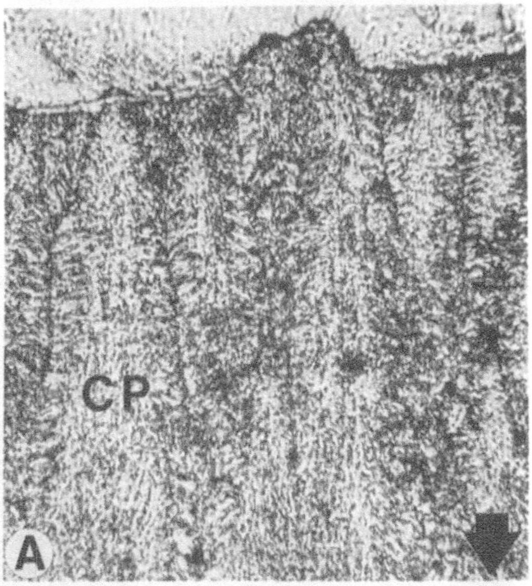

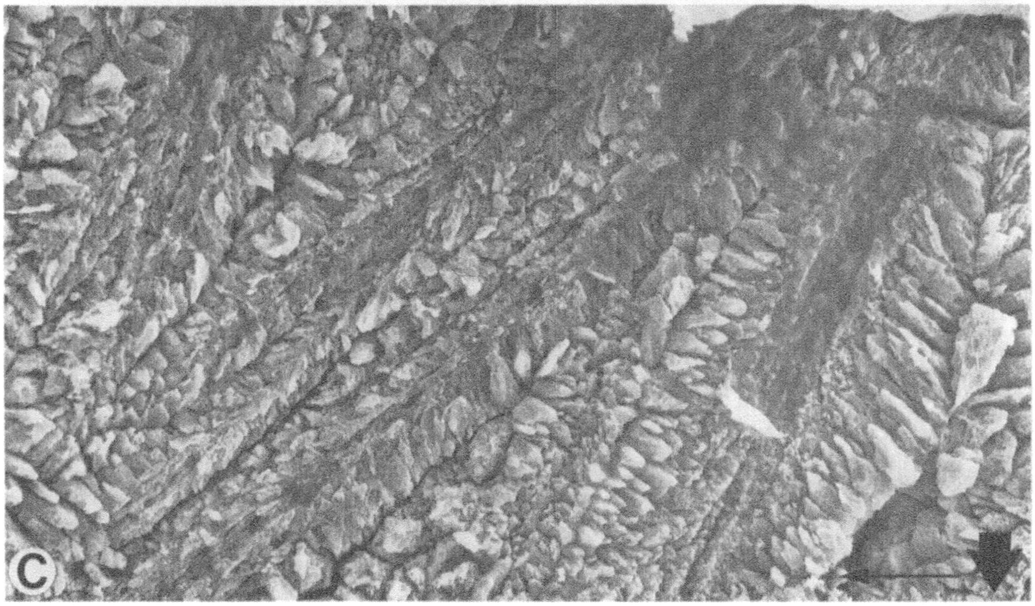

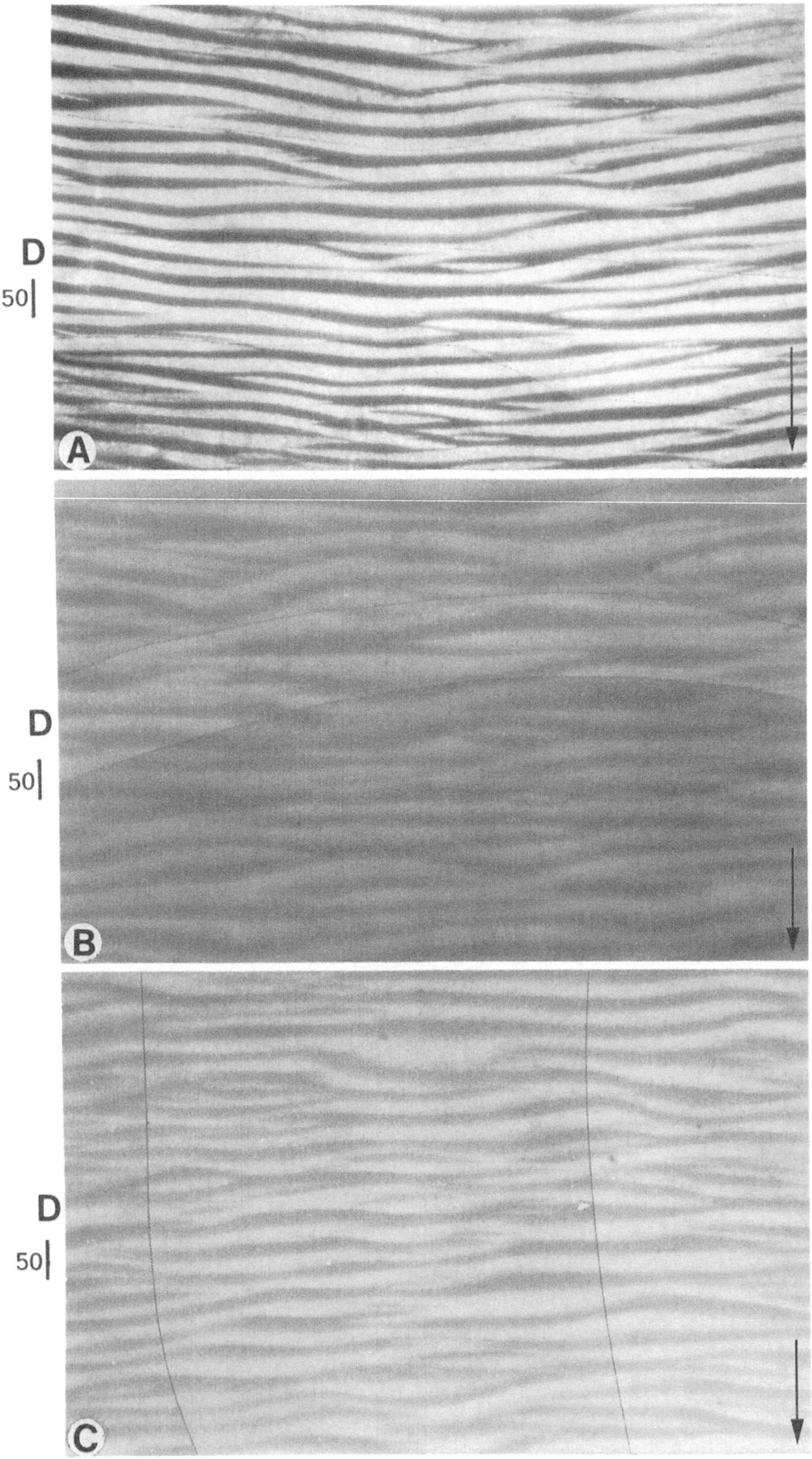

4. Arcidae *Arca*

5. Cucullaeidae *Cucullaea*

6. Noetiidae *Noetia*

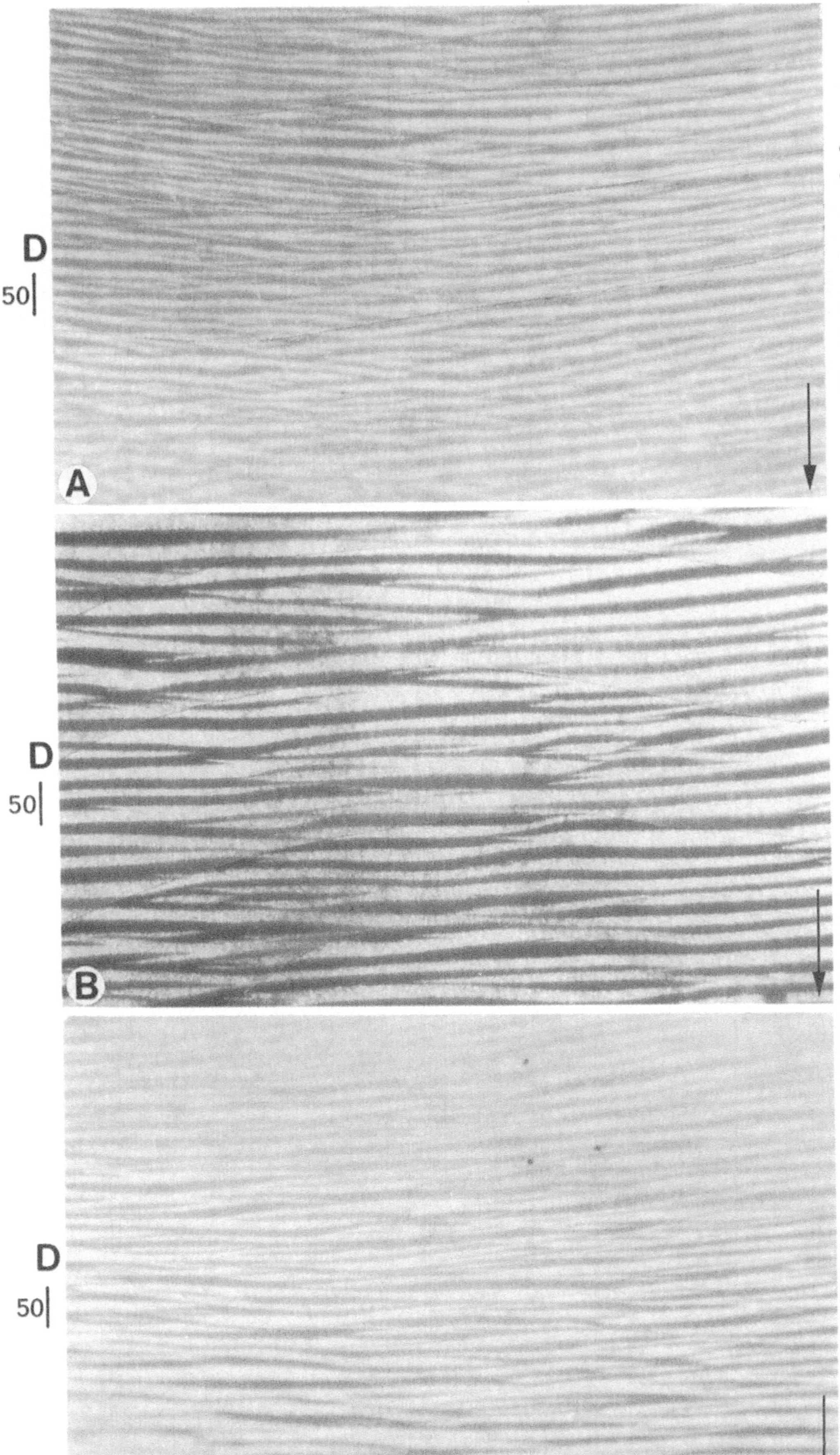

7. Limopsidae *Limopsis*

8. Glycymerididae *Glycymeris*

13. Propeamussiidae *Propeamussium*

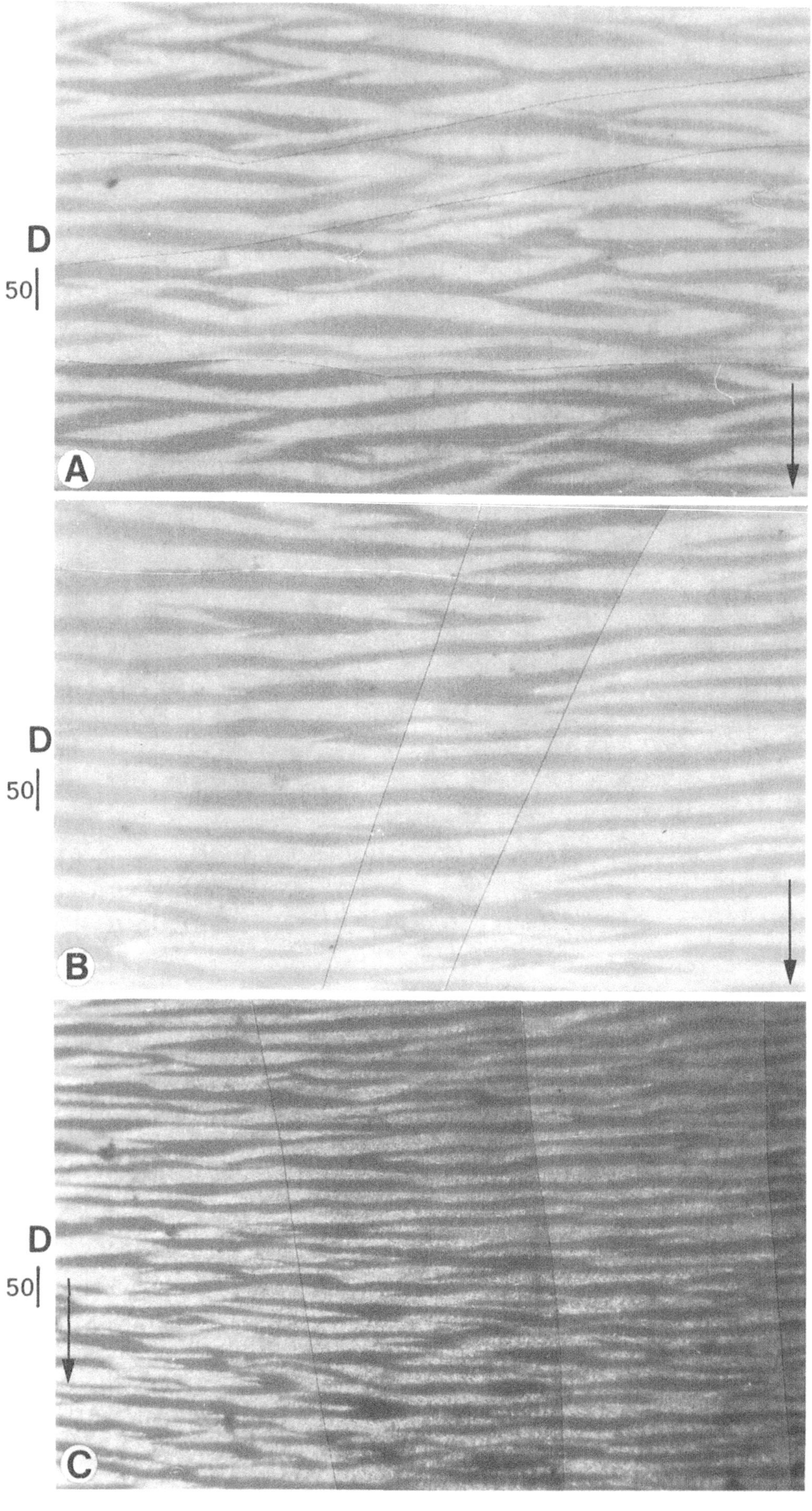

14. Plicatulidae *Plicatula*

15. Spondylidae *Spondylus*

16. Anomiidae *Anomia*

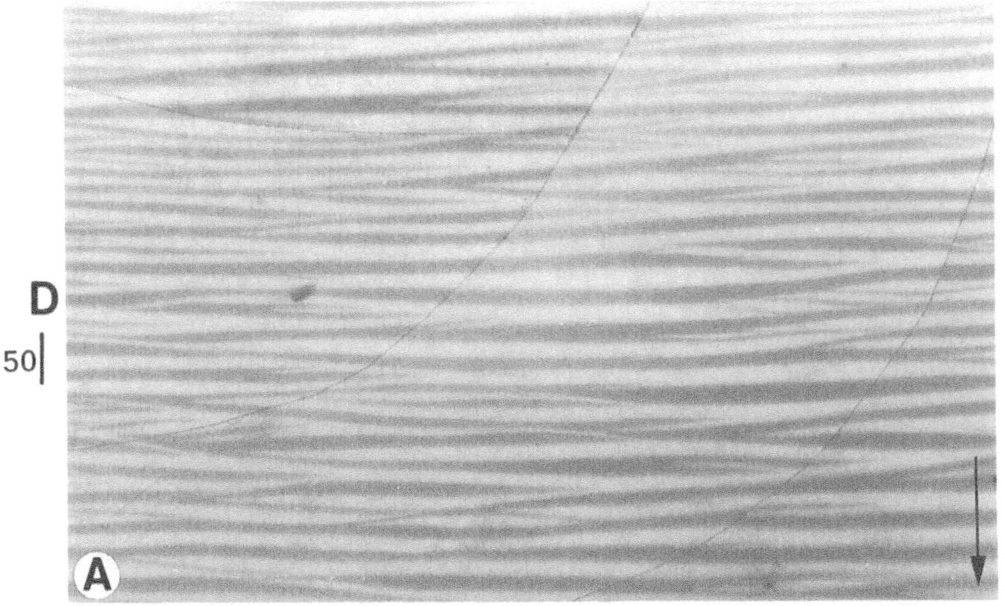

18. Limidae *Ctenoides*

22. Lucinidae *Parvilucina*

23. Lucinidae *Myrtea*

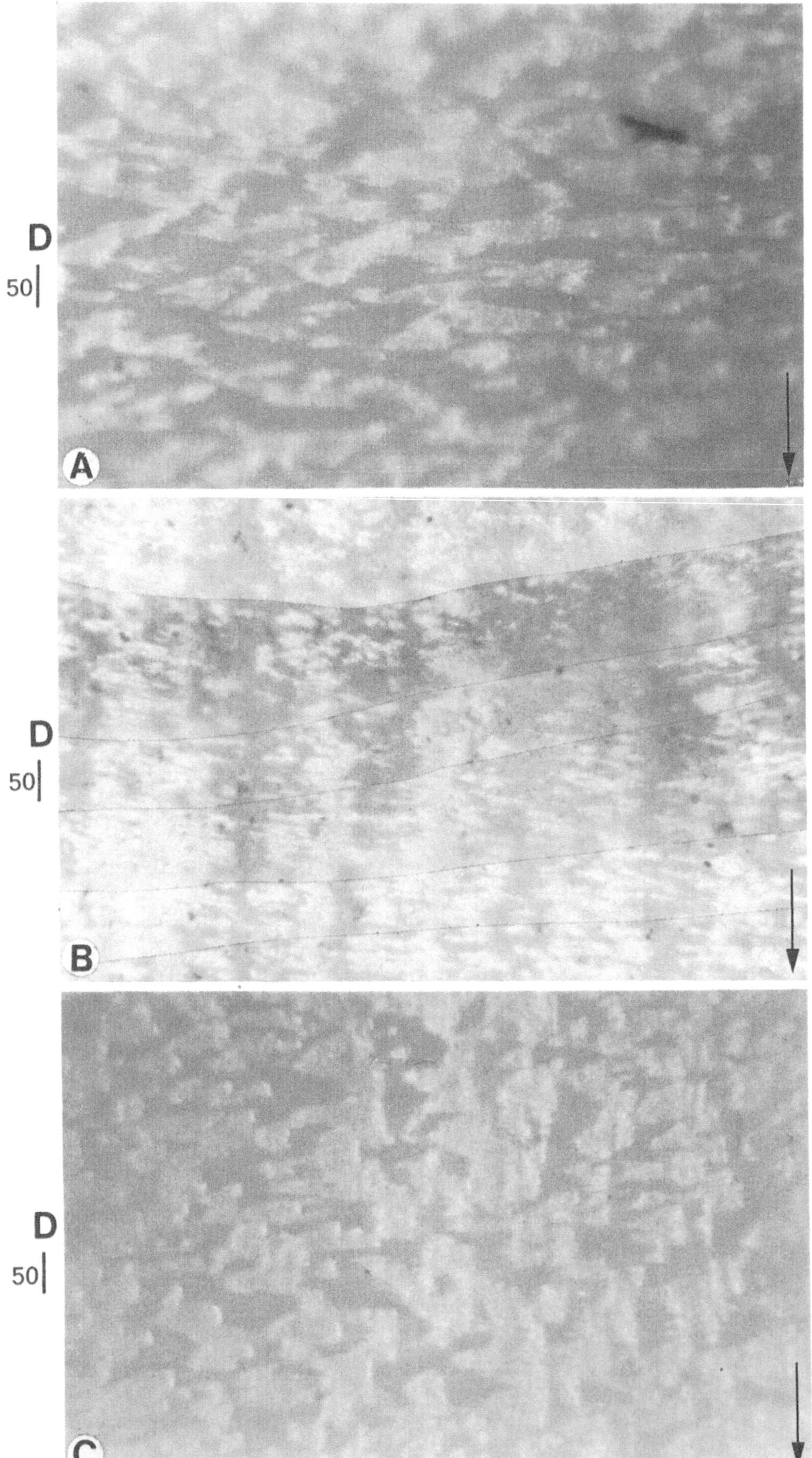

25. Lucinidae *Divalucina*

26. Thyasiridae *Thyasira*

27. Fimbriidae *Fimbria*

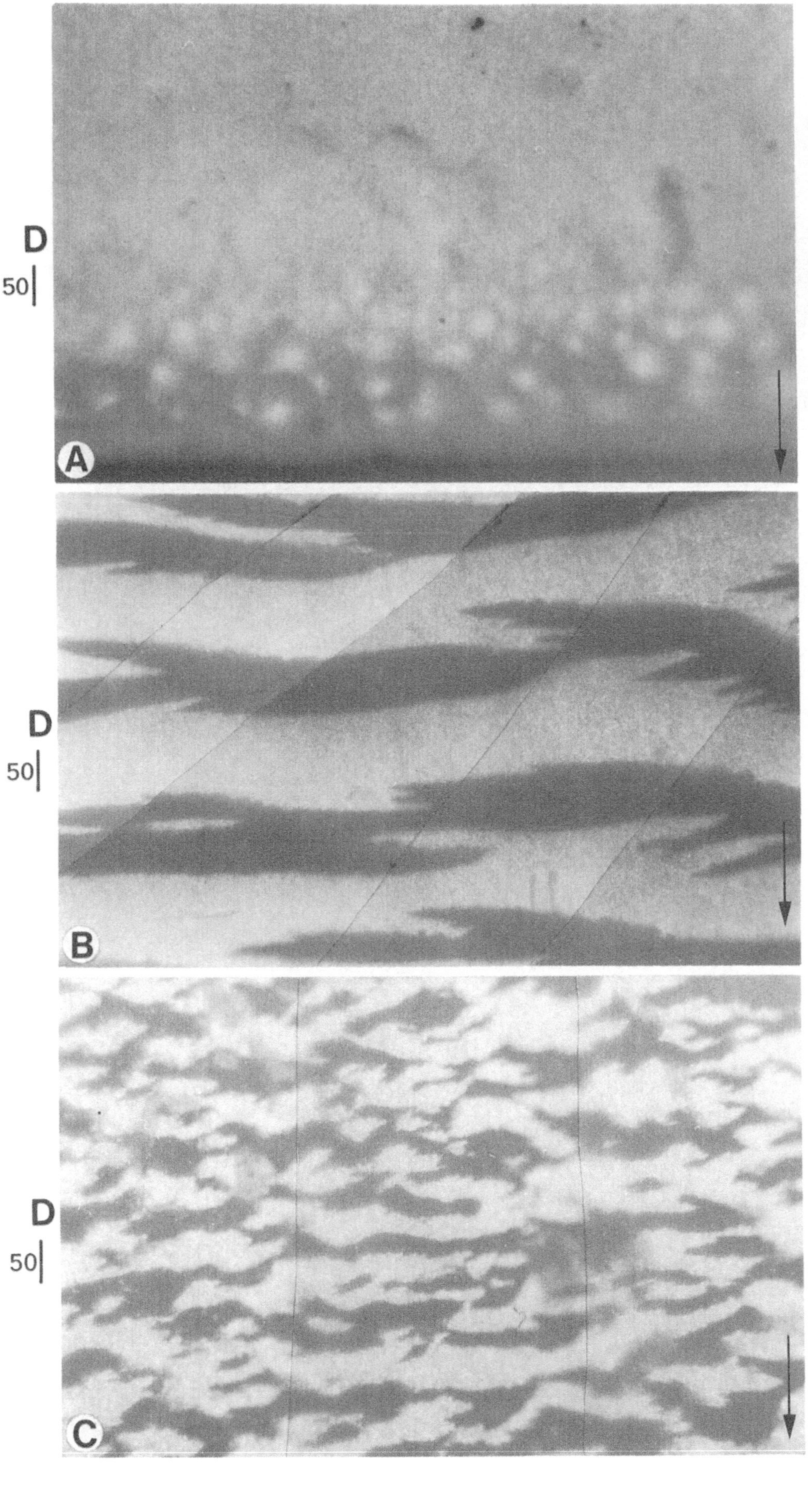

28. Ungulinidae *Felaniella*

29. Chamidae *Chama*

30. Chamidae *Arcinella*

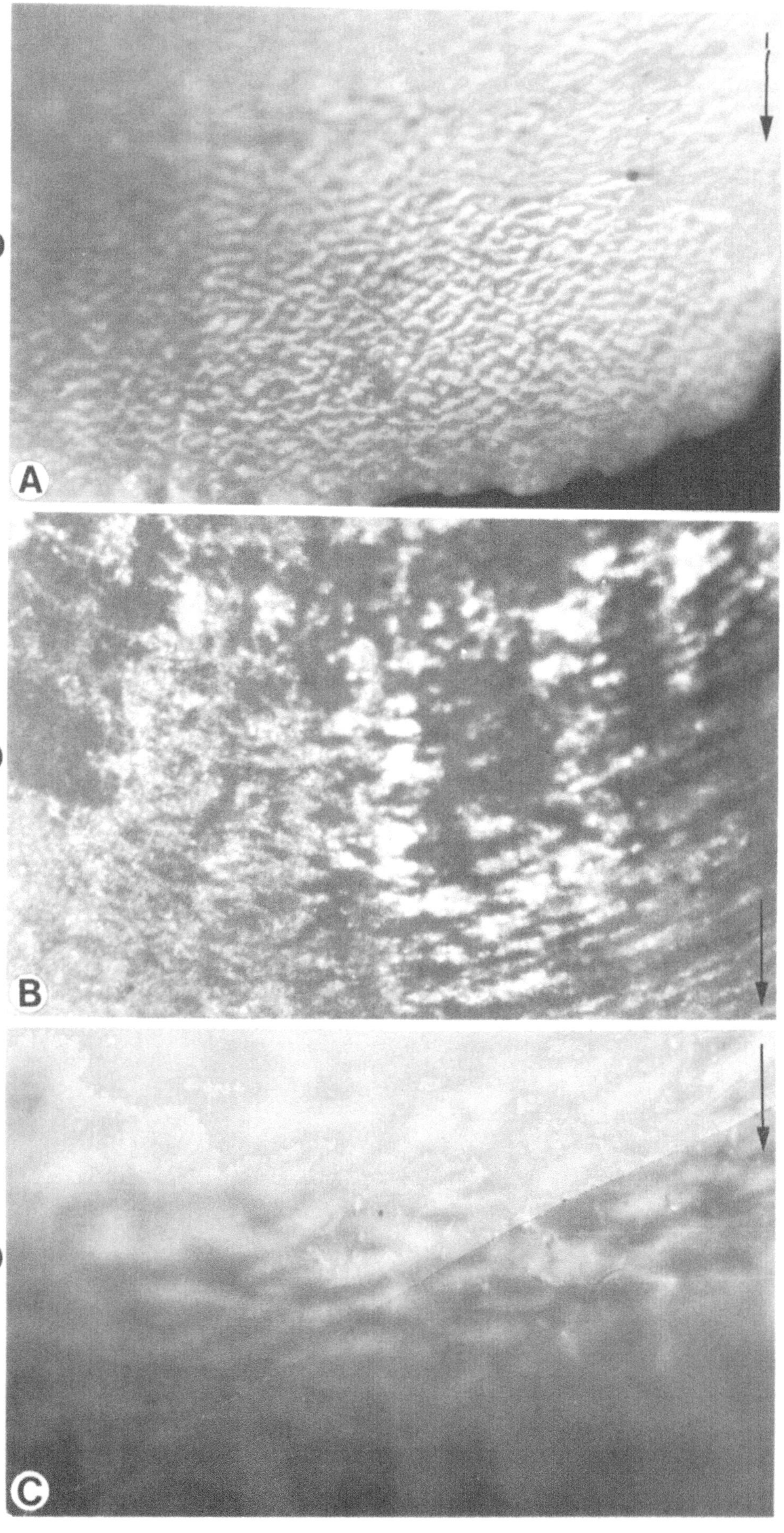

31. Kelliidae *Aligena*

32. Montacutidae *Montacuta*

33. Galeommatidae *Galeomma*

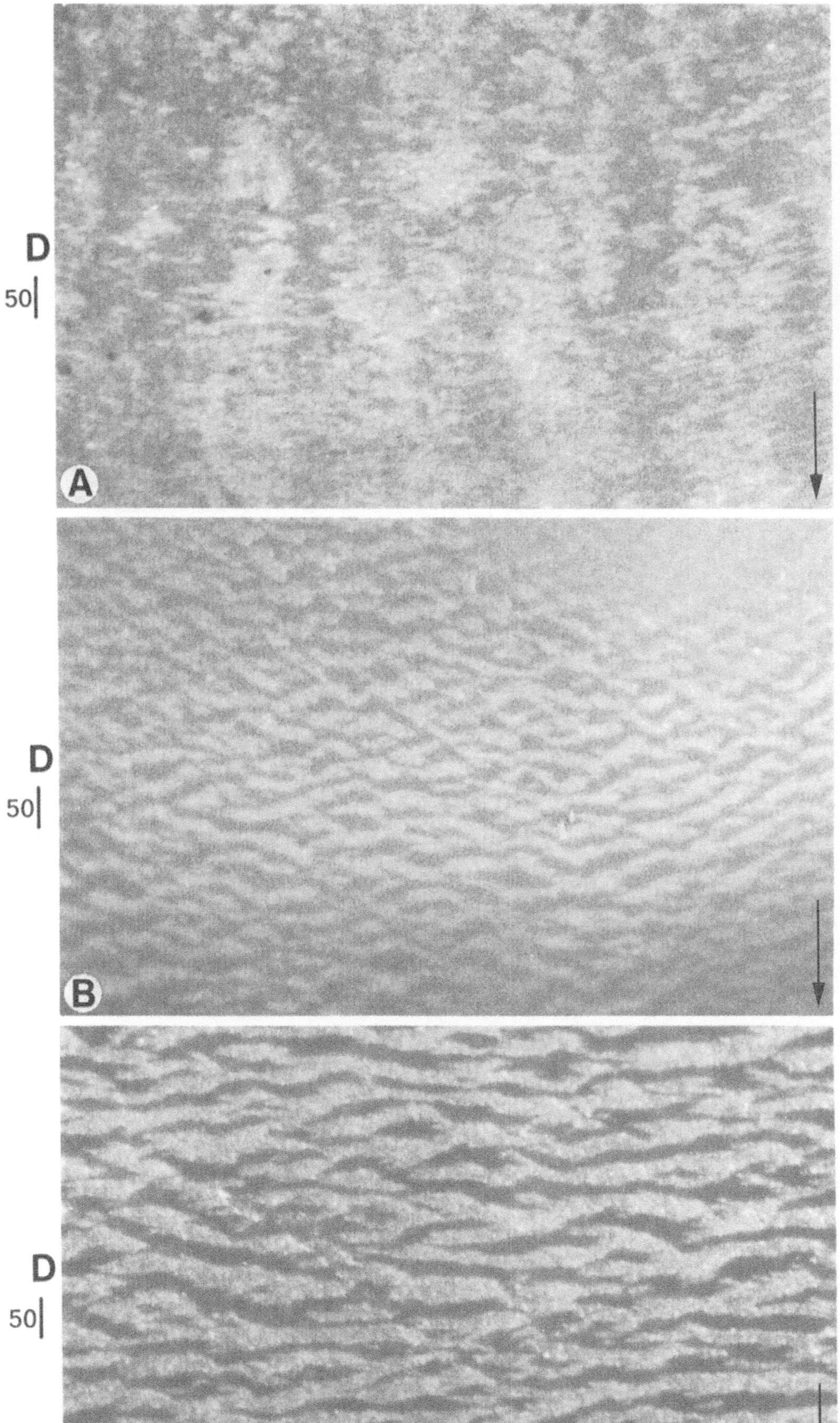

34. Cyamiidae *Cyamium*

35. Sportellidae *Anisodonta*

36. Carditidae *Beguina*

D
50

37. Carditidae *Glans*

D
50

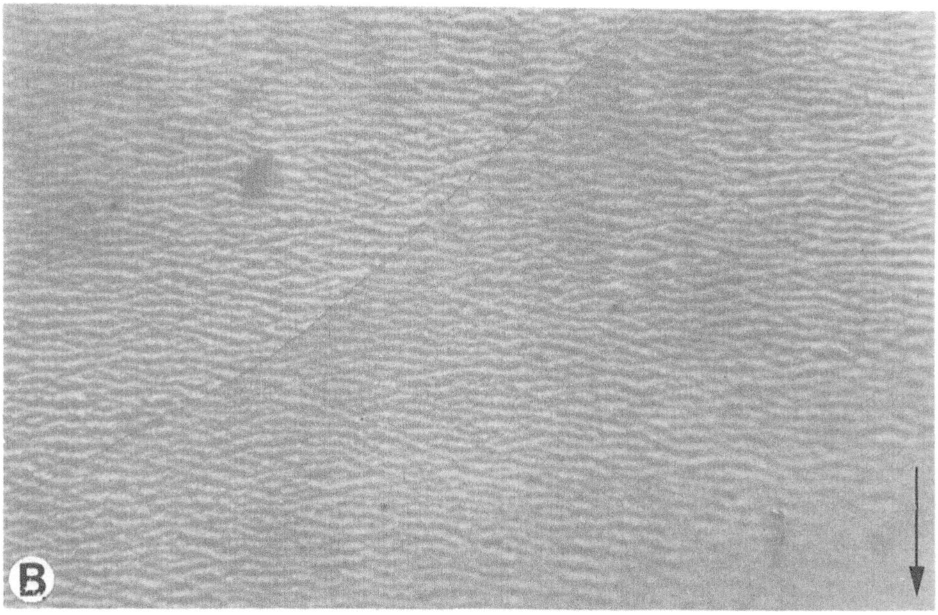

38. Astartidae *Astarte*

D
50

39. Crassatellidae *Eucrassatella*

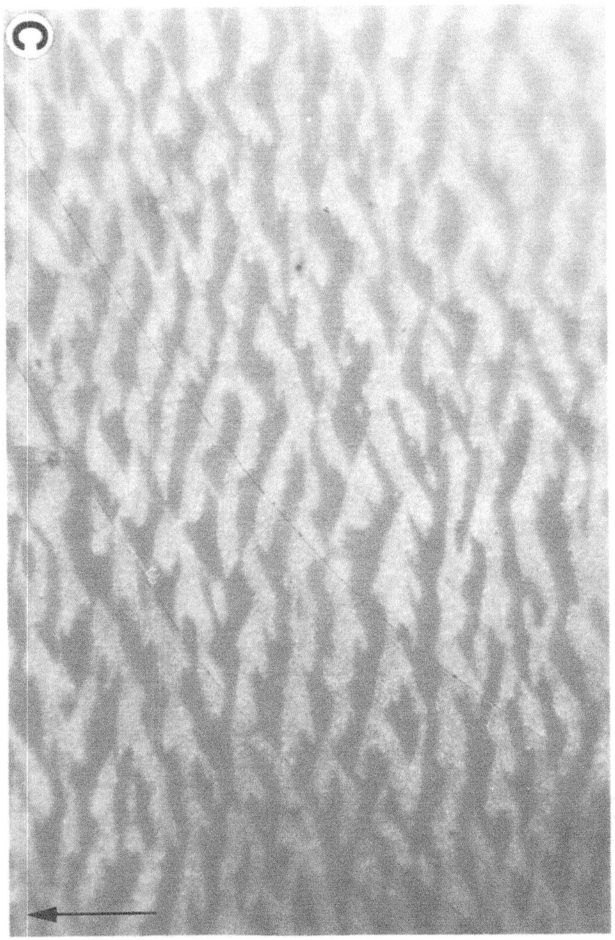

45. Mactridae
Lutraria

44. Tridacnidae
Hippopus

43. Cardiidae
Acrosterigma

46. Mesodesmatidae *Anapella*

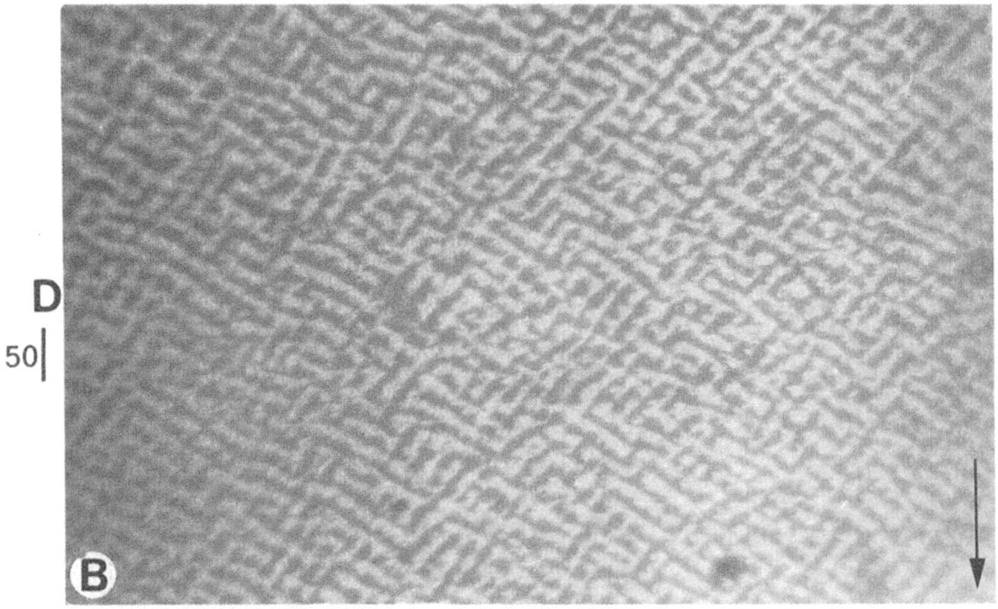

47. Cultellidae *Ensis*

48. Cultellidae *Ensis*

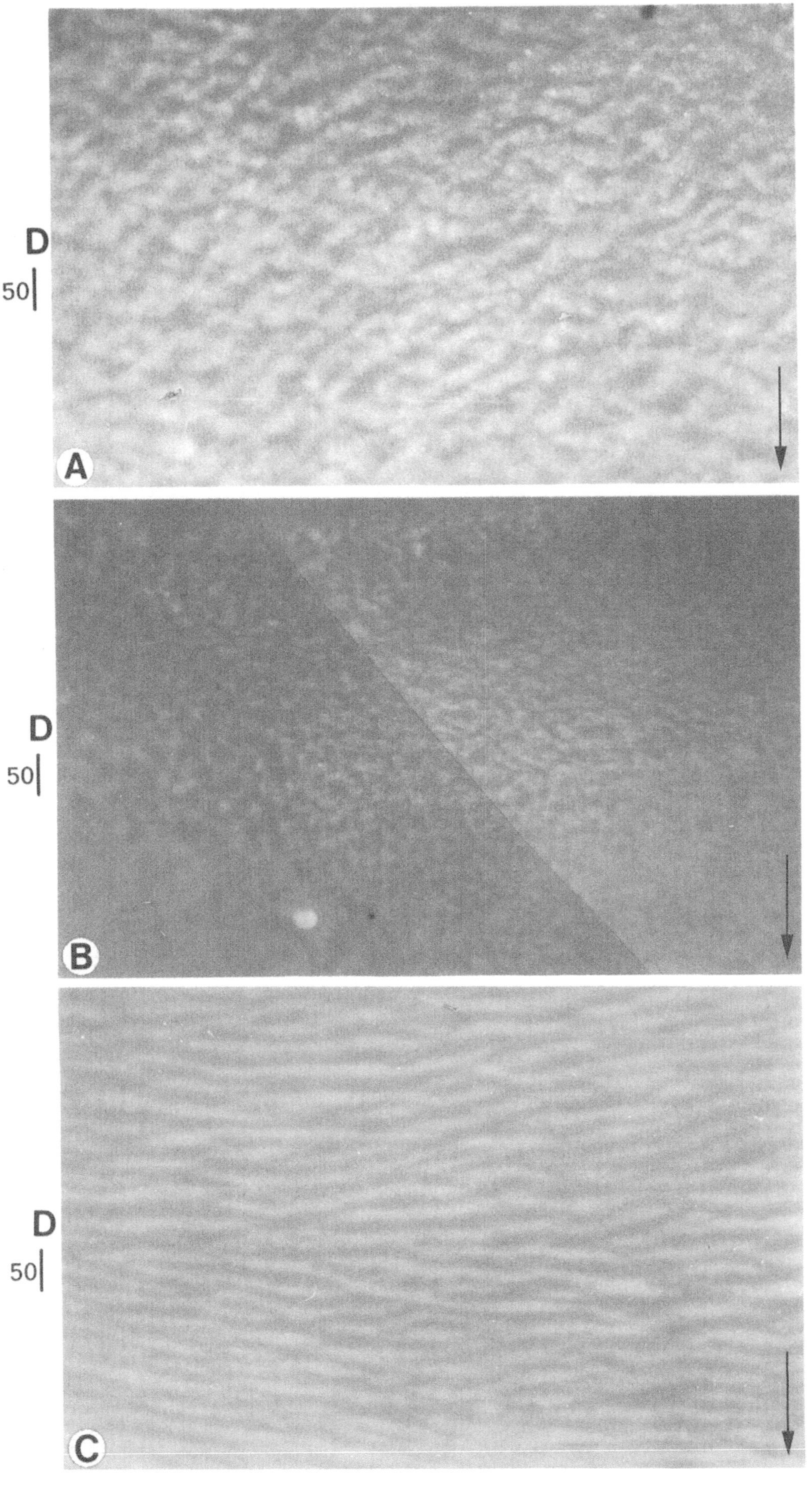

49. Cultellidae *Siliqua*

50. Tellinidae *Tellina*

51. Tellinidae *Florimetis*

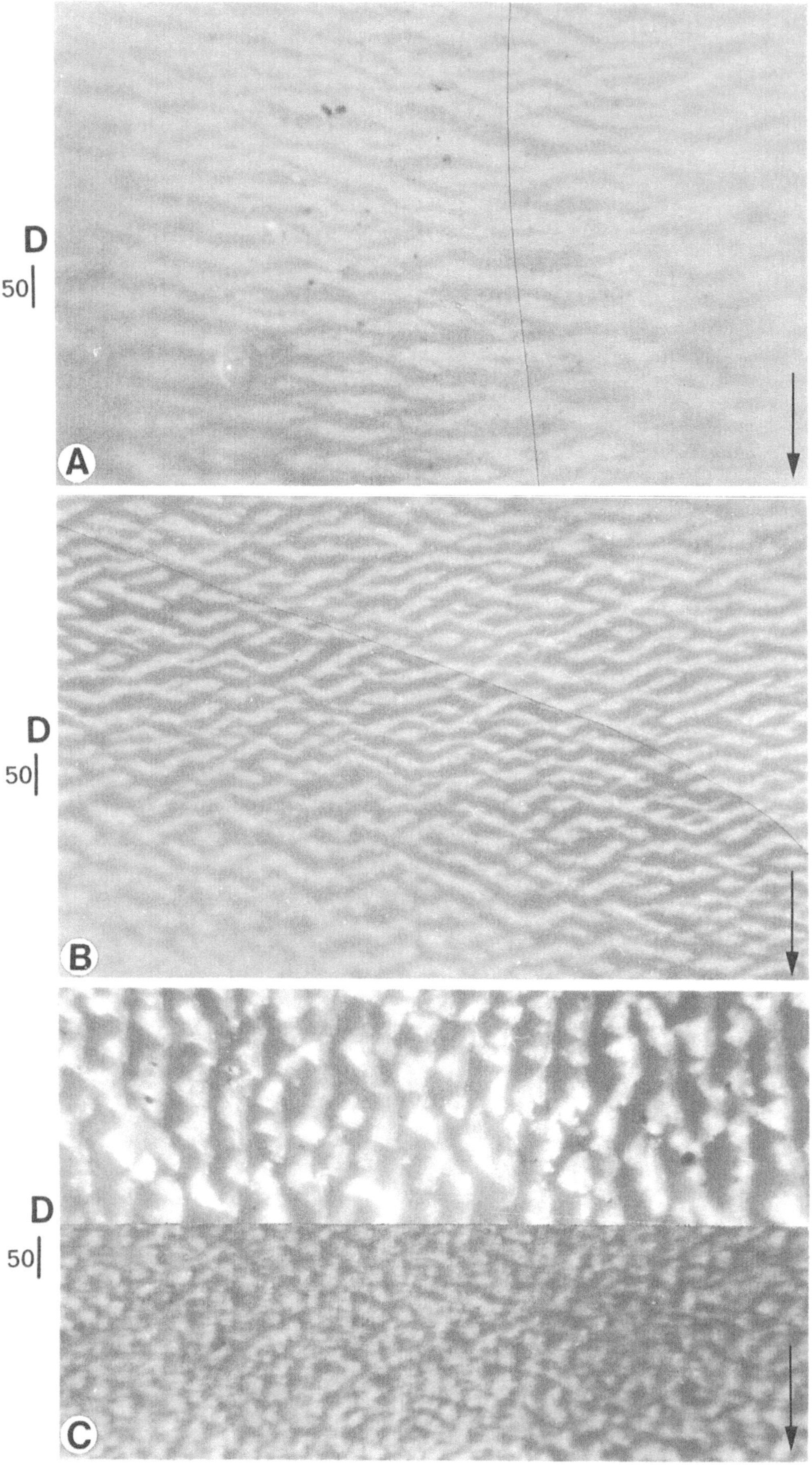

52. Hemidonacidae *Hemidonax*

54. Psammobiidae *Sanguinolaria*

55. Psammobiidae *Sanguinolaria*

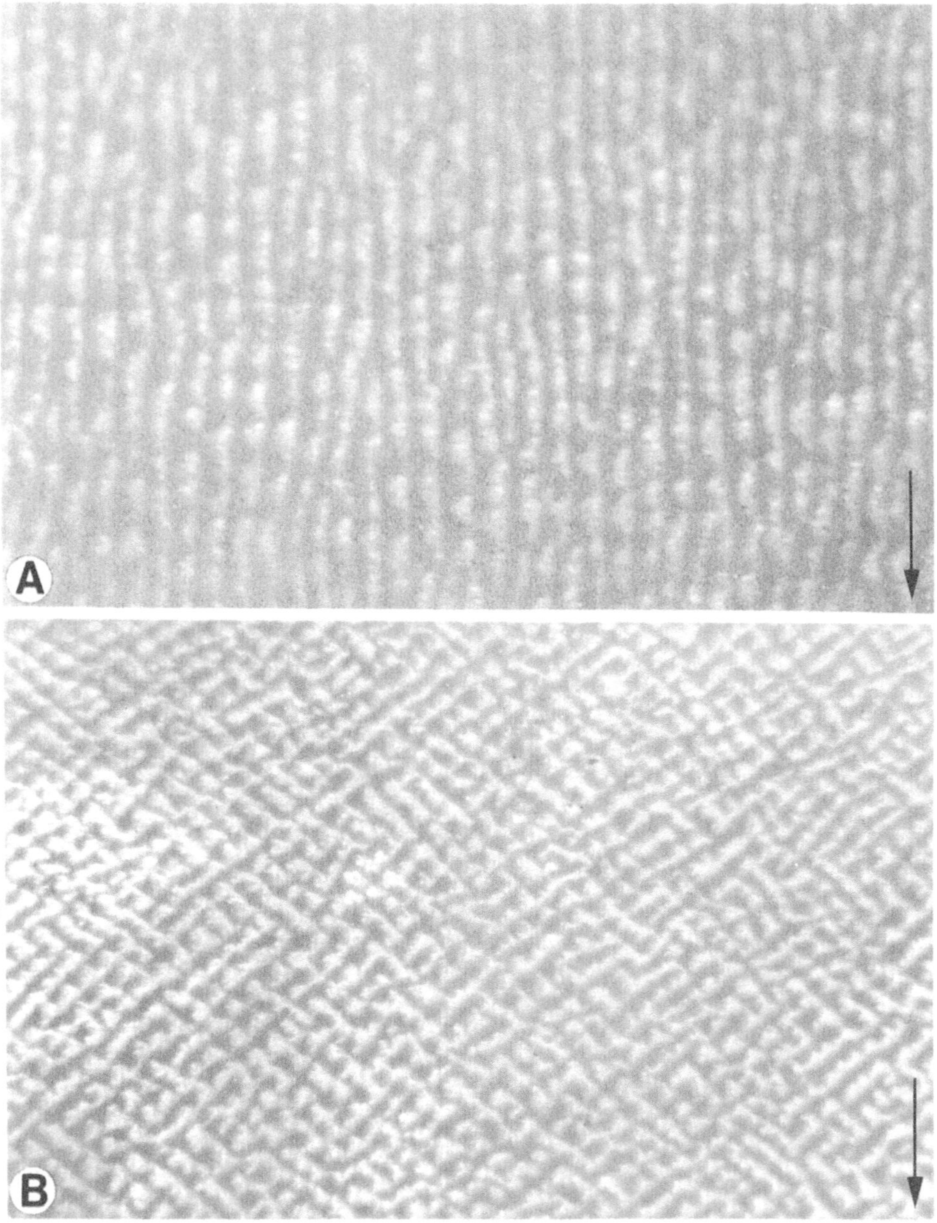

D
50

58. Solecurtidae *Solecurtus*

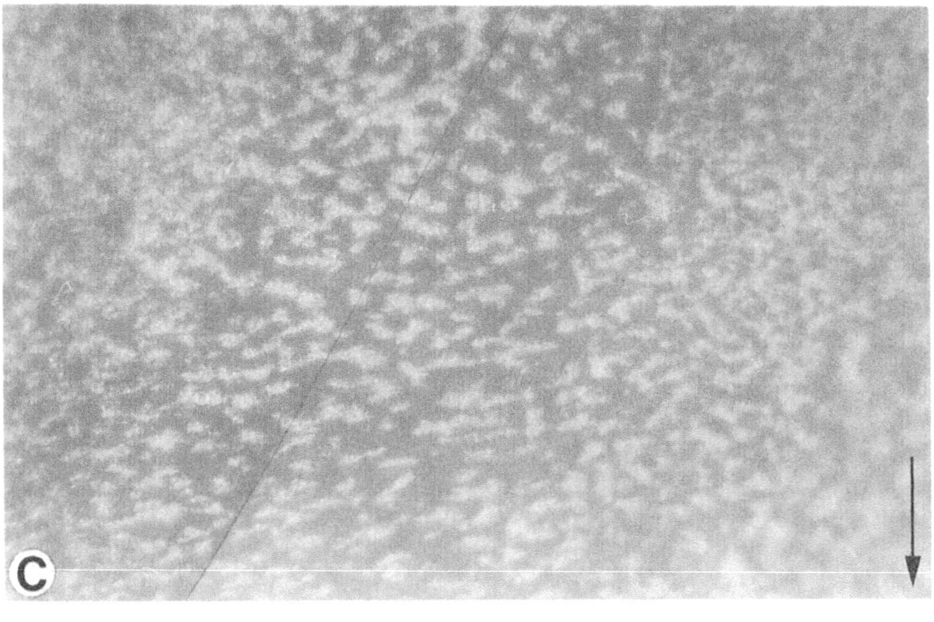

D
50

59. Solecurtidae *Tagelus*

D
50

D
50

D
50

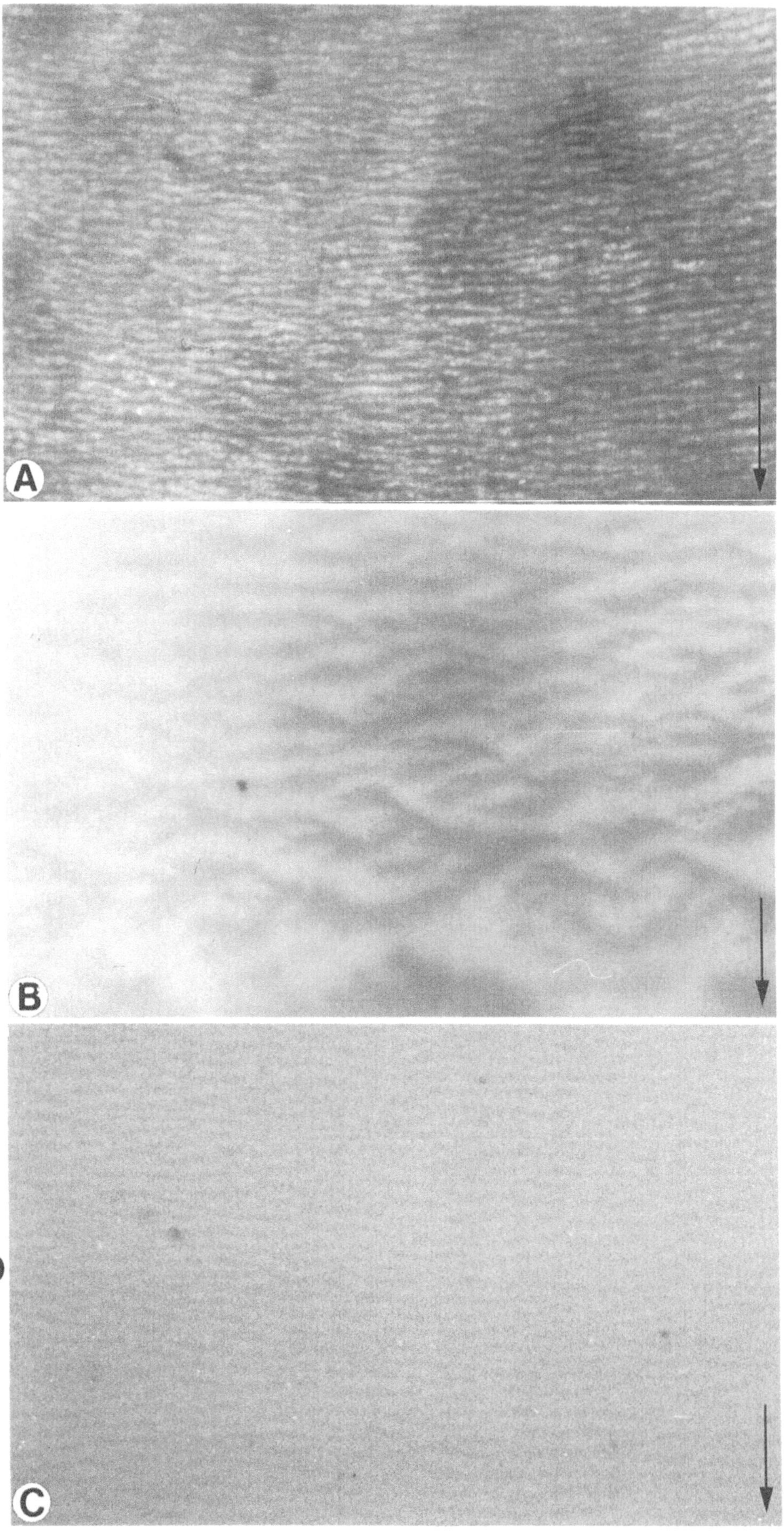

60. Dreissenidae *Mytilopsis*

62. Trapeziidae *Trapezium*

D
50

63. Trapeziidae
Coralliophaga

D
50

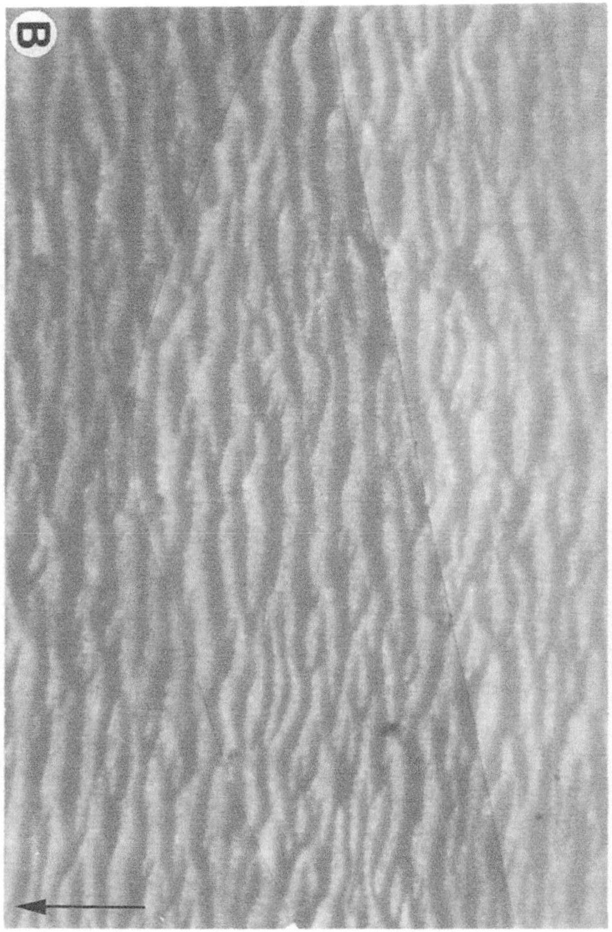

64. Glossidae
Glossus

D
50

65. Corbiculidae
Batissa

66. Veneridae *Dosinia*

67. Veneridae *Cyclina*

68. Veneridae *Circe*

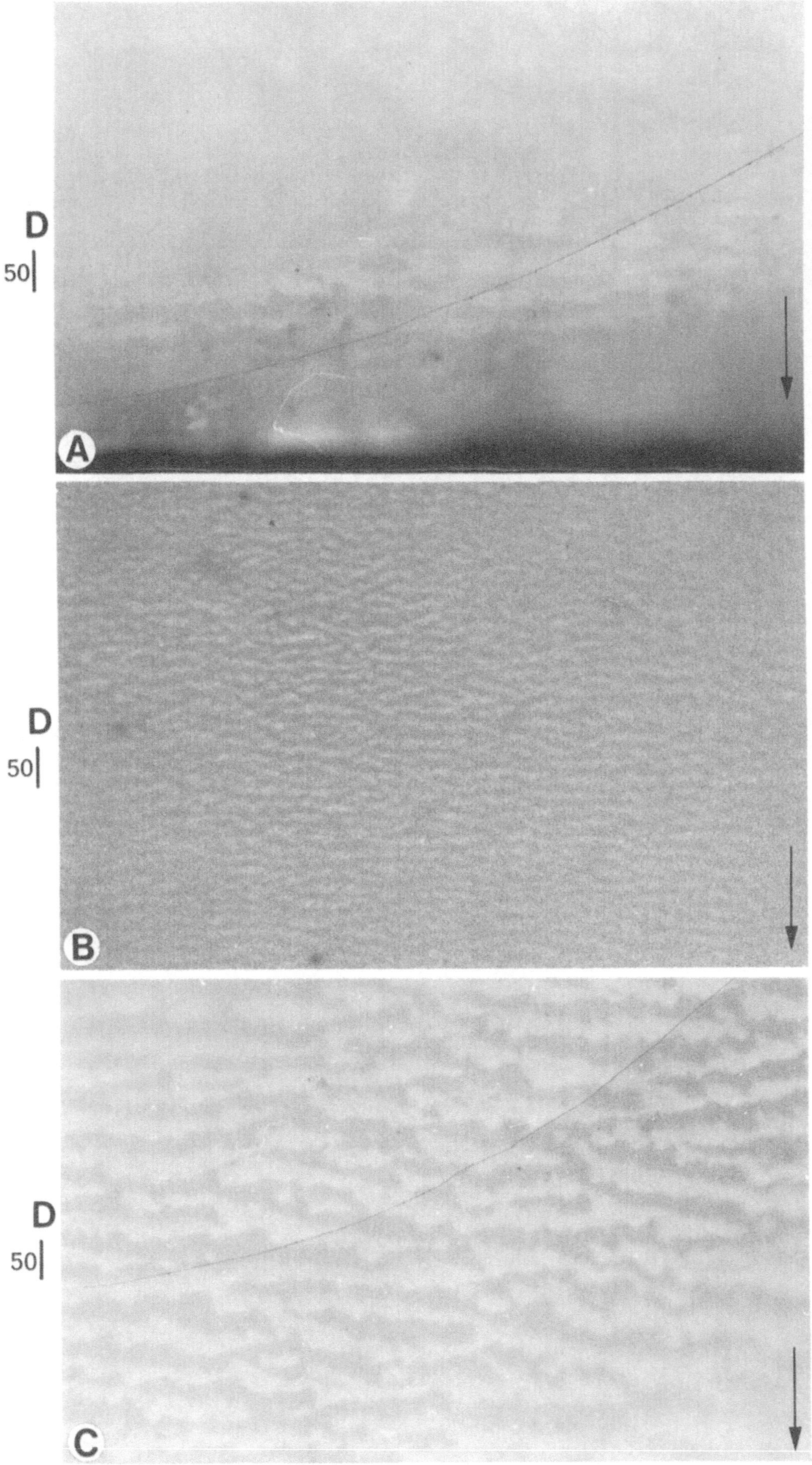

72. Veneridae *Meretrix*

73. Veneridae *Tivela*

74. Veneridae *Saxidomus*

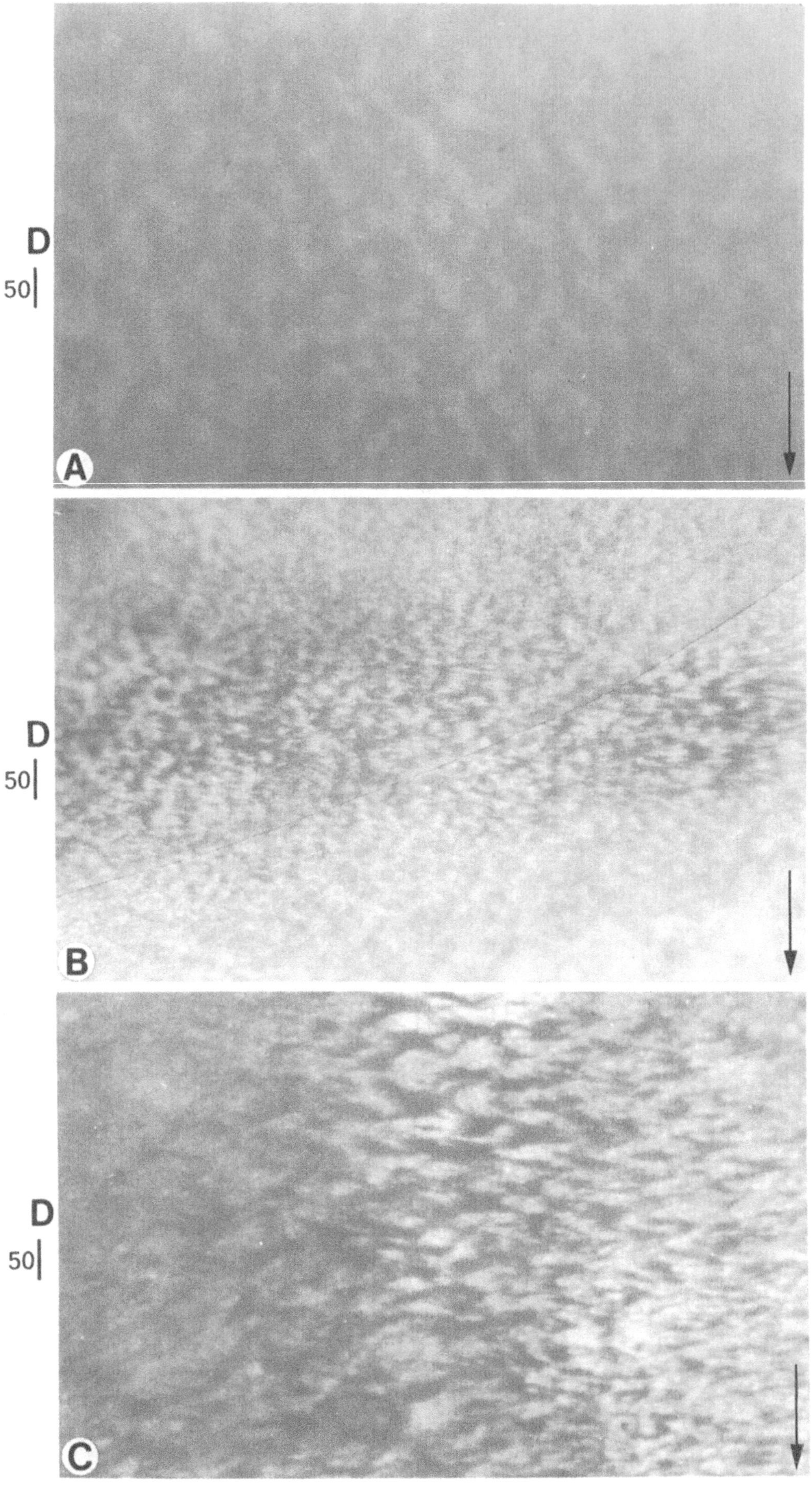

76. Veneridae *Chione*

77. Cooperellidae *Cooperella*

78. Glauconomidae *Glauconome*

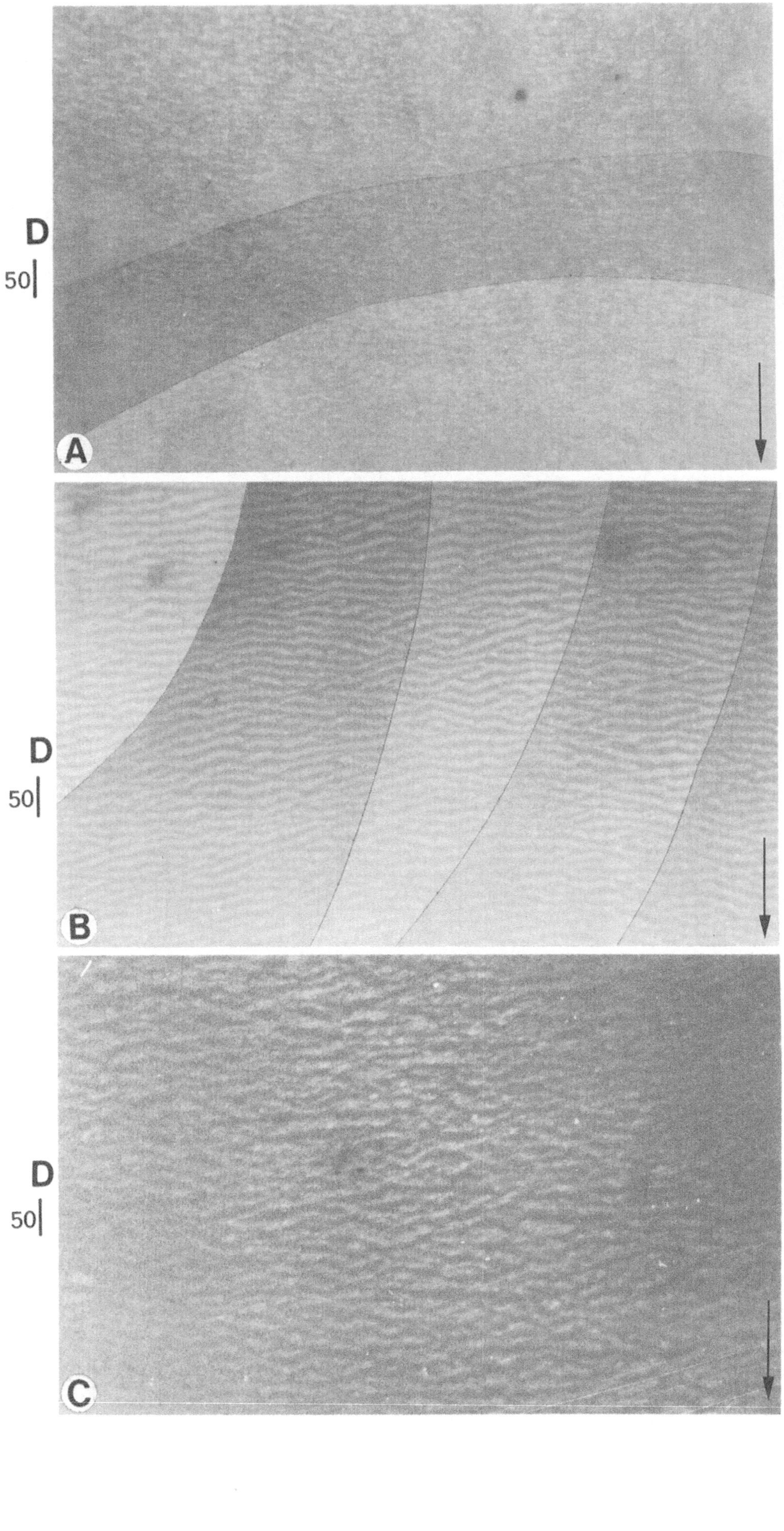

79. Myidae
Mya

81. Myidae
Cryptomya

83. Myidae *Platyodon*

84. Corbulidae *Corbula*

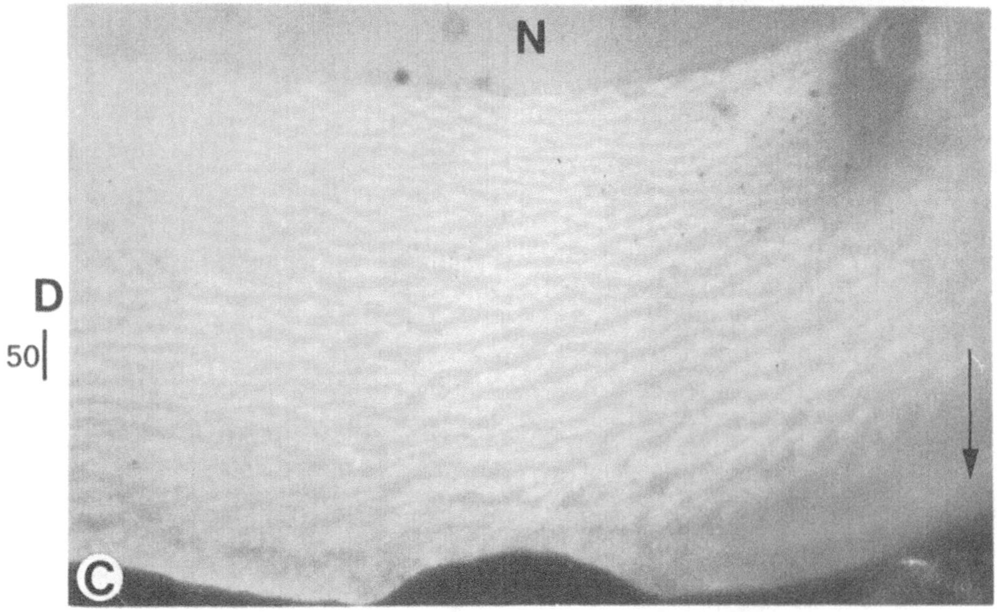

86. Corbulidae *Corbula*

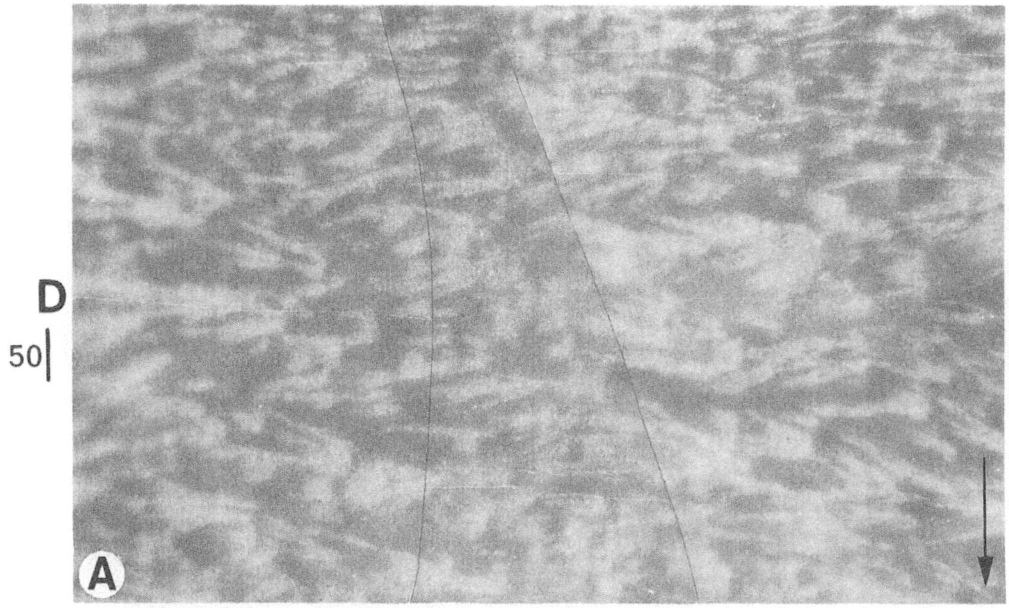

87. Gastrochaenidae *Rocellaria*

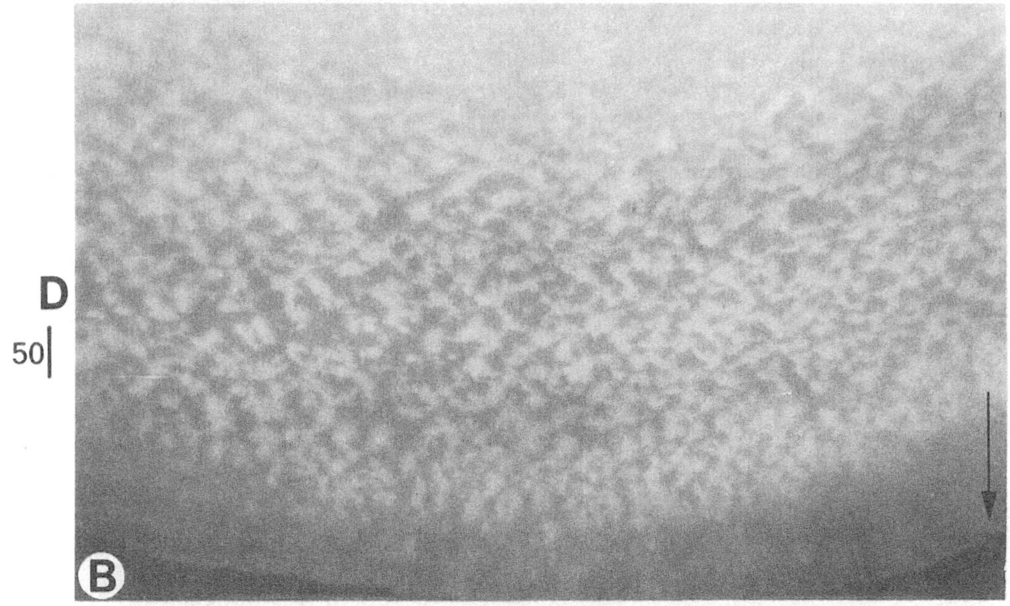

88. Gastrochaenidae *Gastrochaena*

89. Gastrochaenidae *Spengleria*

D
50

90. Hiatellidae
Hiatella

D
50

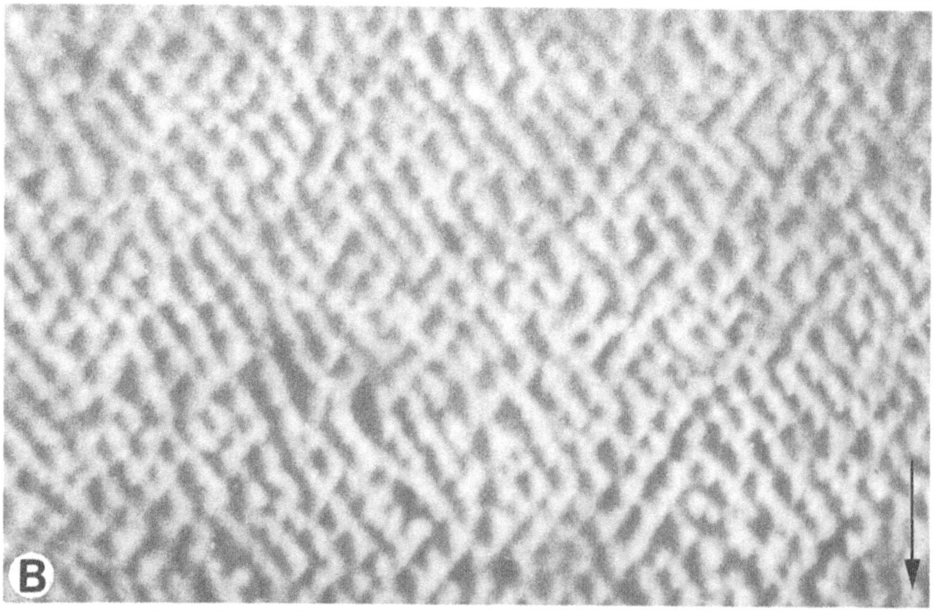

93. Pholadidae
Barnea

D
50

94. Pholadidae
Zirfaea

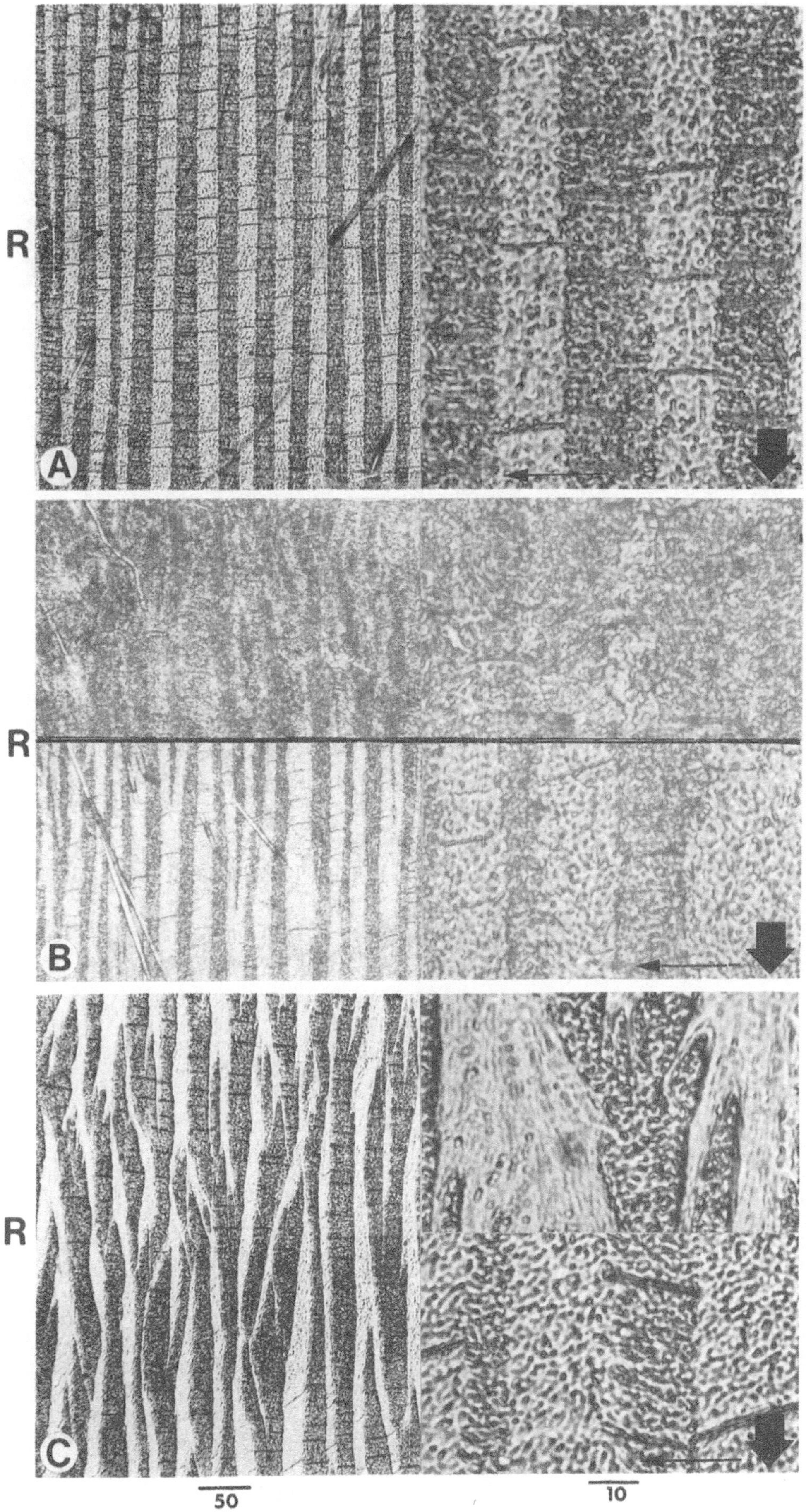

4. Arcidae *Arca*

5. Cucullaeidae *Cucullaea*

6. Noetiidae *Noetia*

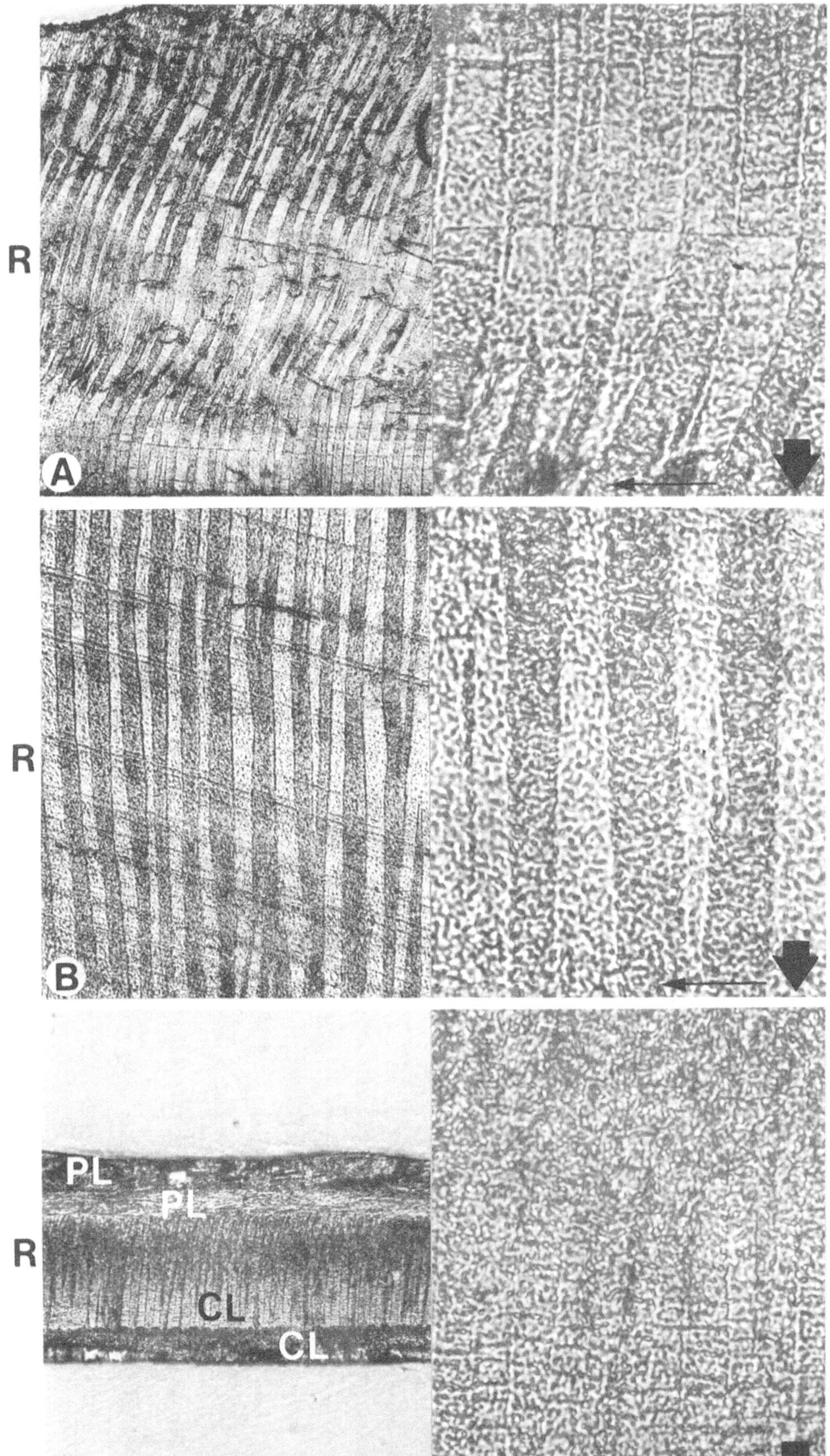

7. Limopsidae *Limopsis*

8. Glycymerididae *Glycymeris*

13. Propeamussiidae *Propeamussium*

16.
Anomiidae
Anomia

25. Lucinidae *Divalucina*

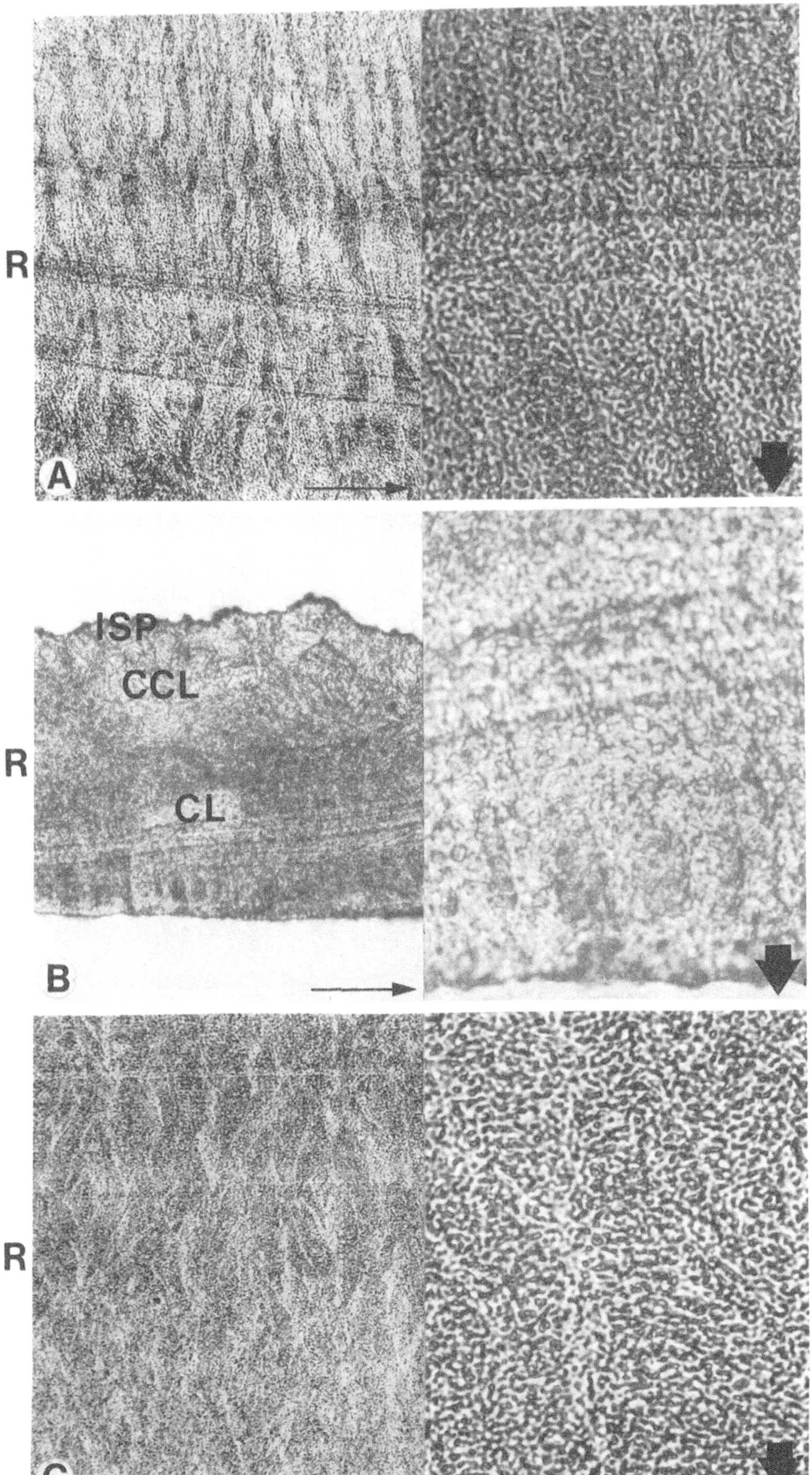

25. Lucinidae *Divalucina*

26. Thyasiridae *Thyasira*

27. Fimbriidae *Fimbria*

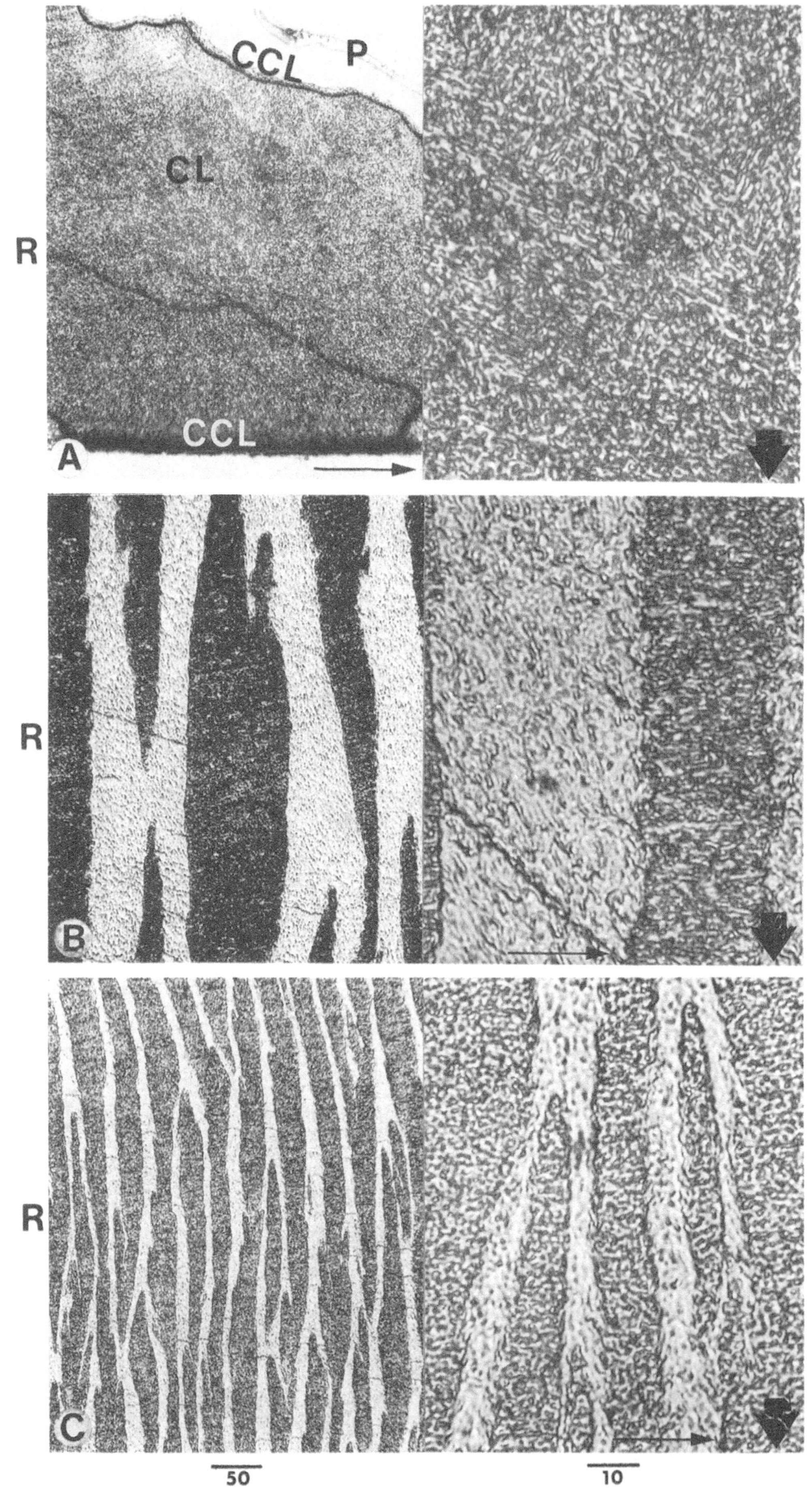

28. Ungulinidae *Felaniella*

29. Chamidae *Chama*

30. Chamidae *Arcinella*

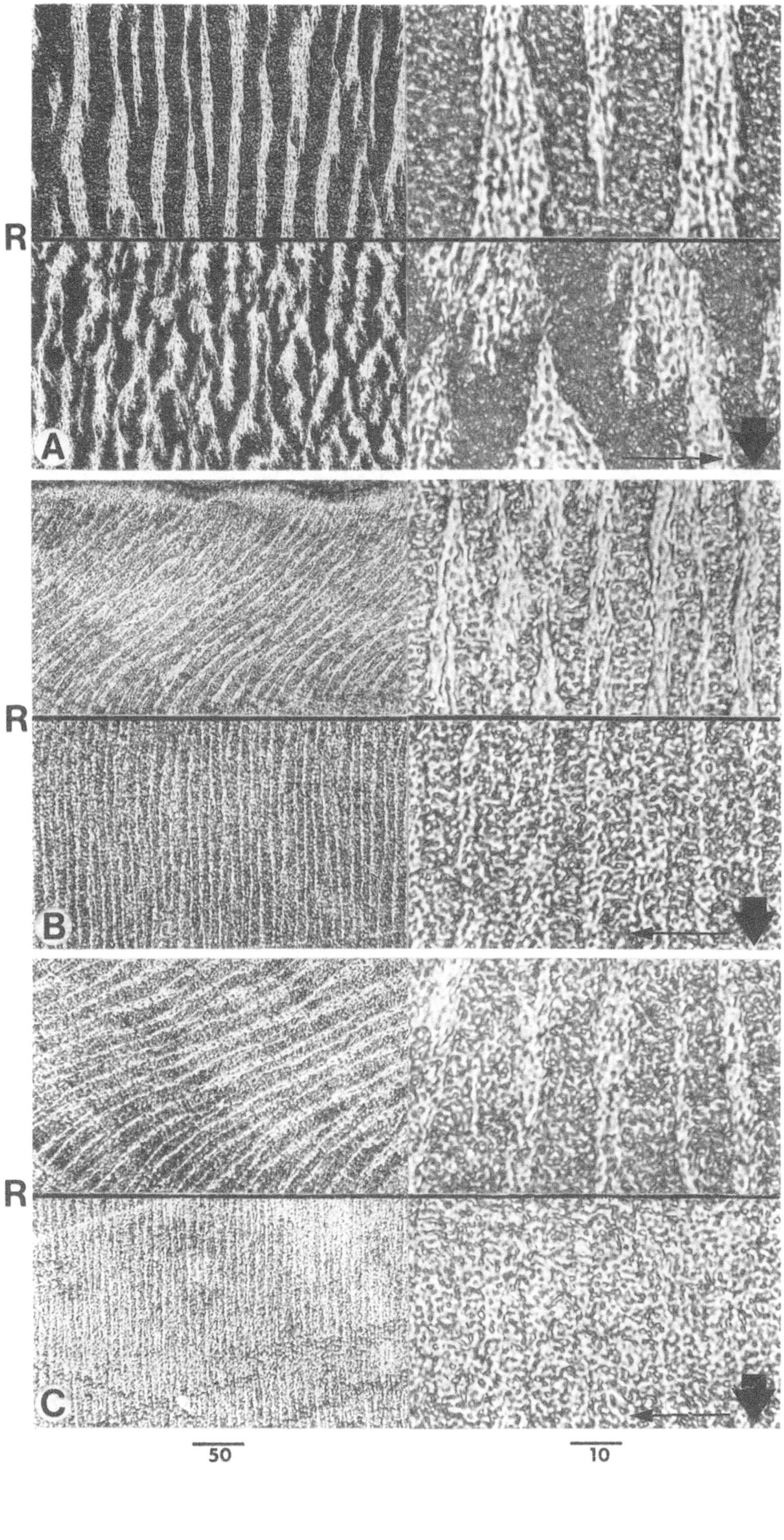

37. Carditidae *Glans*

38. Astartidae *Astarte*

40. Crassatellidae *Eucrassatella*

41.
Cardiidae
Fragum

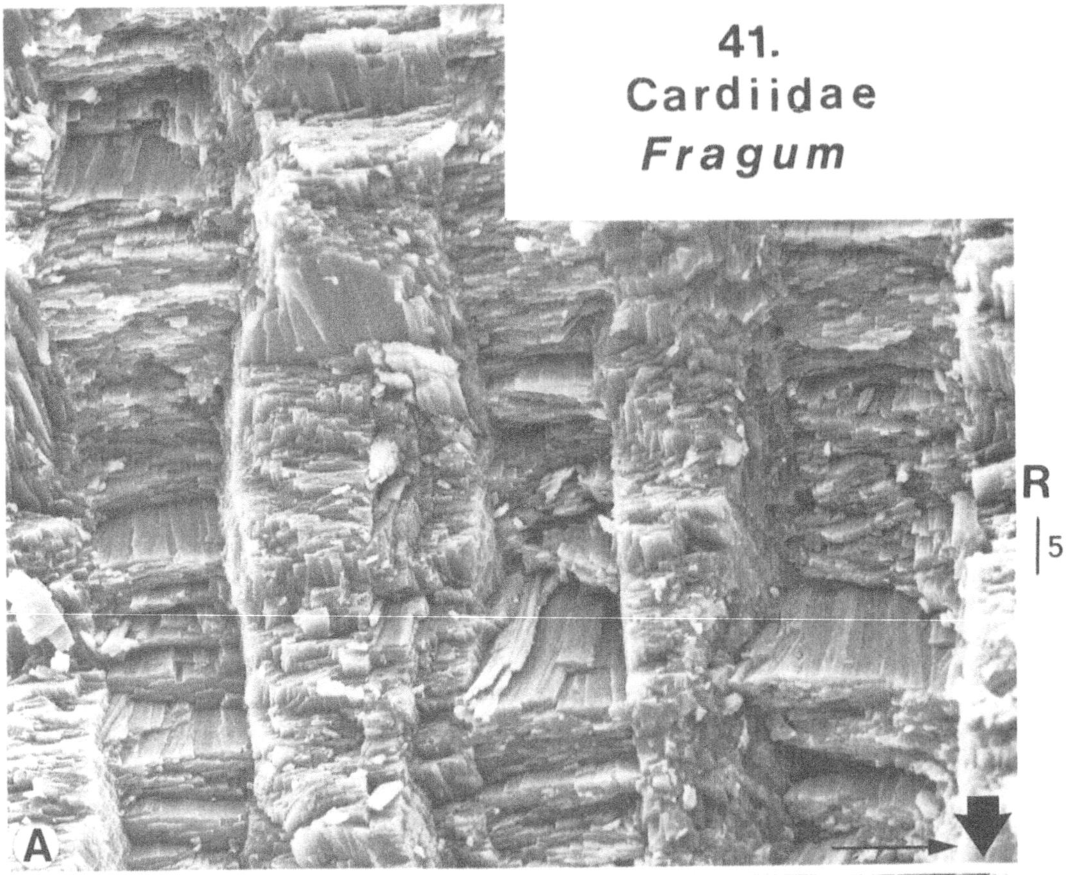

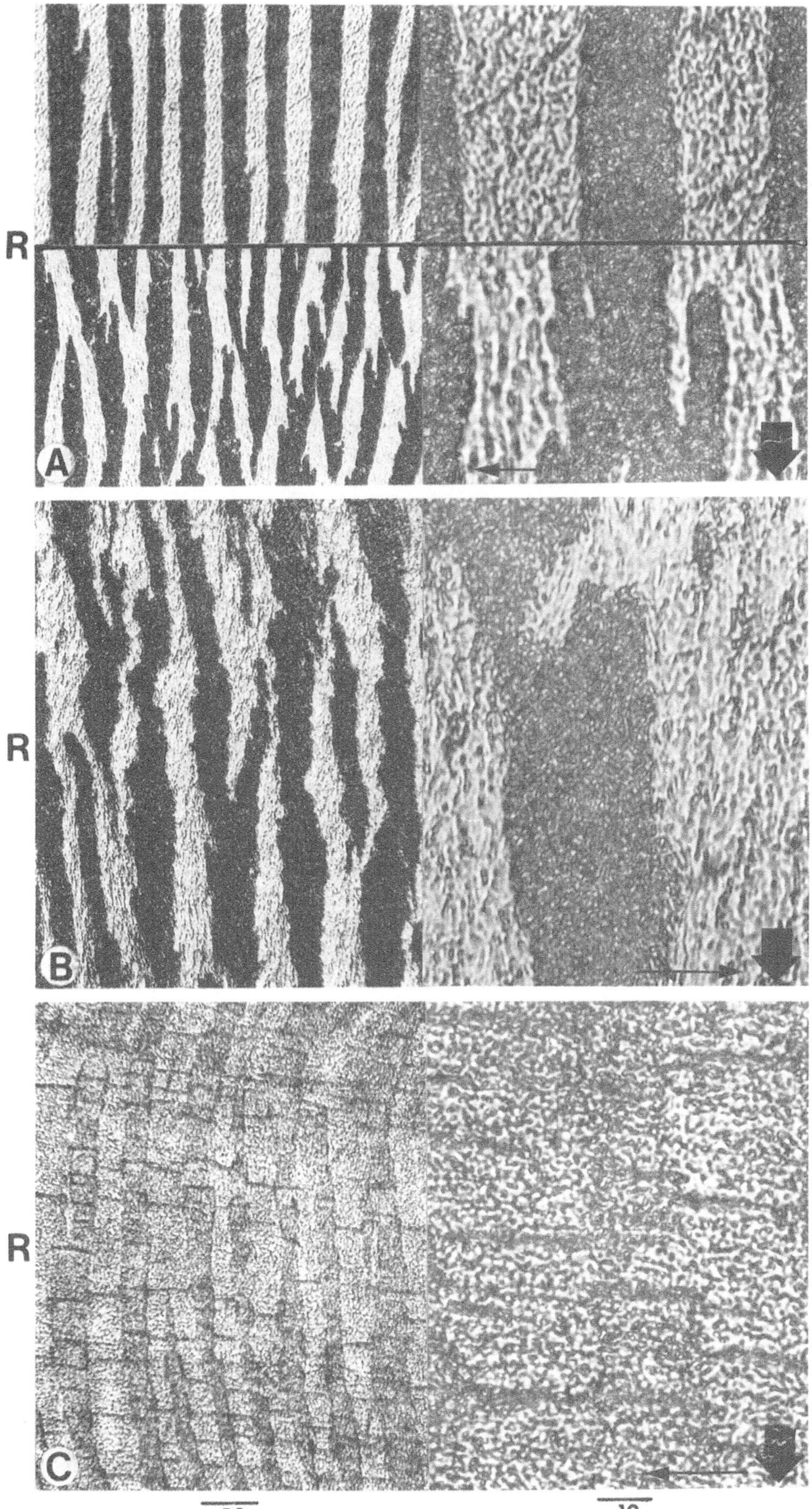

43. Cardiidae *Acrosterigma*

44. Tridacnidae *Hippopus*

45. Mactridae *Lutraria*

47. Cultellidae *Ensis*

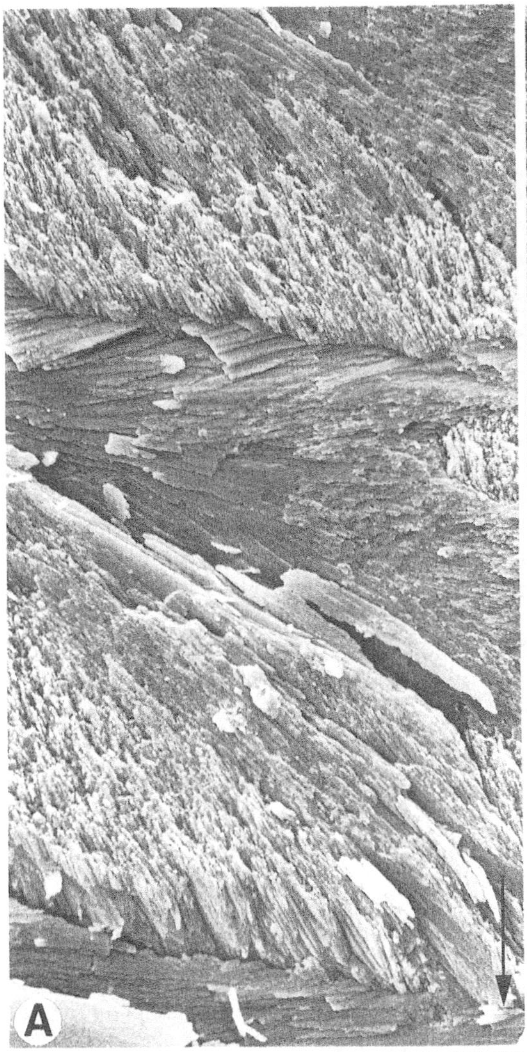

52.
Hemidonacidae
Hemidonax

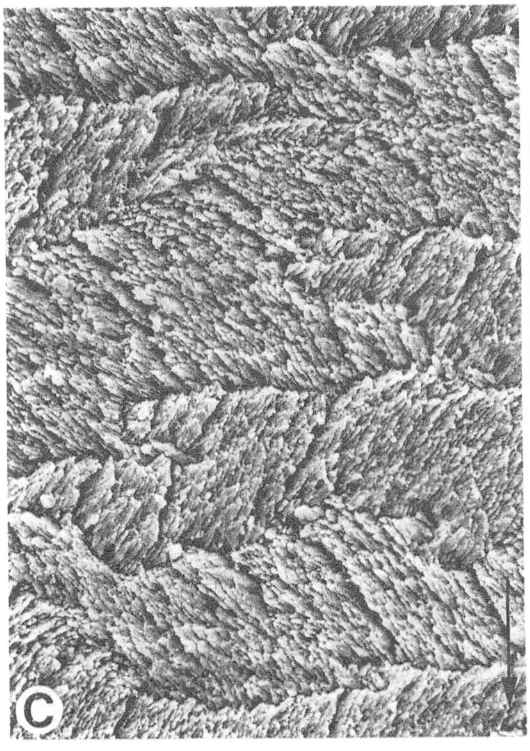

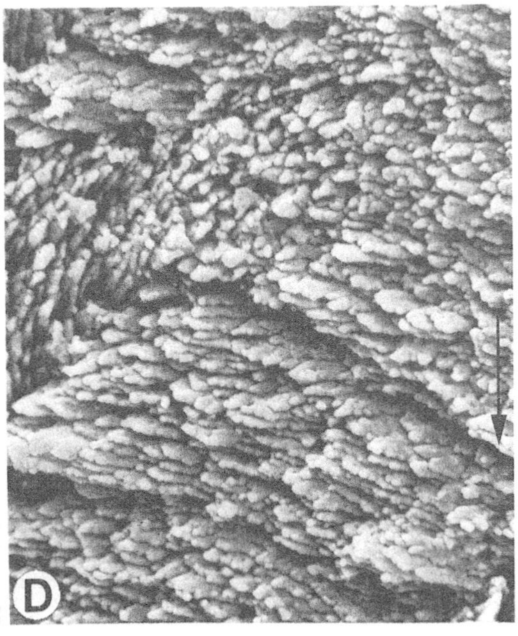

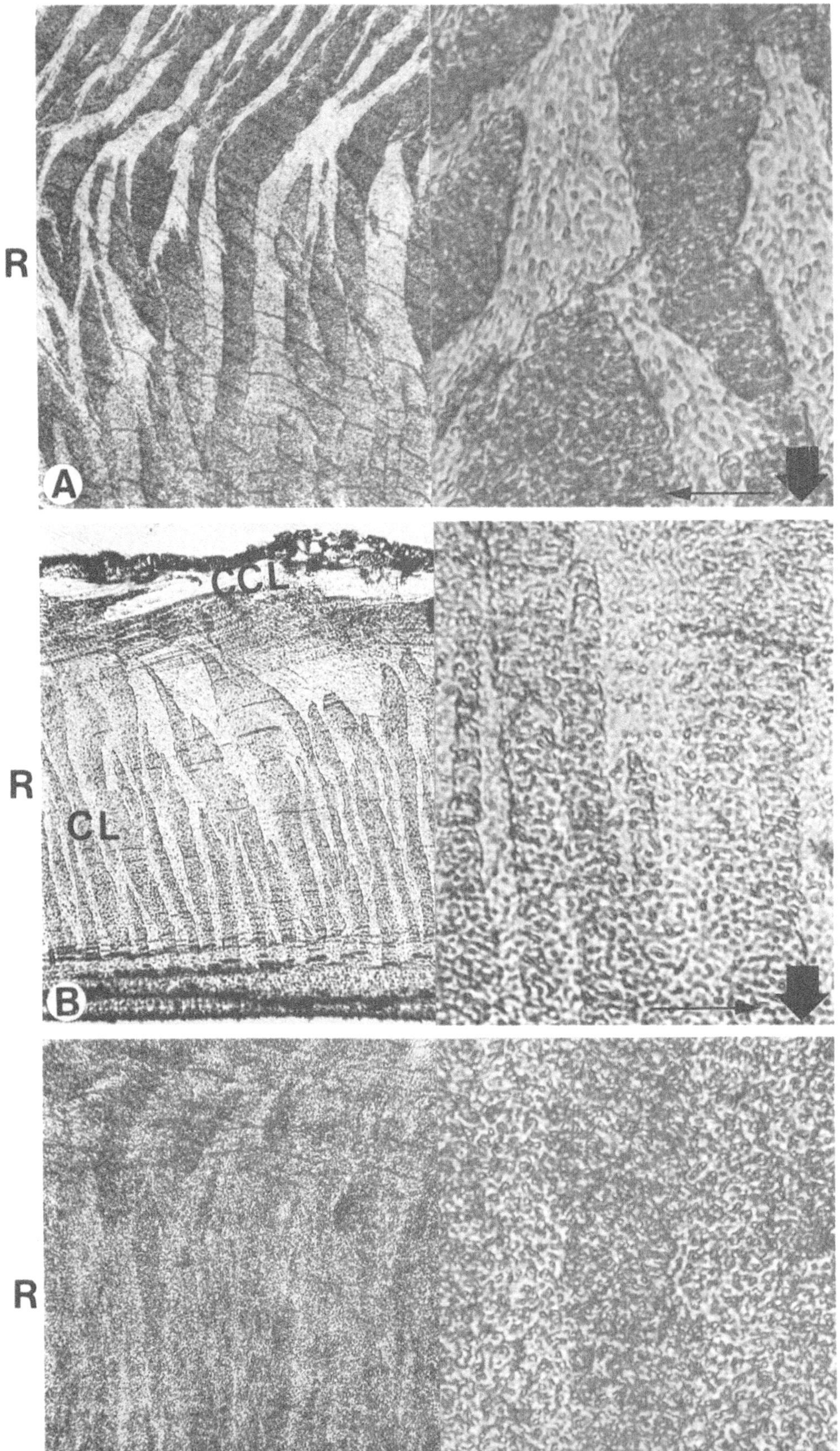

52. Hemidonacidae *Hemidonax*

54. Psammobiidae *Sanguinolaria*

55. Psammobiidae *Sanguinolaria*

T 5

58.
Solecurtidae
Solecurtus

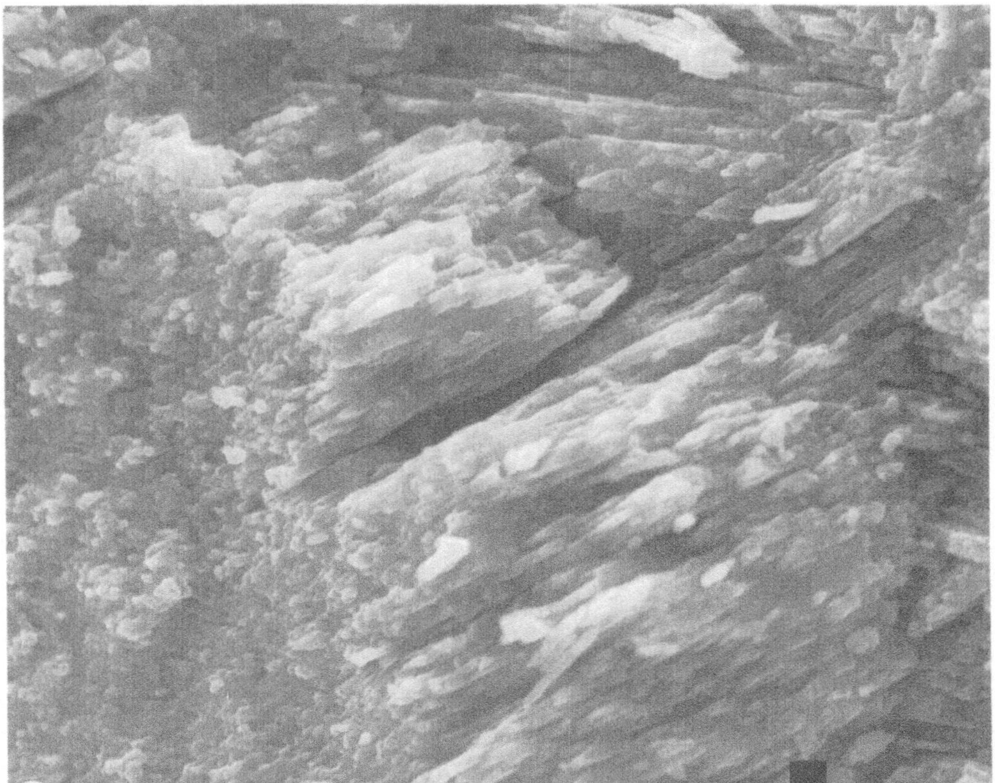

T 1

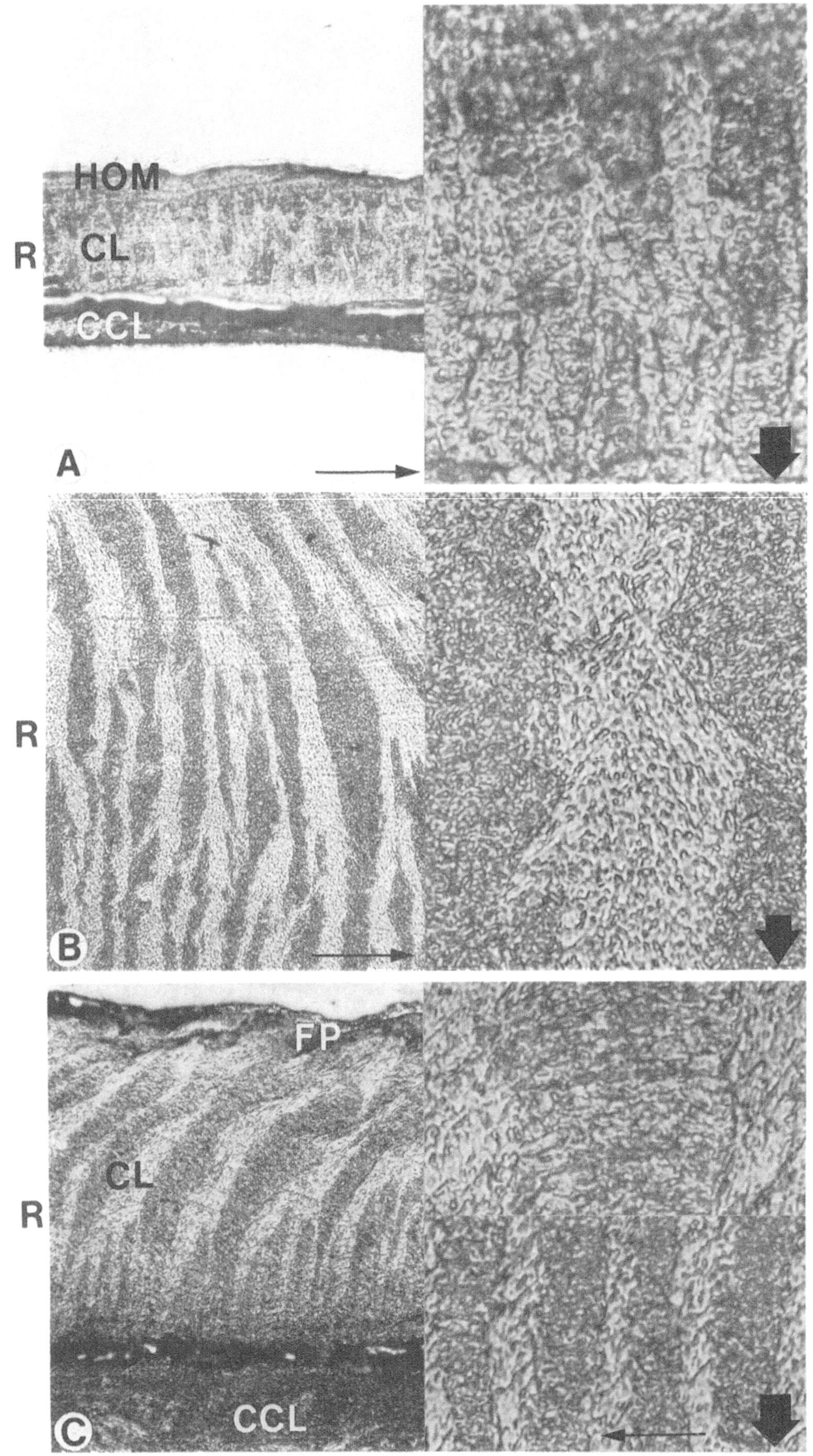

60. Dreissenidae *Mytilopsis*

61. Trapeziidae *Trapezium*

62. Trapeziidae *Trapezium*

62.
Trapeziidae
Trapezium

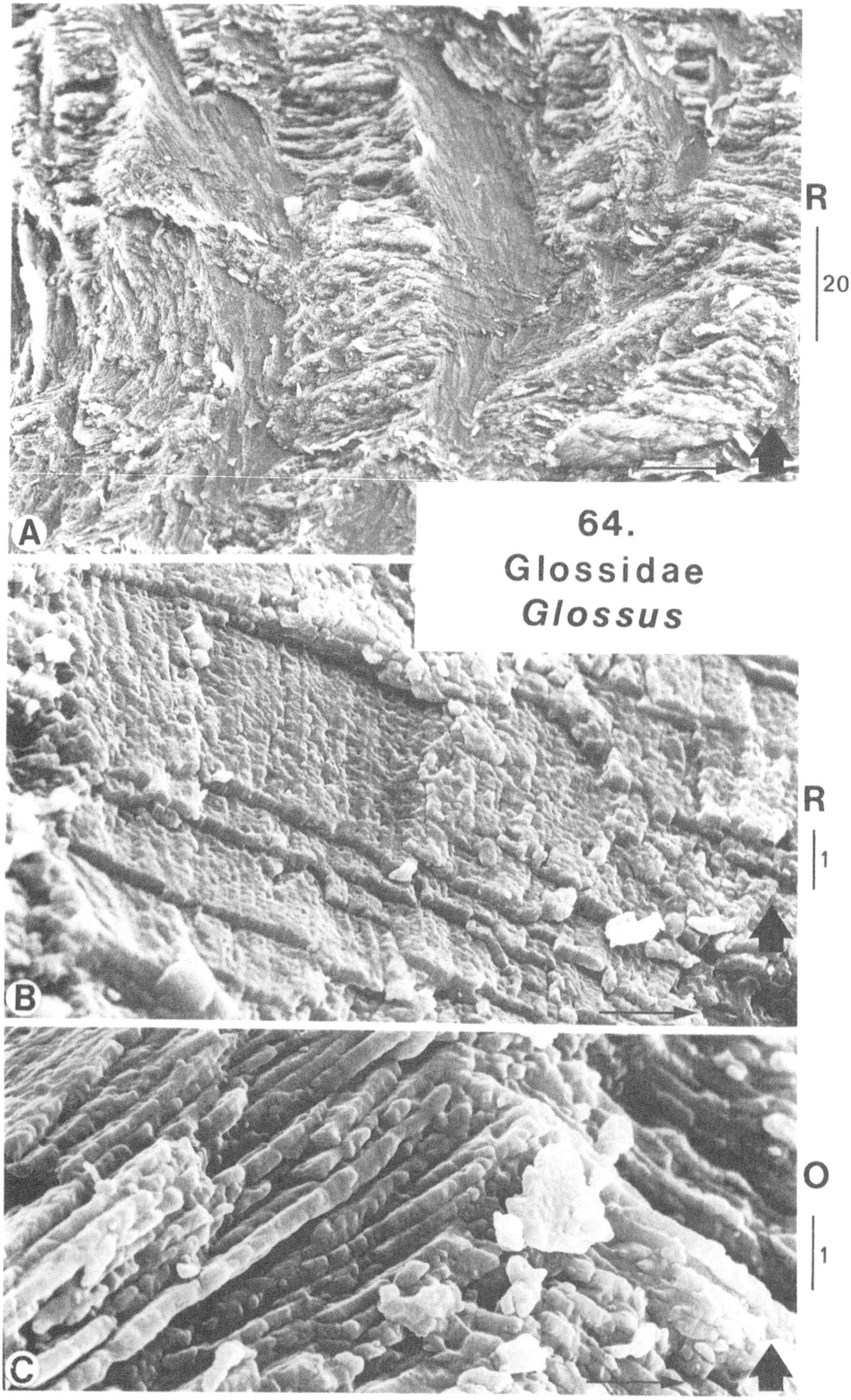

64.
Glossidae
Glossus

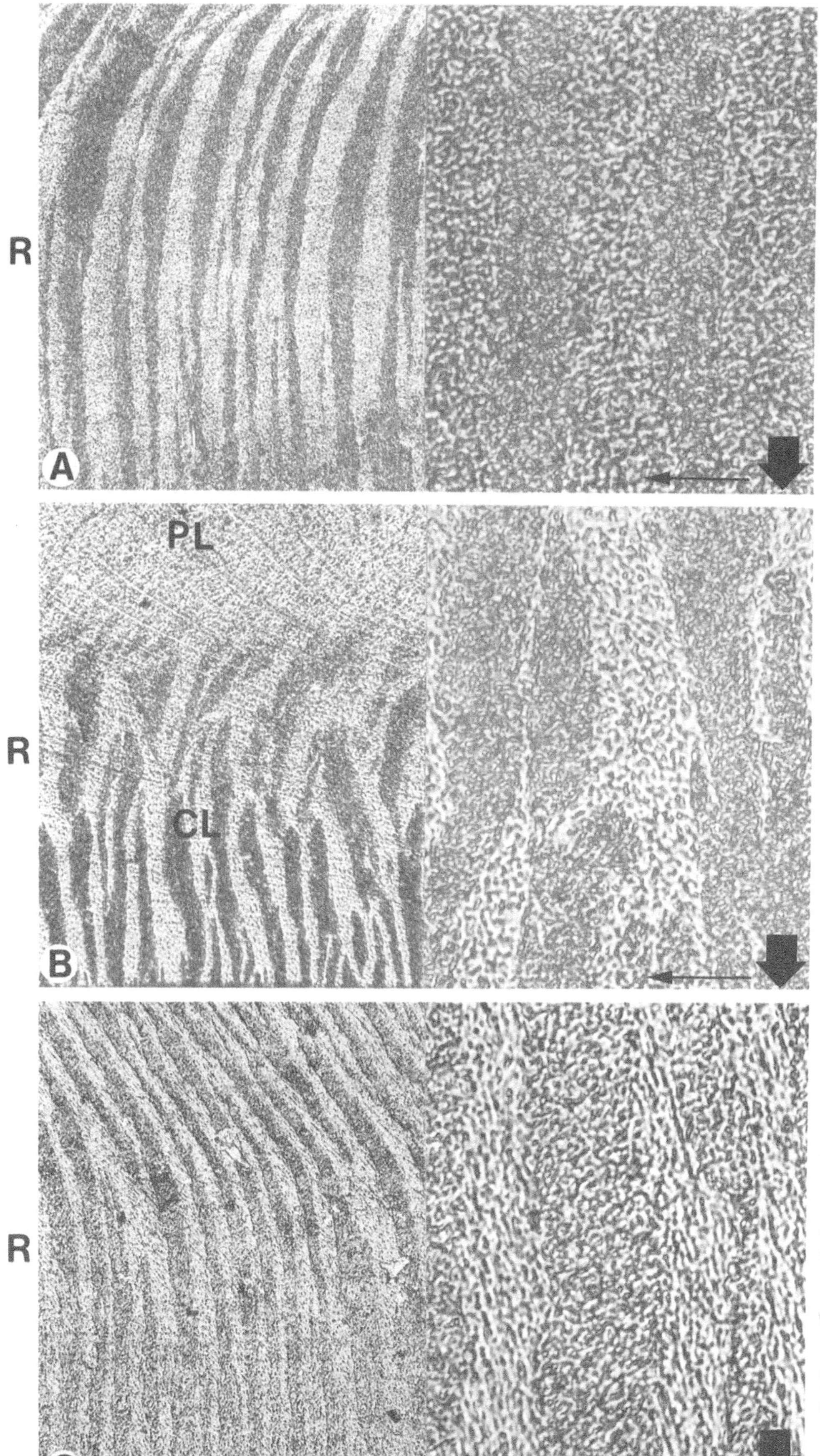

63. Trapeziidae *Coralliophaga*

64. Glossidae *Glossus*

65. Corbiculidae *Batissa*

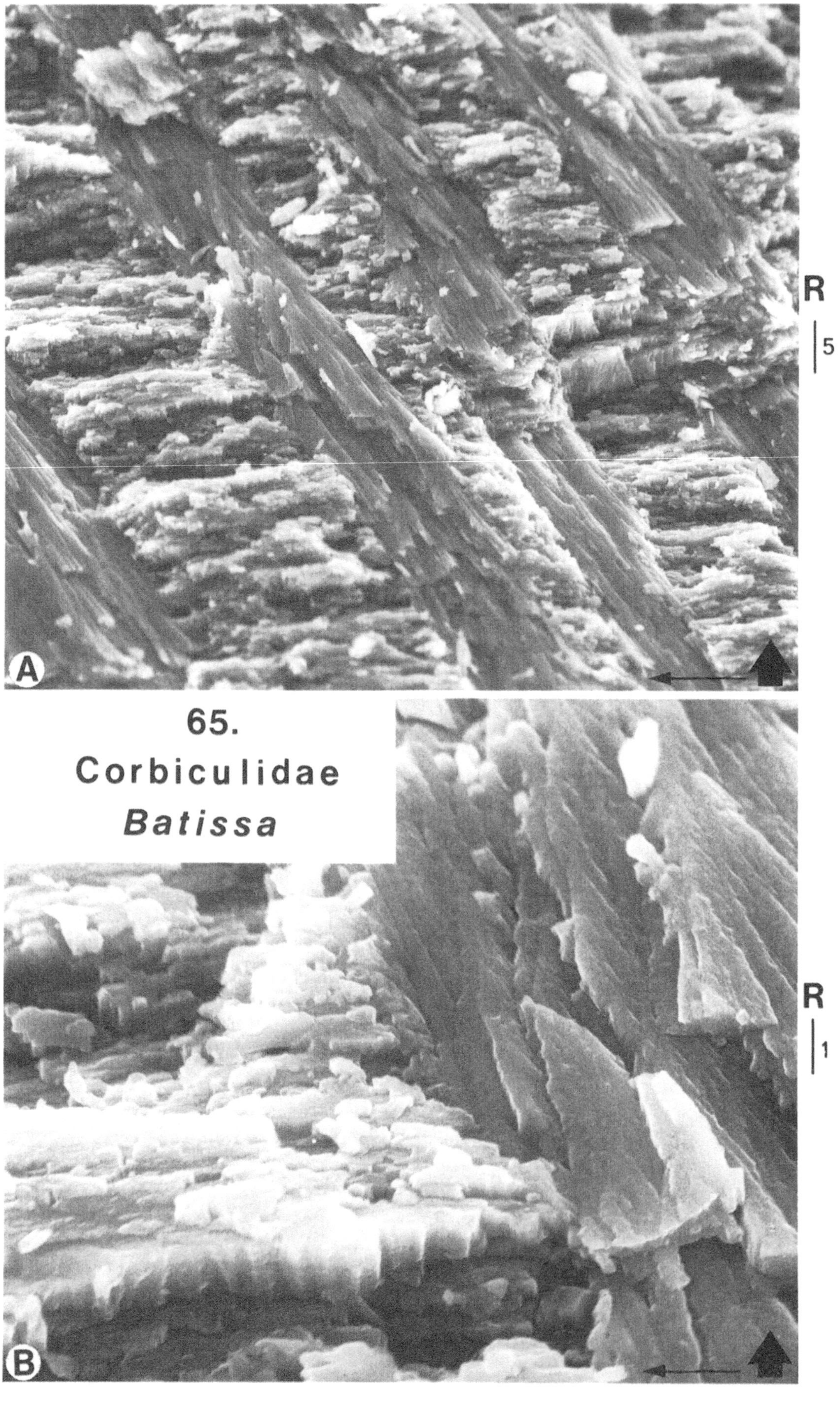

65.
Corbiculidae
Batissa

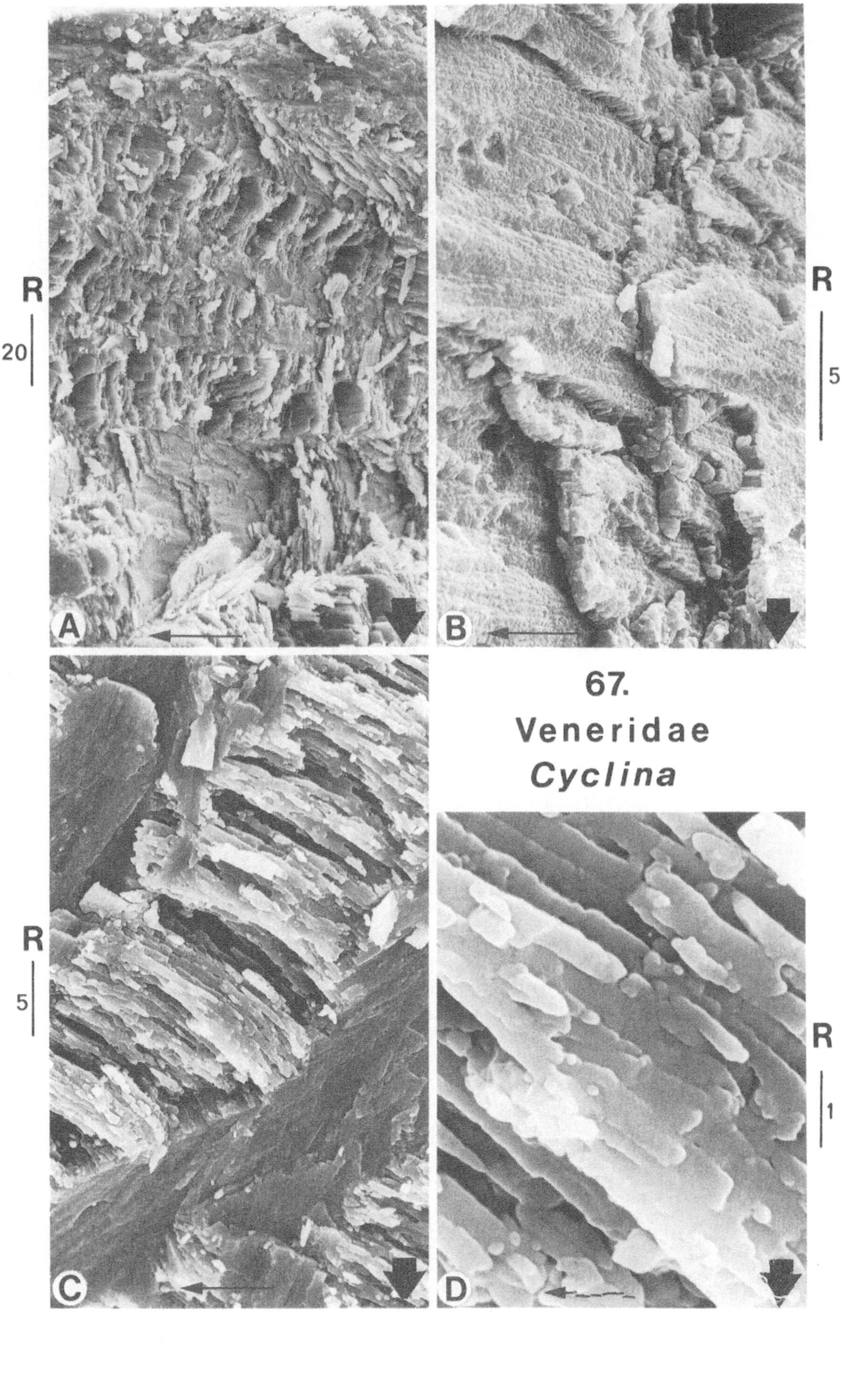

67.
Veneridae
Cyclina

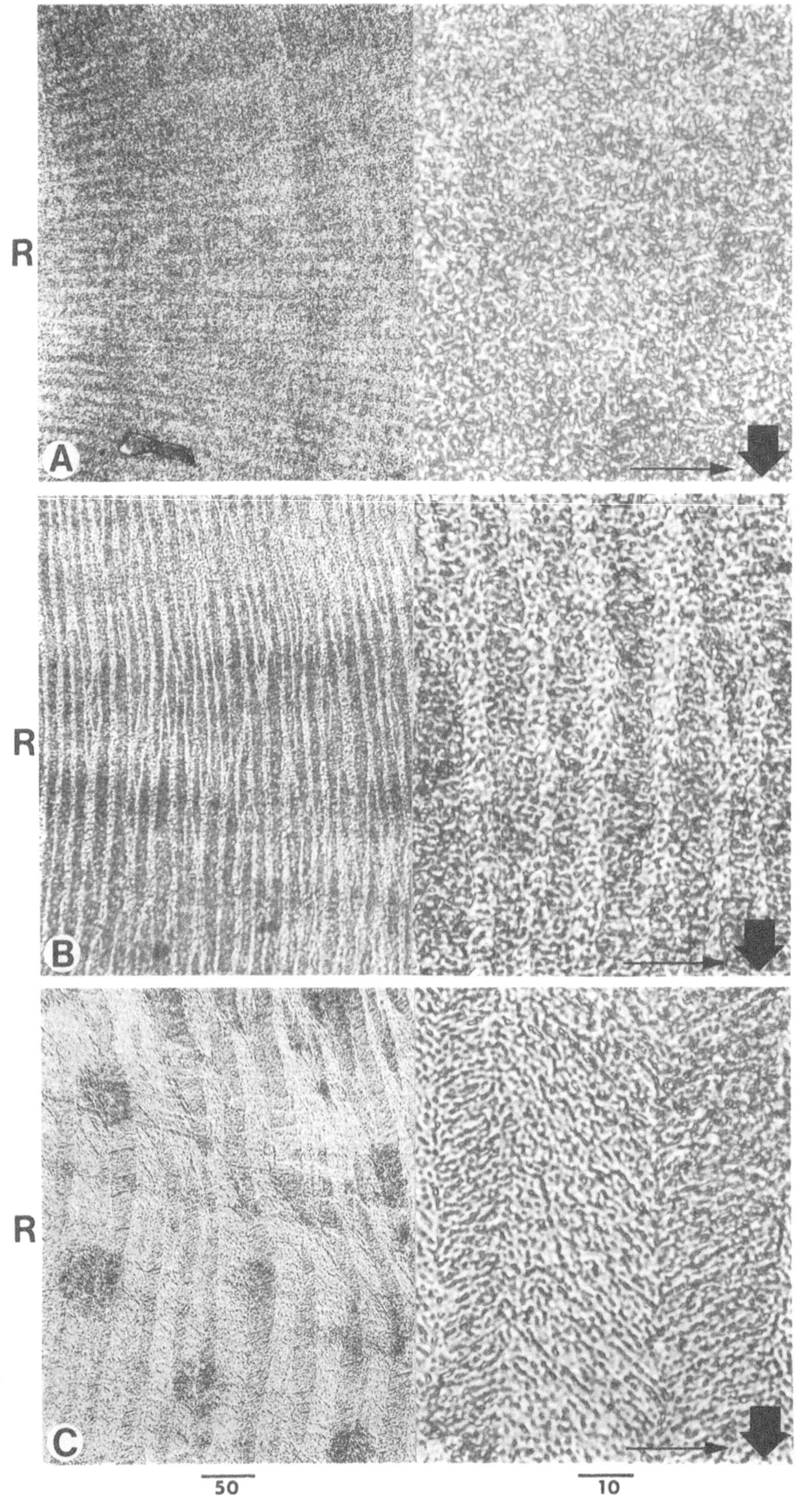

66. Veneridae *Dosinia*

67. Veneridae *Cyclina*

68. Veneridae *Circe*

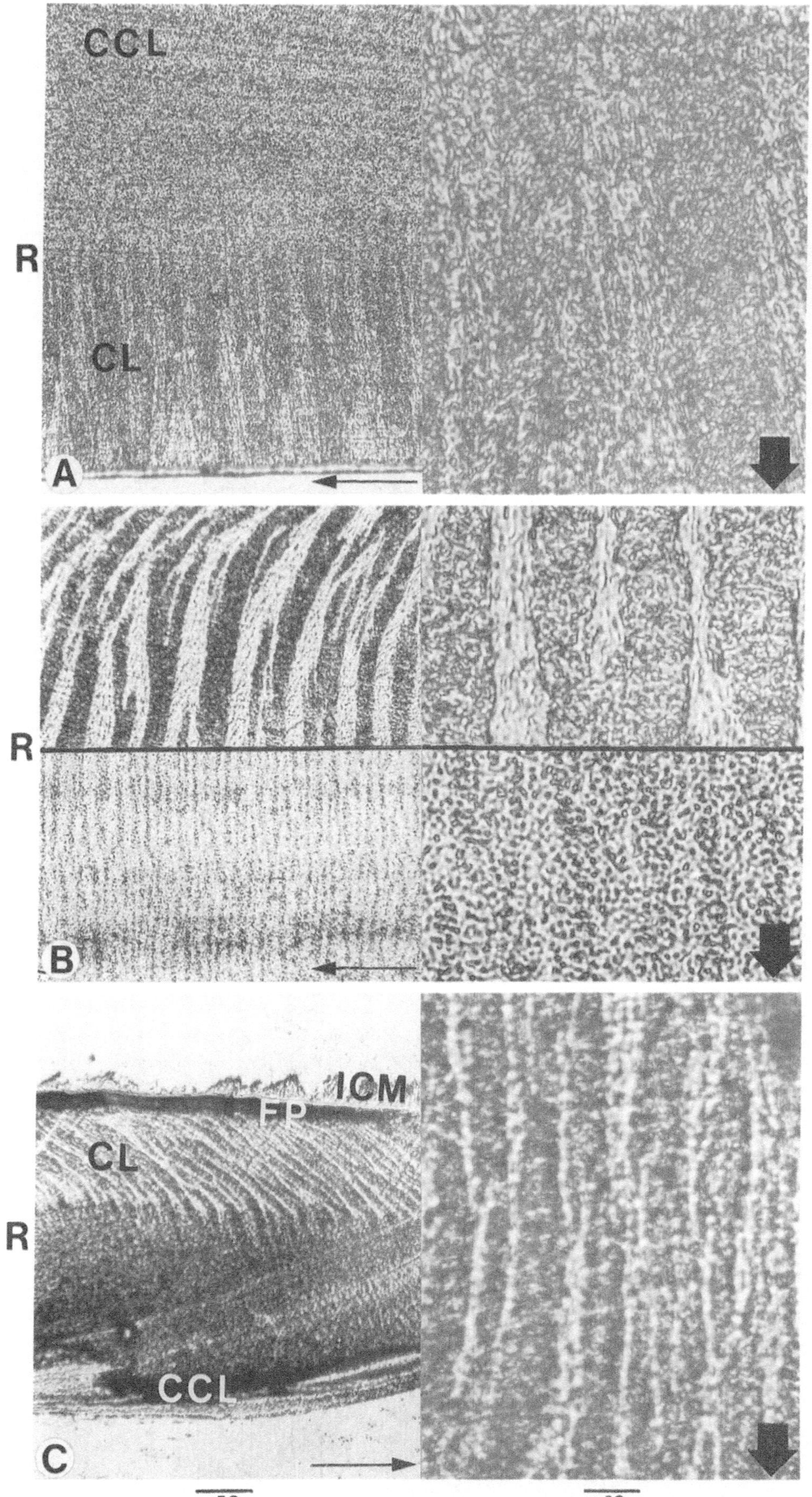

69. Veneridae *Compsomyax*

70. Veneridae *Paphia*

71. Veneridae *Tivela*

79.
Myidae
Mya

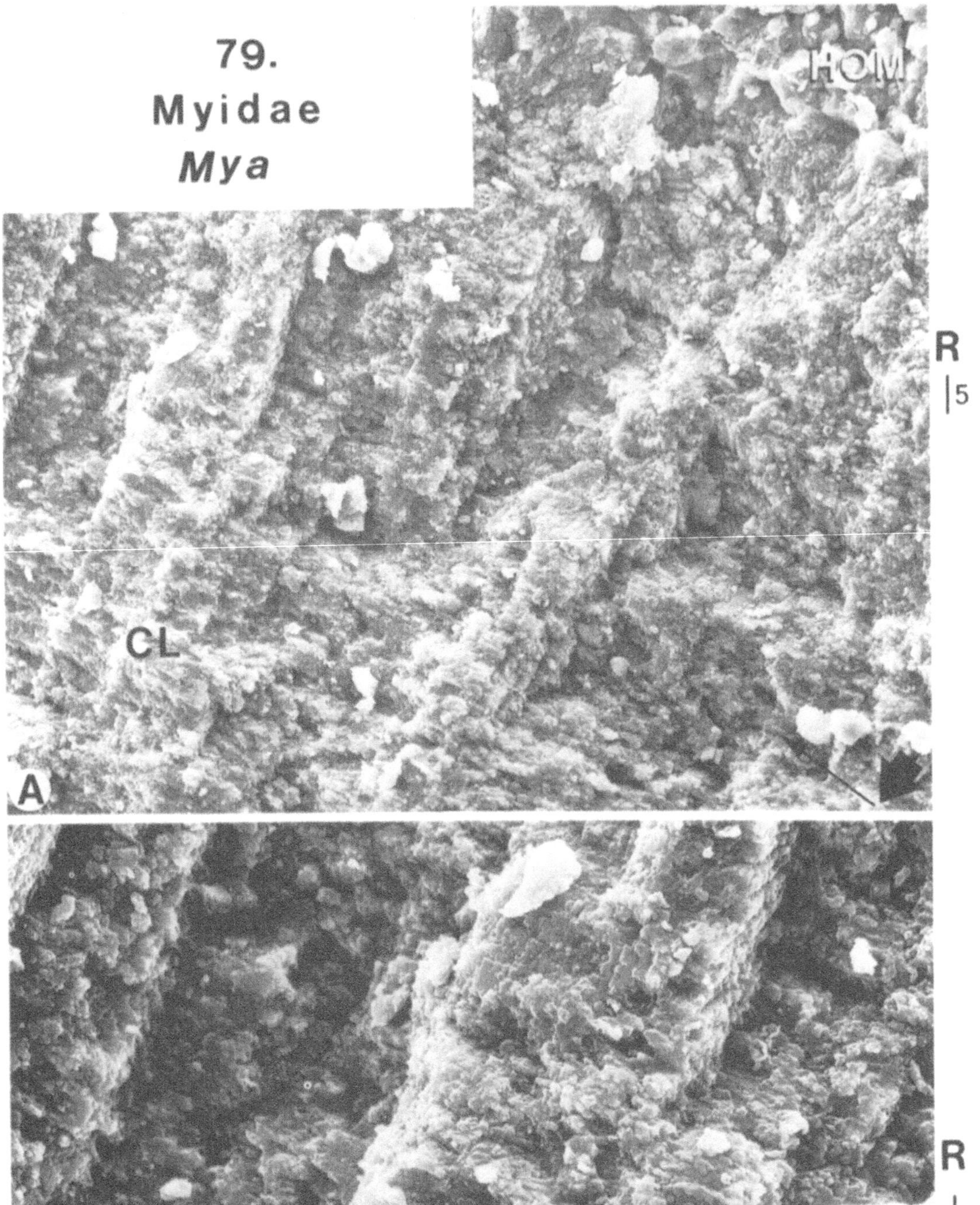

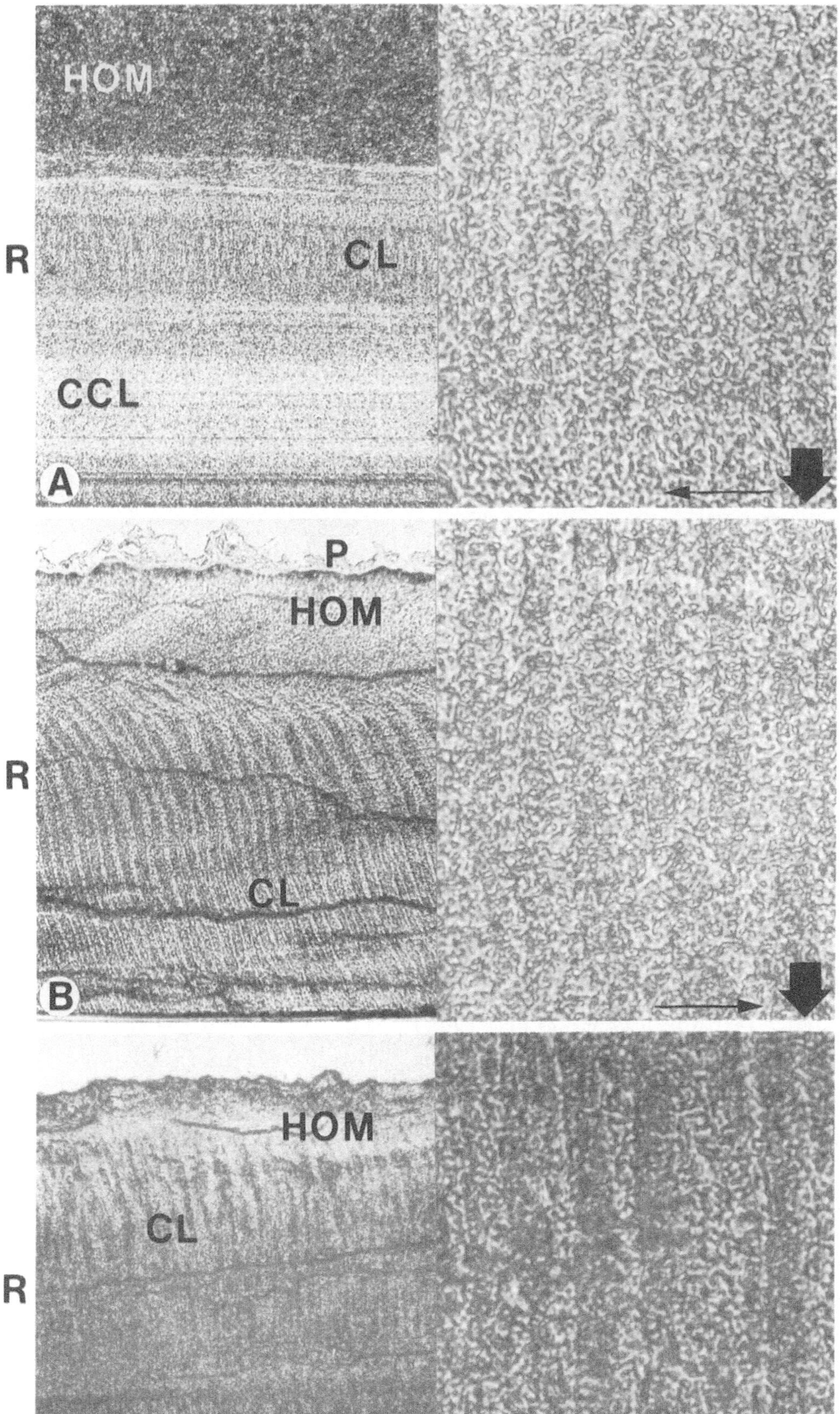

80. Myidae *Mya*

81. Myidae *Cryptomya*

82. Myidae *Paramya*

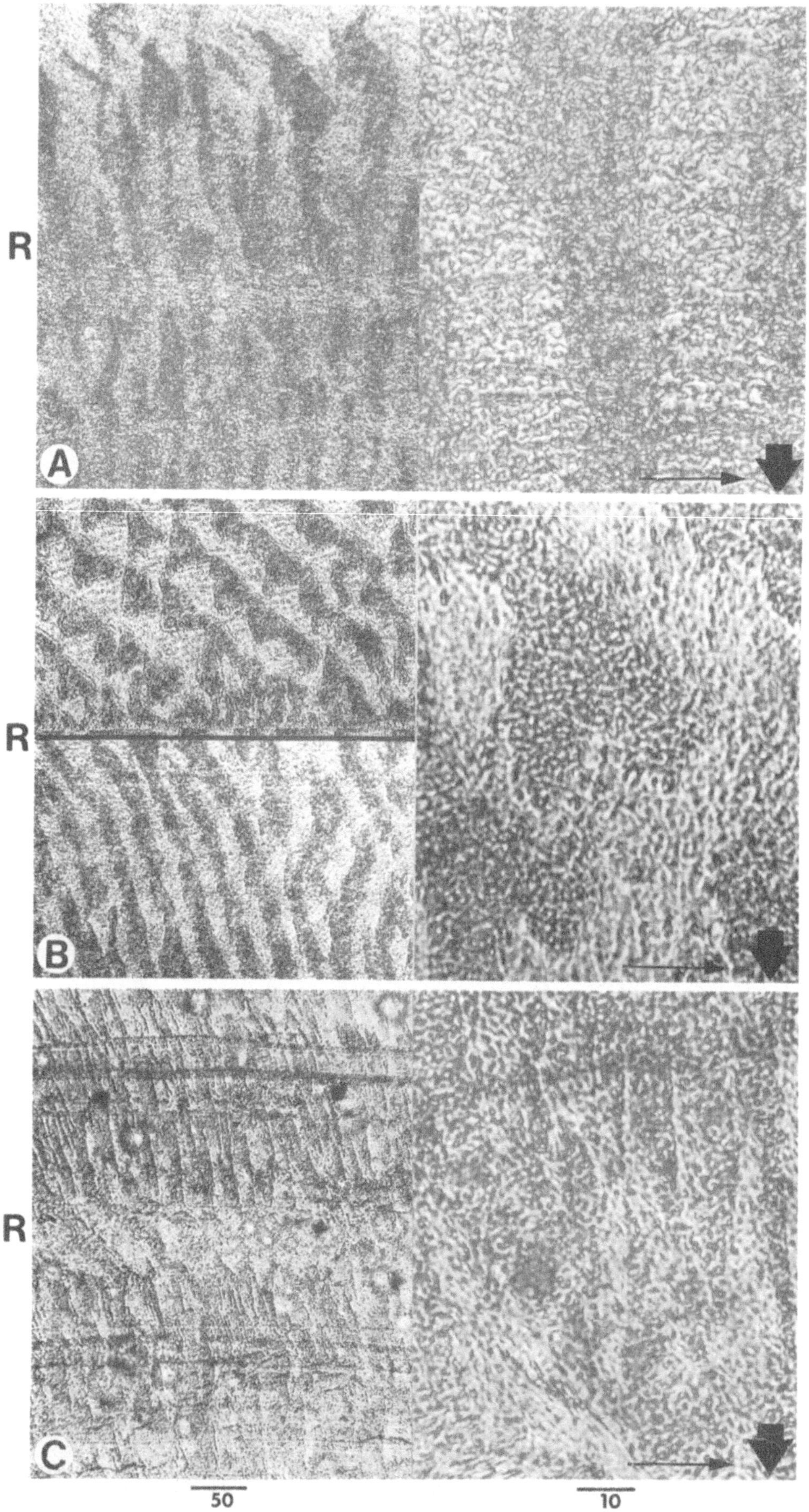

95. Pholadidae *Pholas*

96. Pholadidae *Cyrtopleura*

97. Pholadidae *Zirfaea*

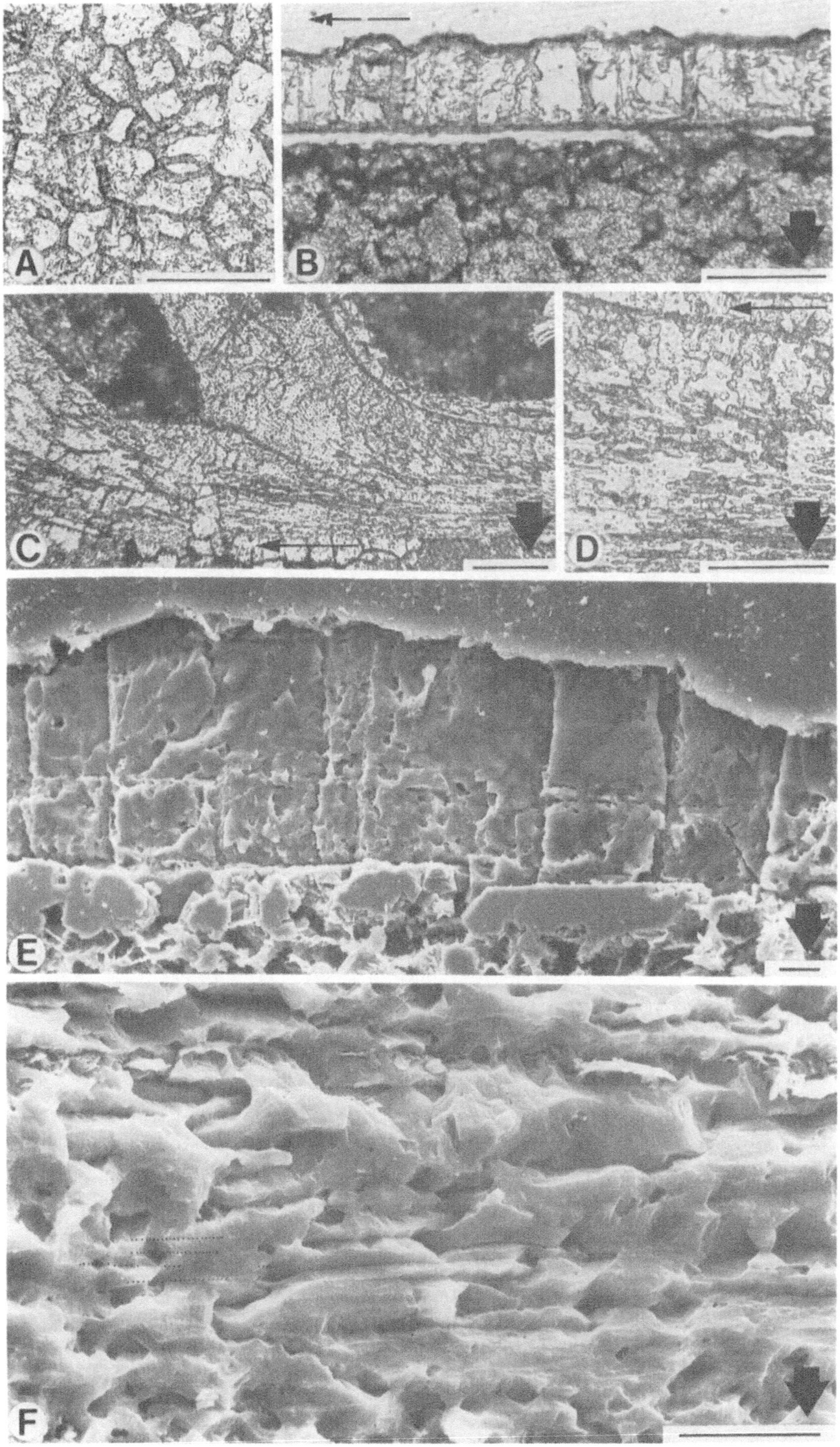
A
B
C
D
E
F

PLATE 102

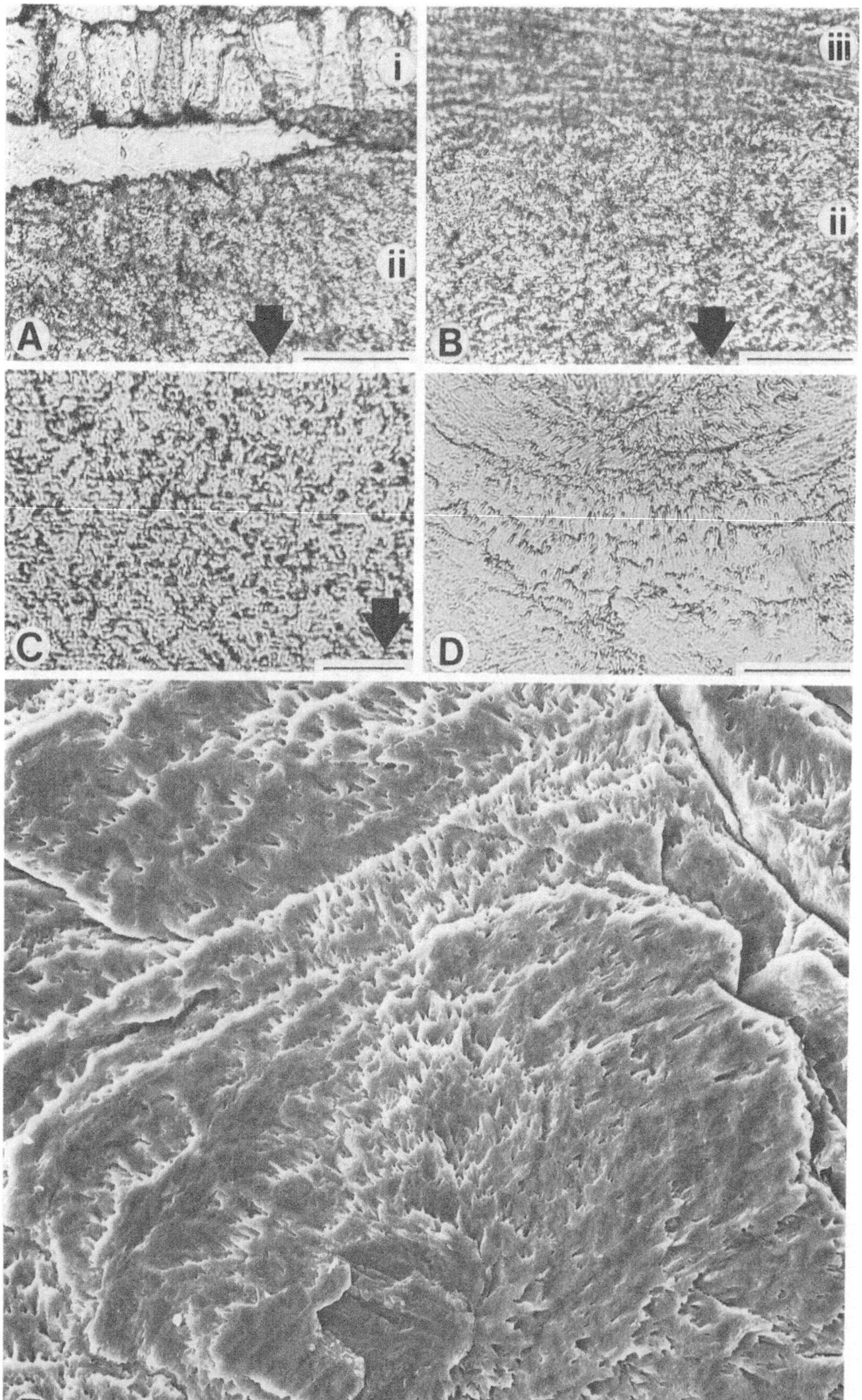

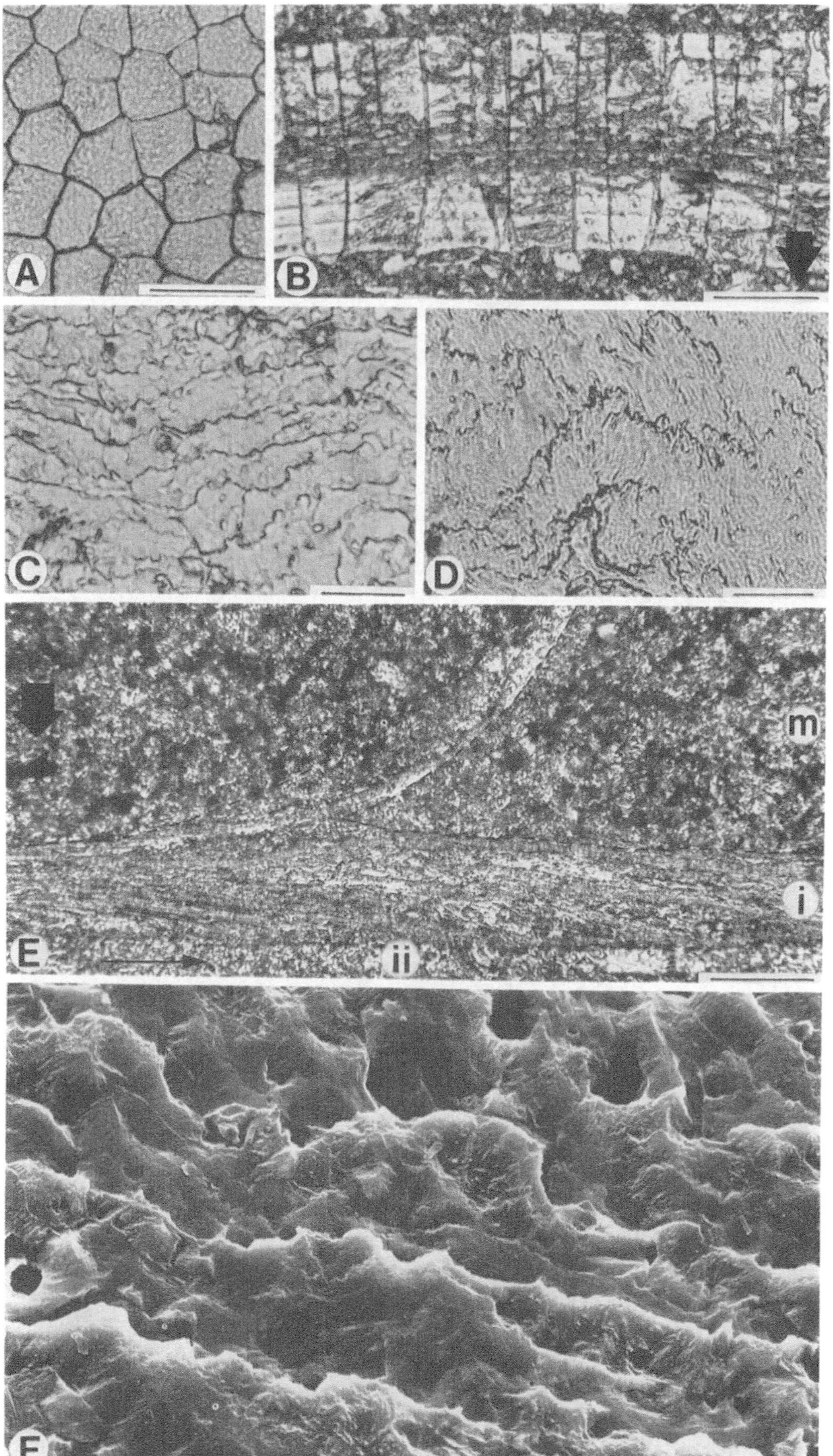
A
B
C
D
E
m
i
ii
F

PLATE 104

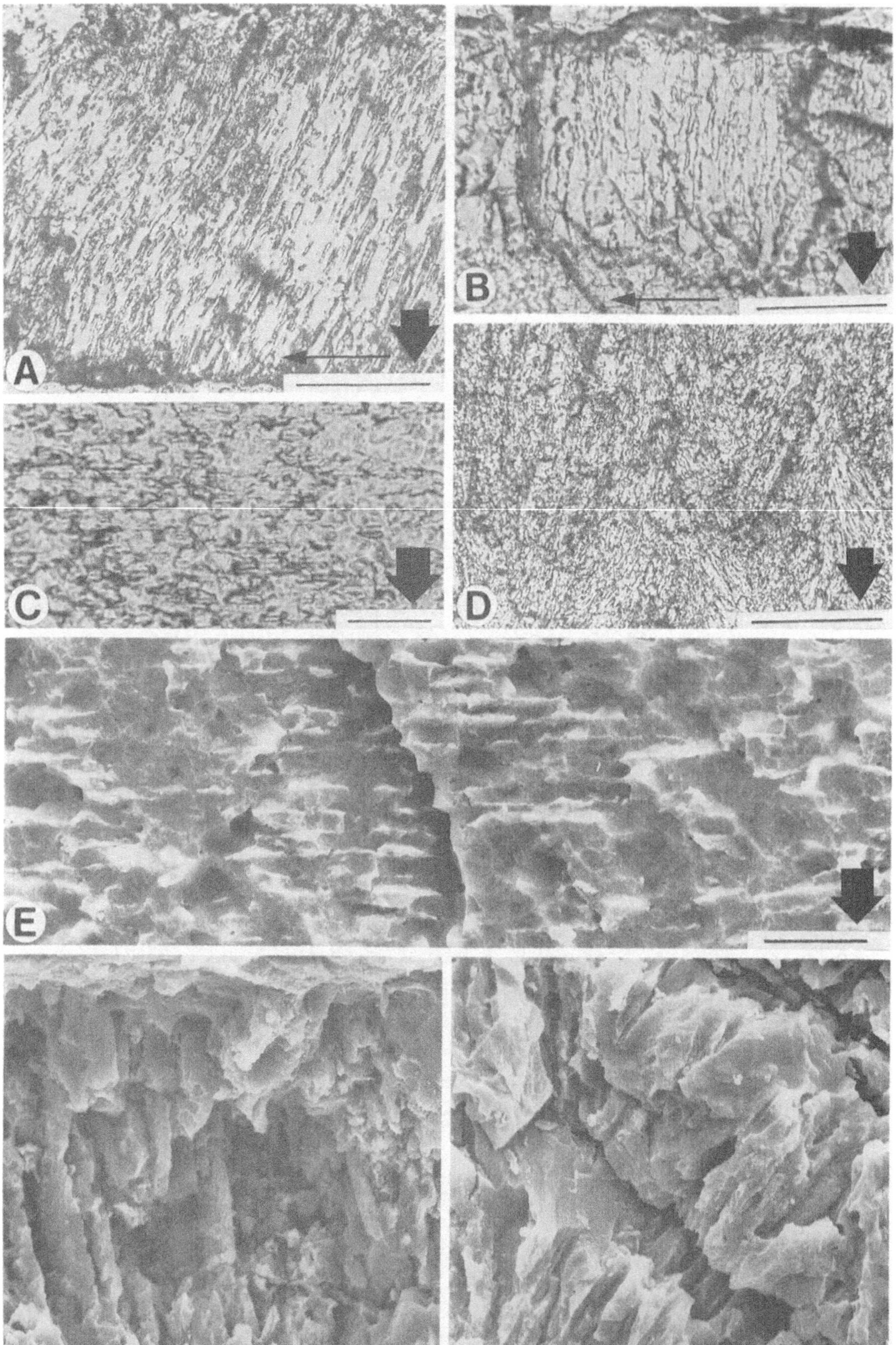

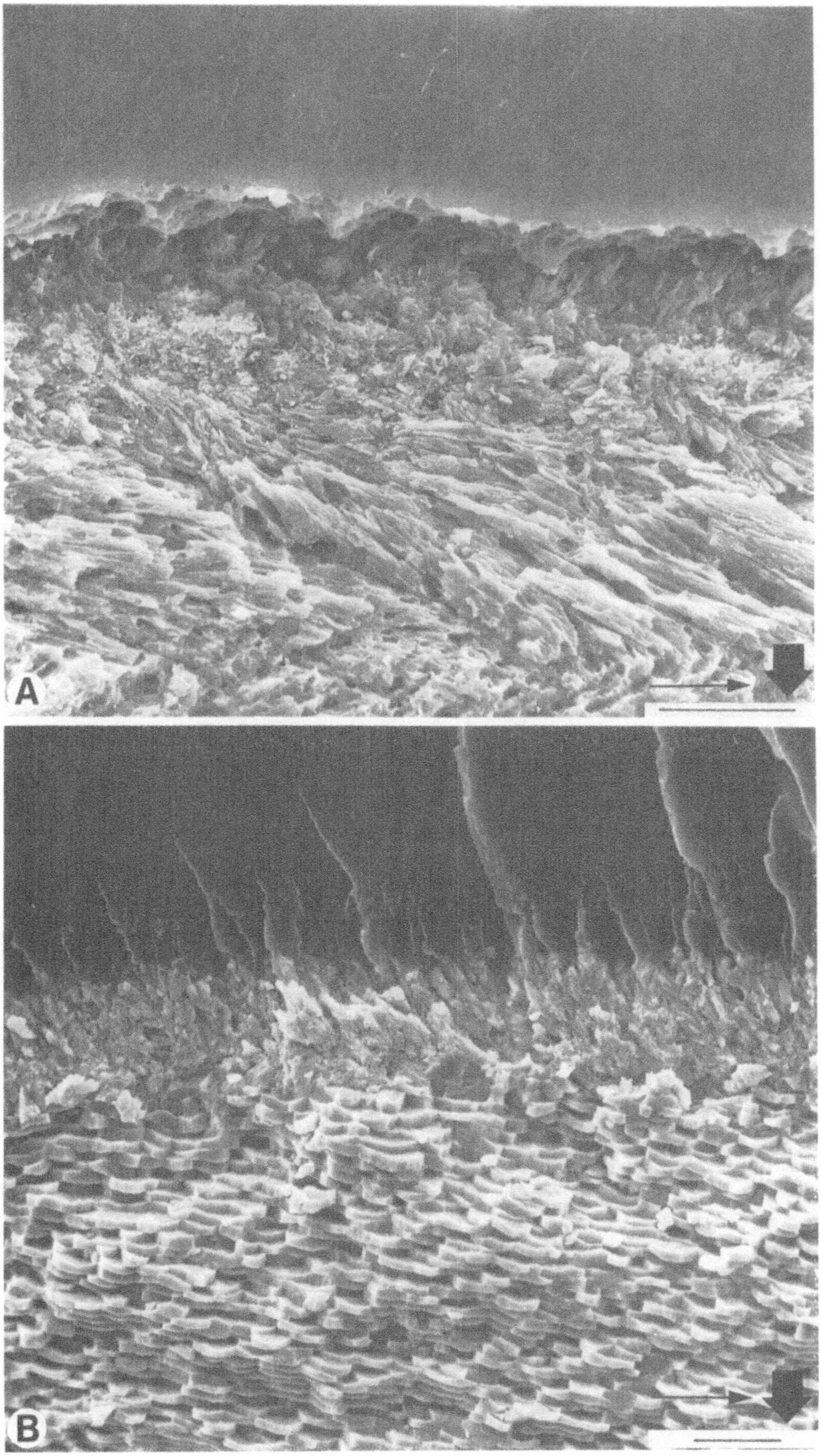
A
B

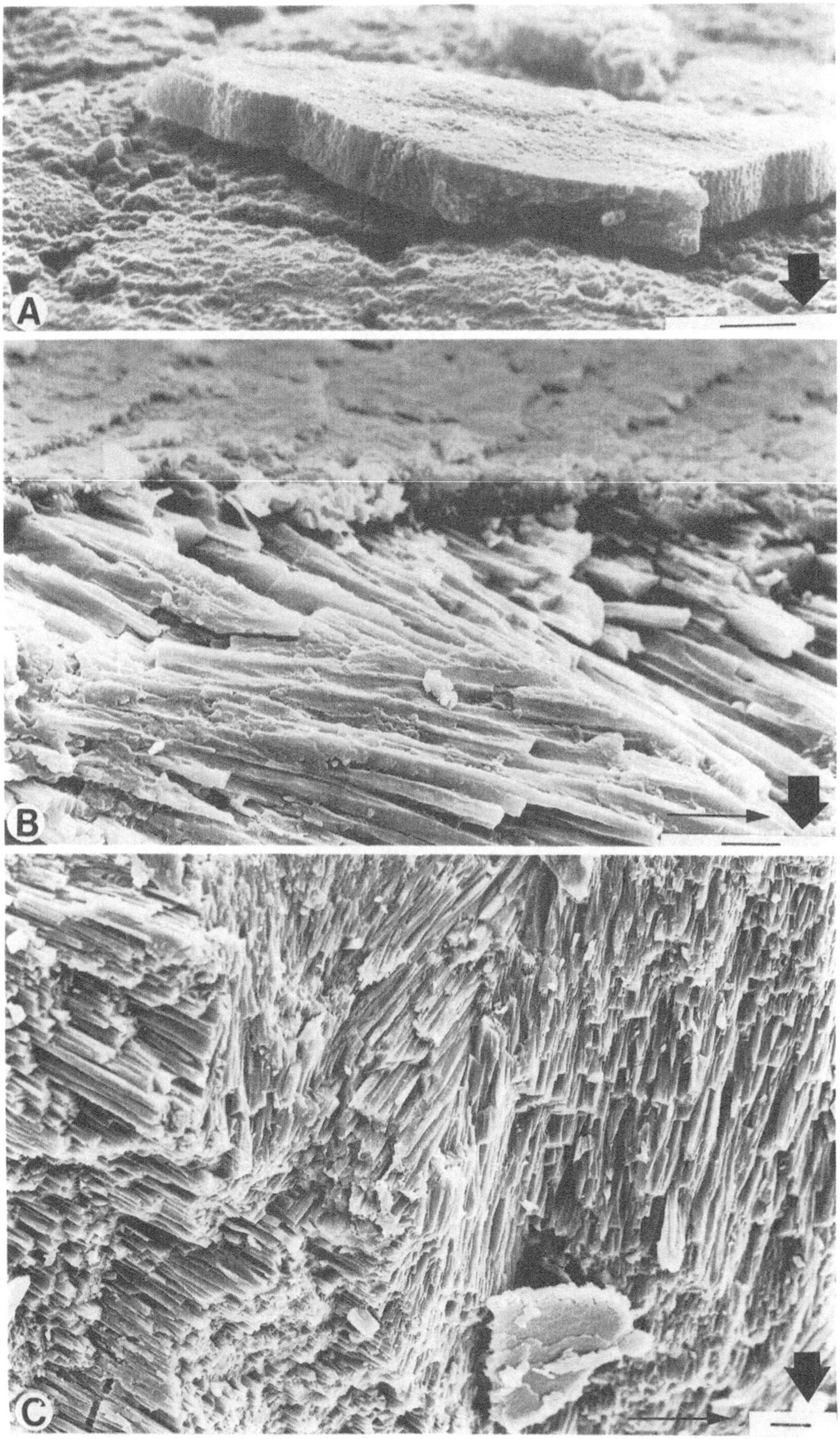
A
B
C

A
B
C

A
B
C

A
B
C

PLATE 110

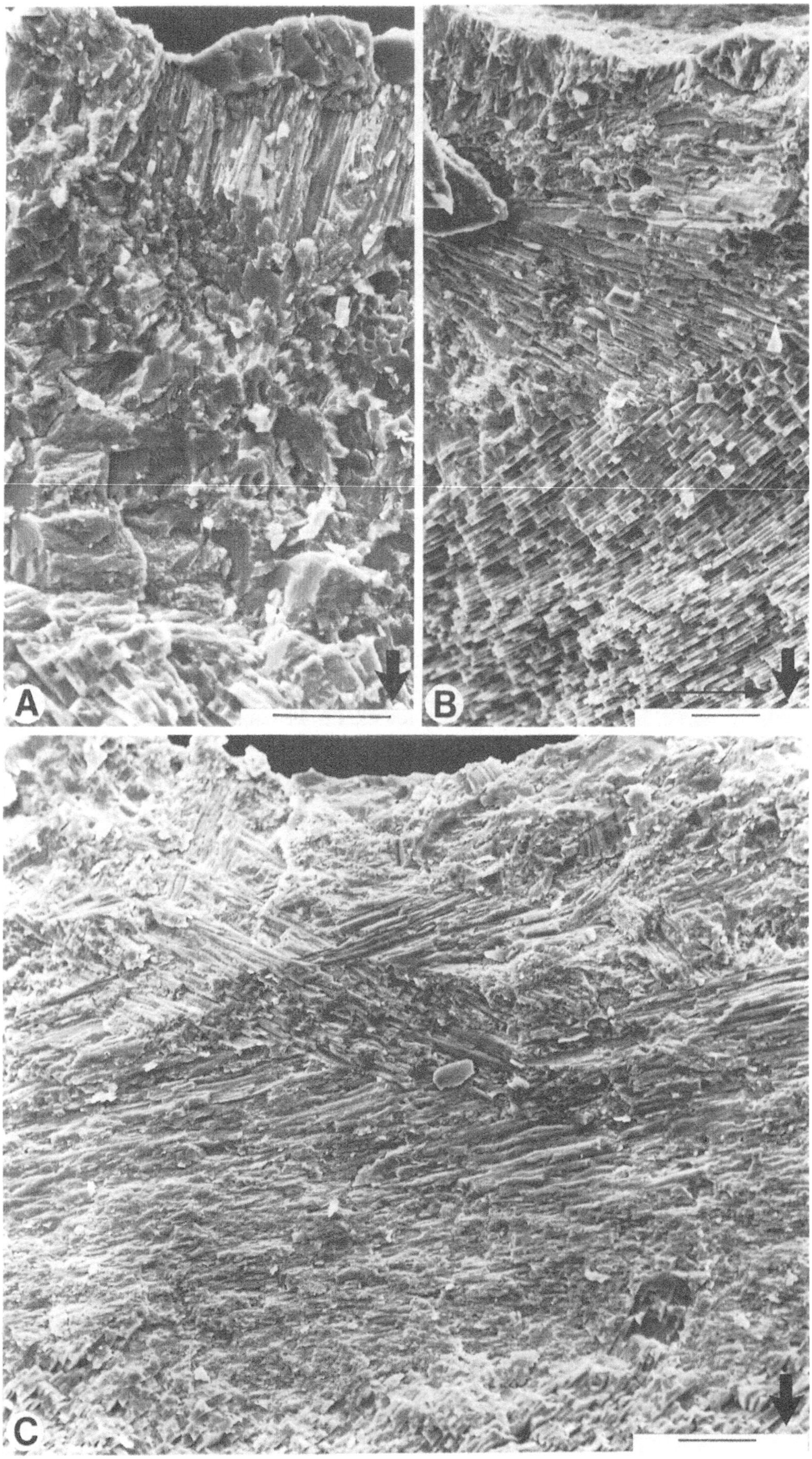

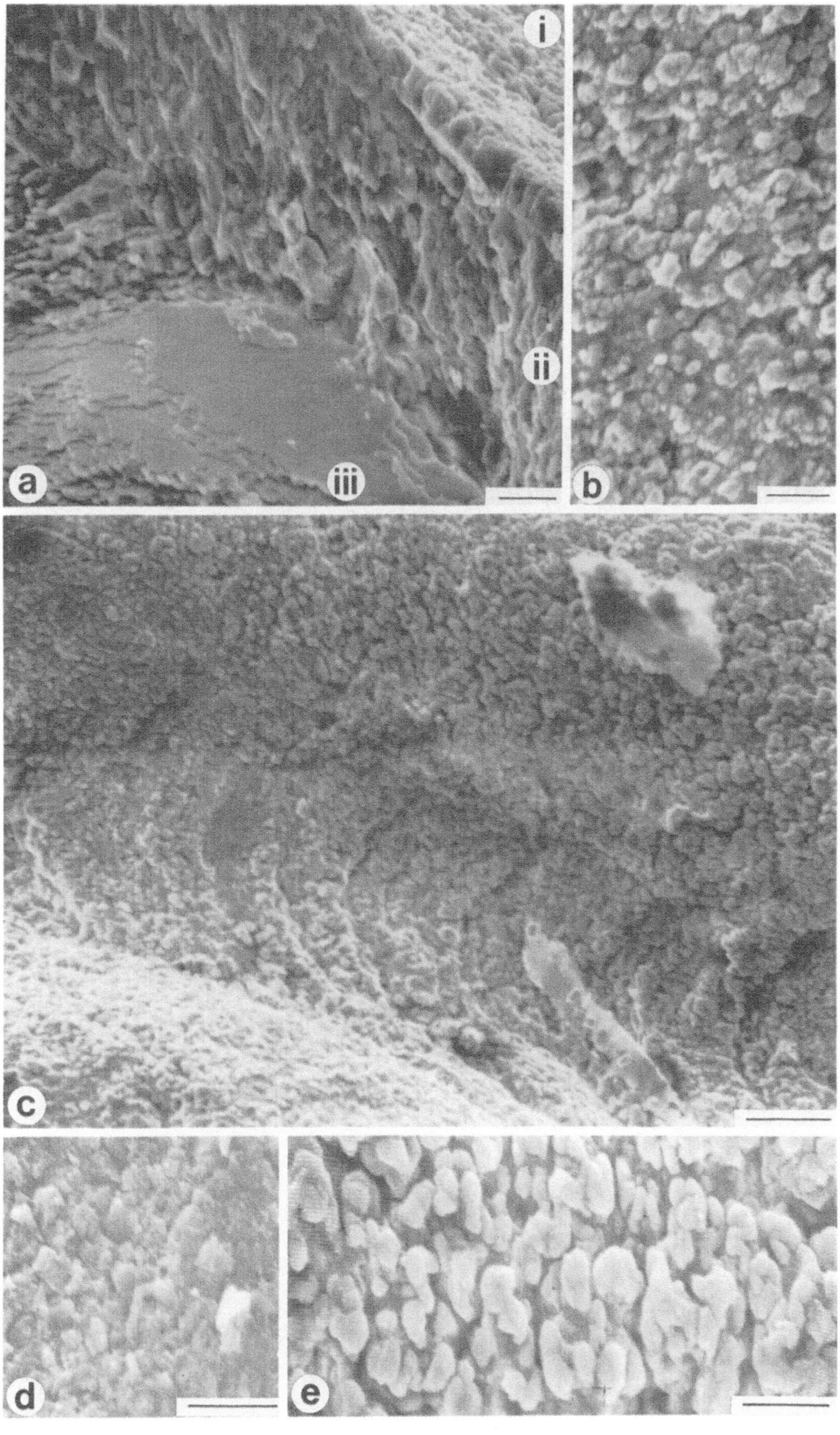
i
ii
iii
a
b
c
d
e

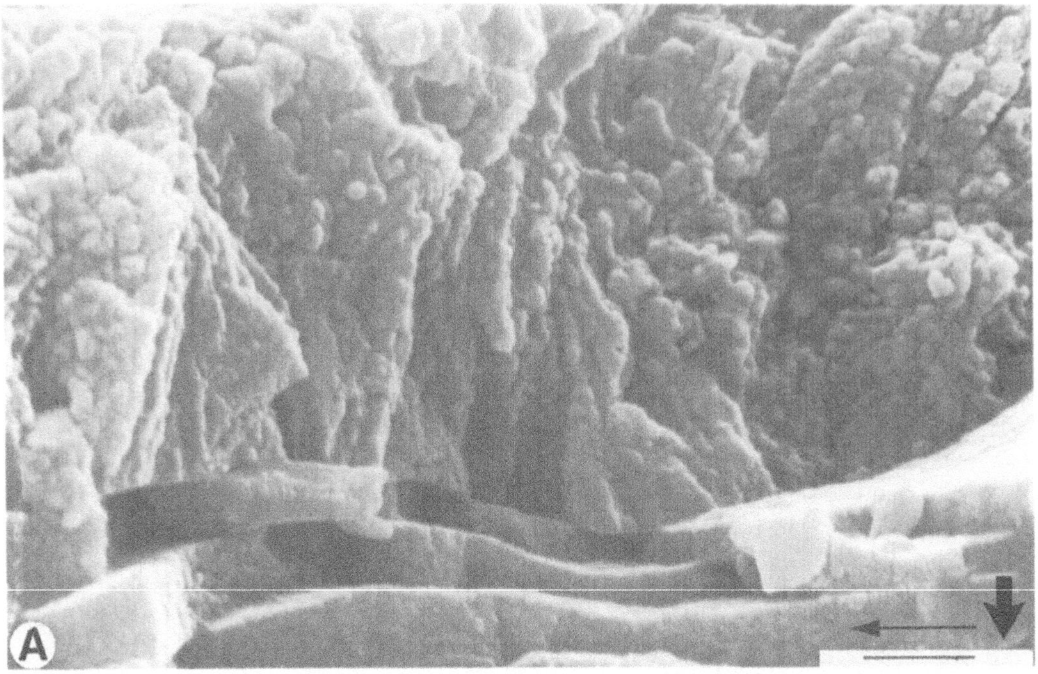
A

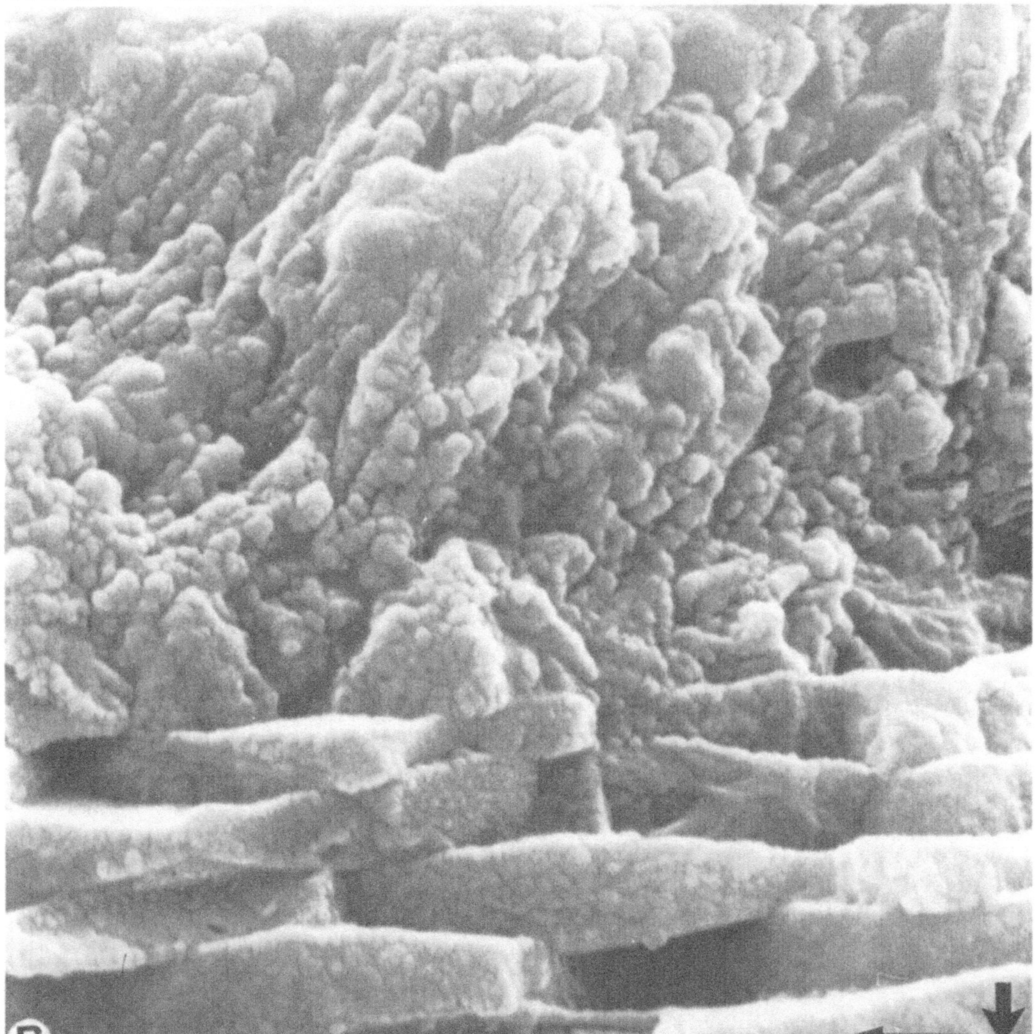
B

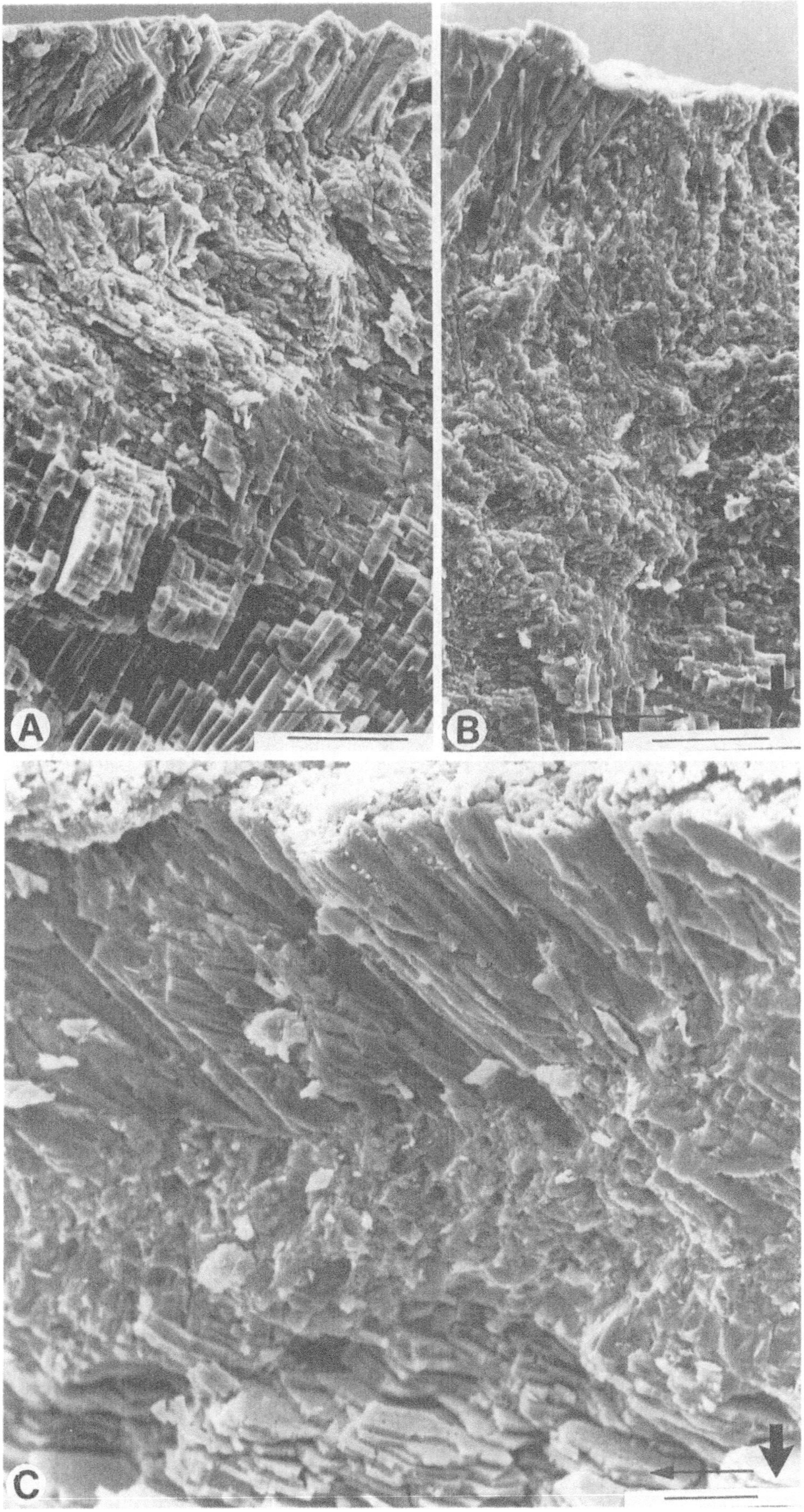

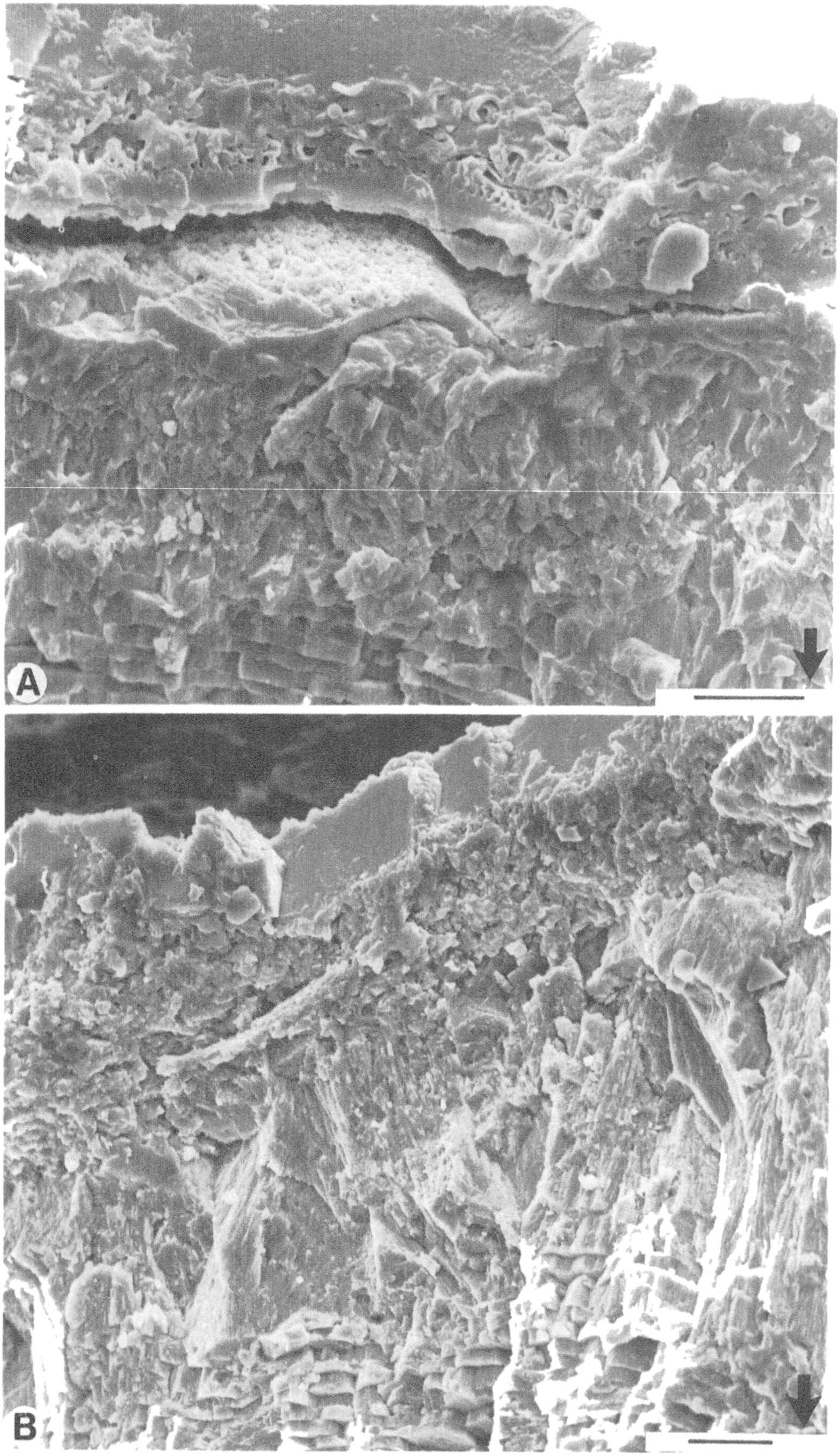
A
B

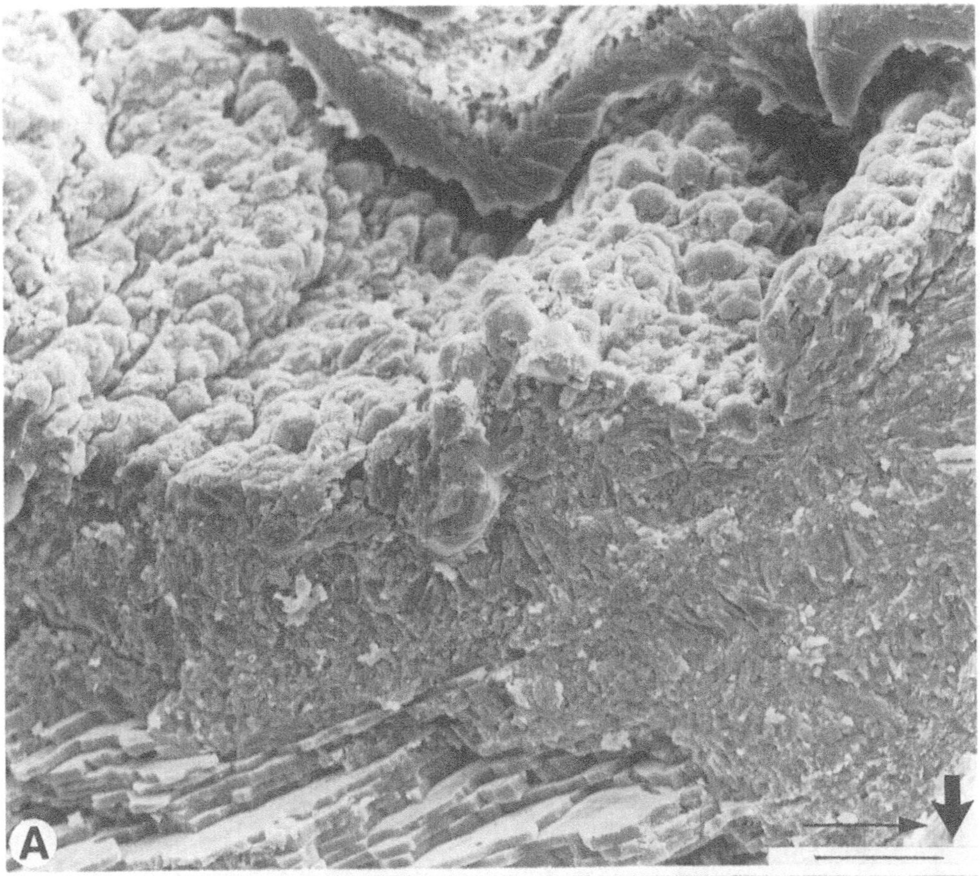
A

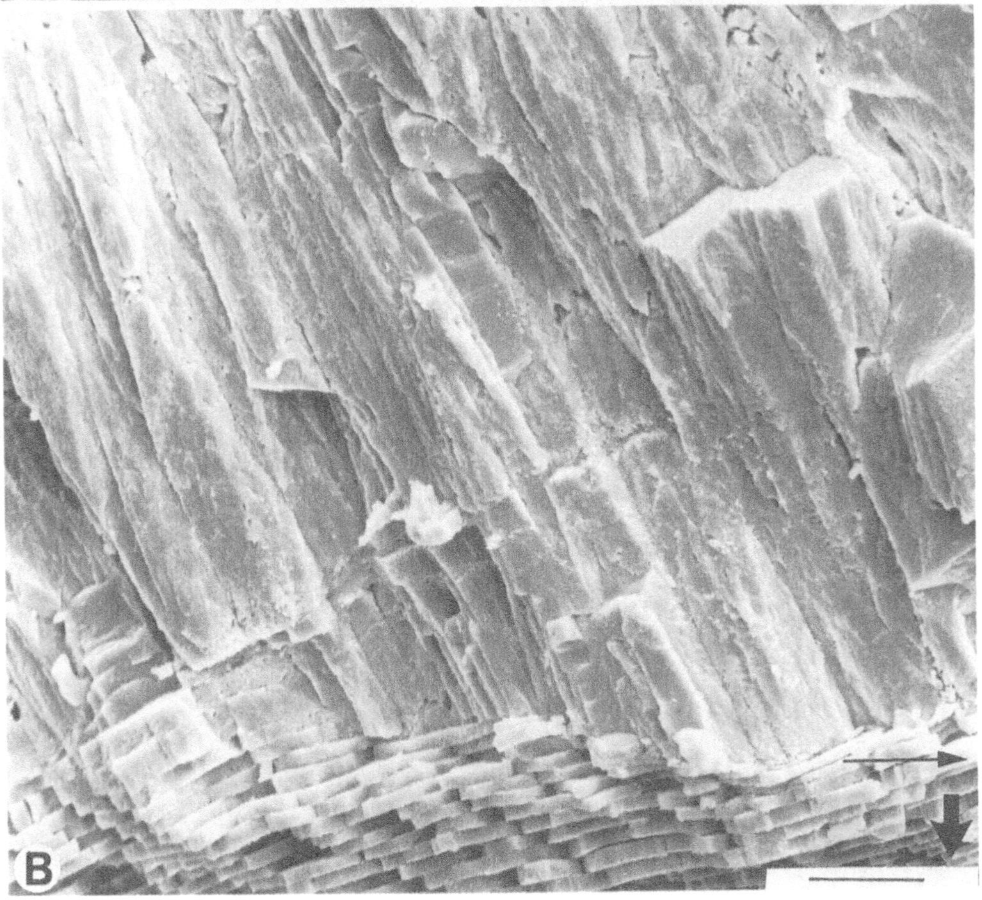
B

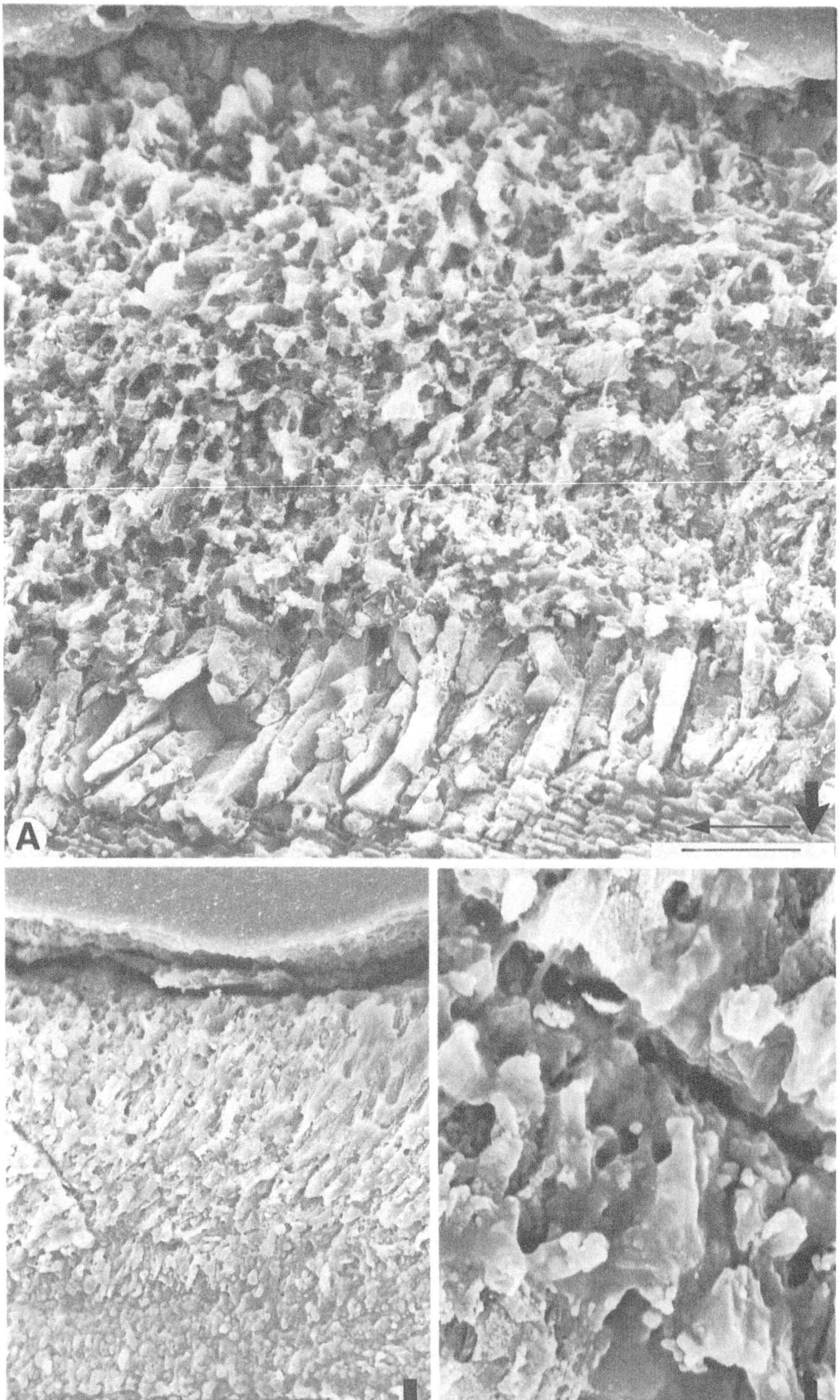
A
B
C

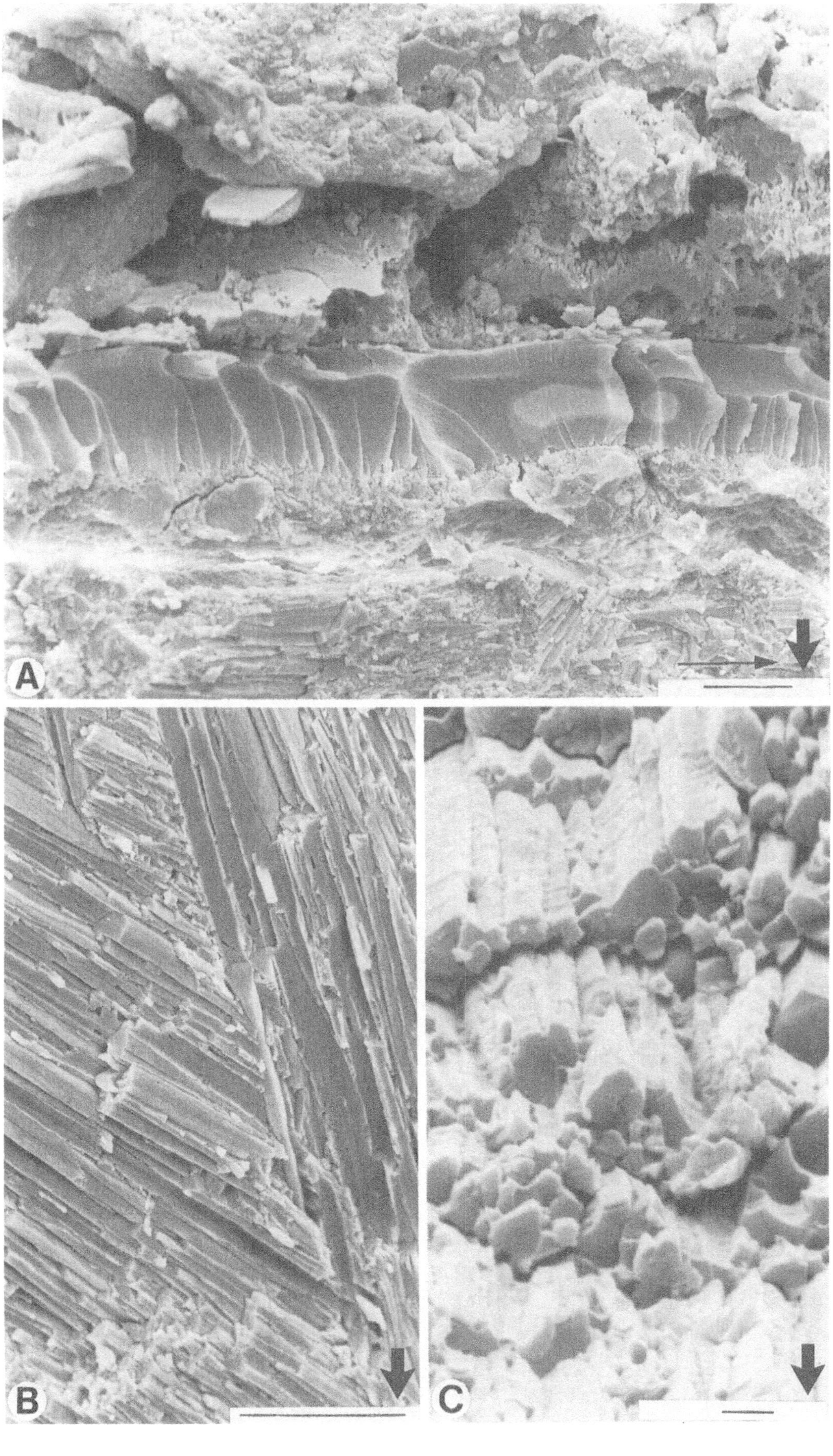
A
B
C

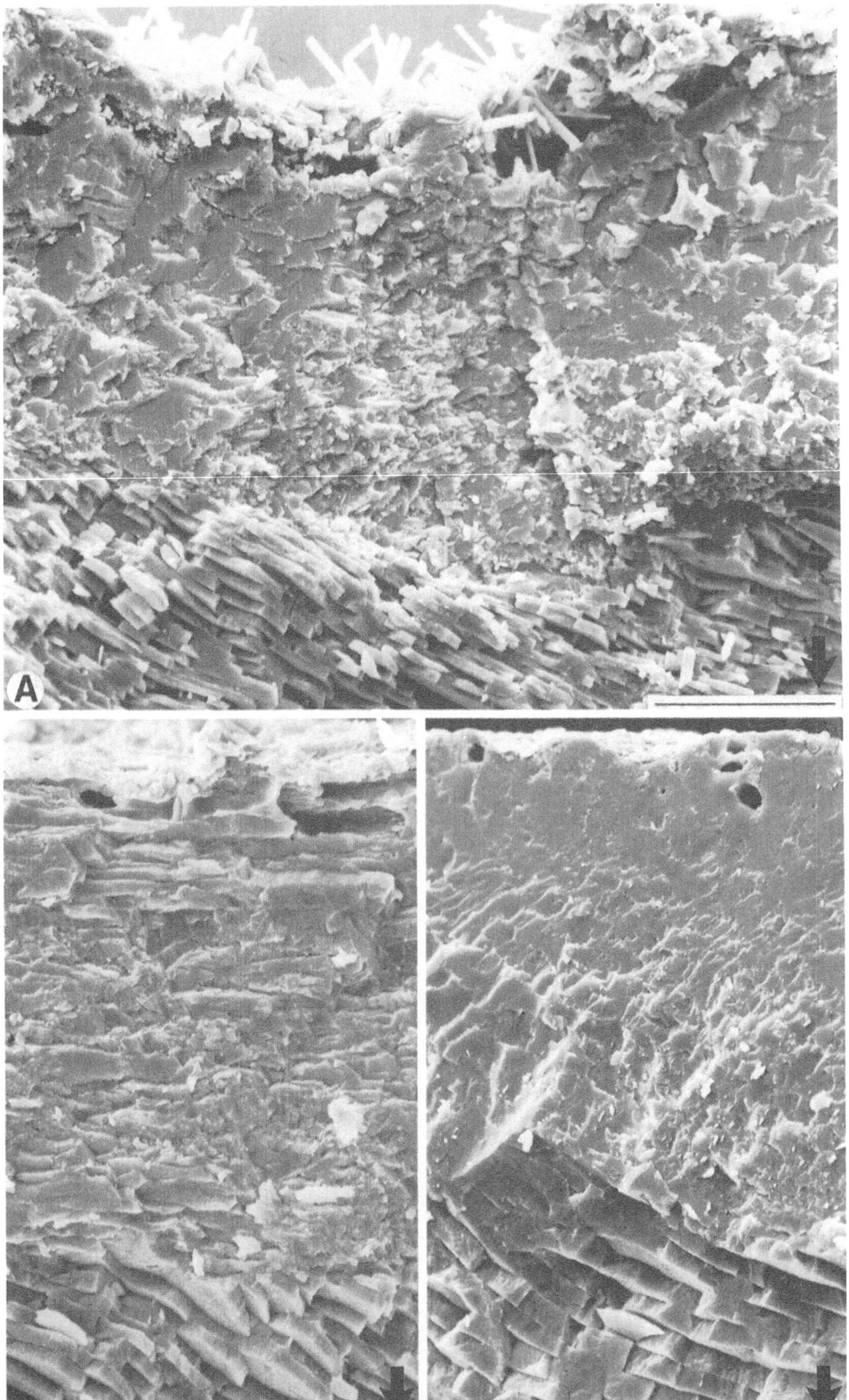
A
B
C

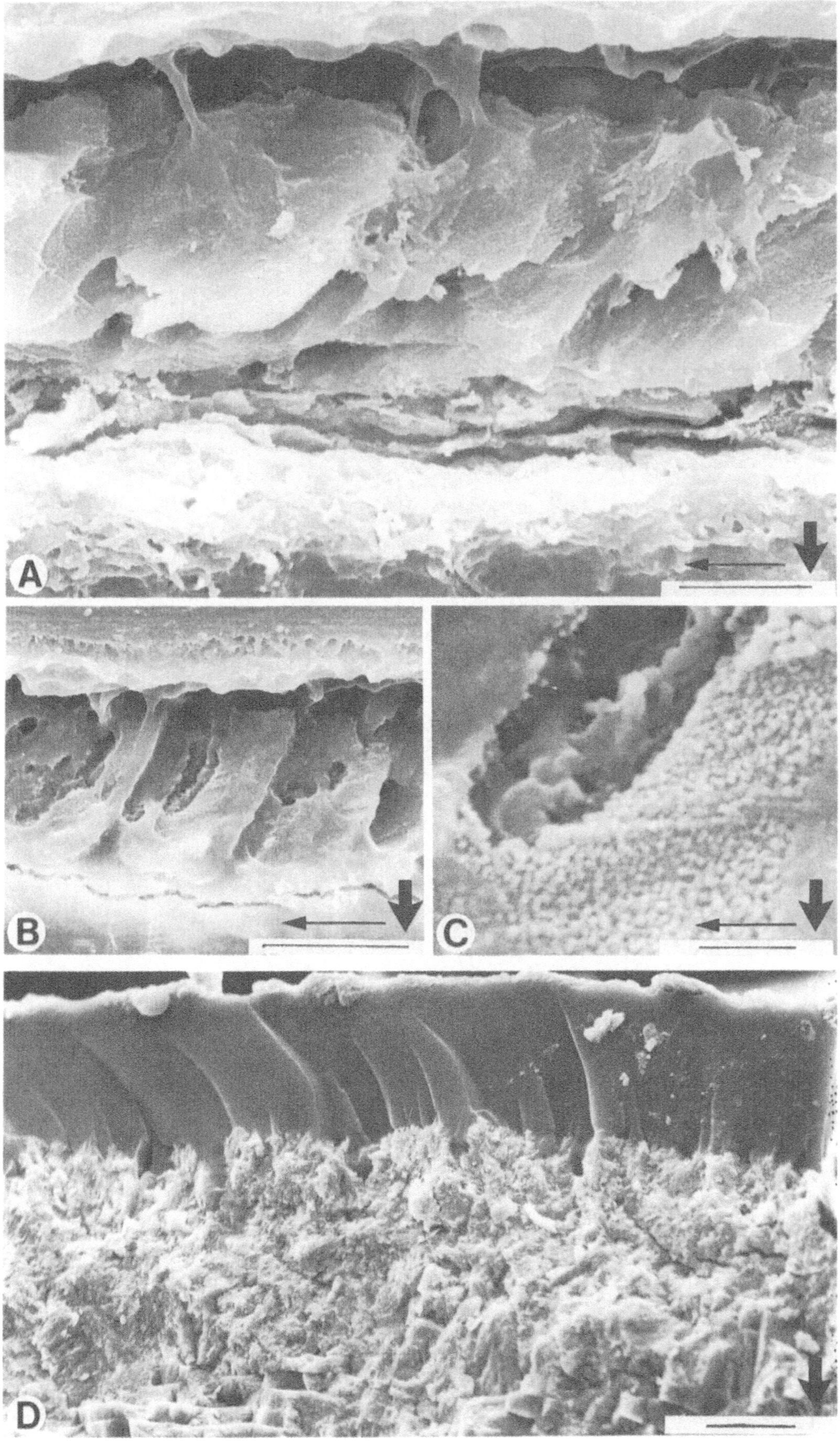
A
B
C
D

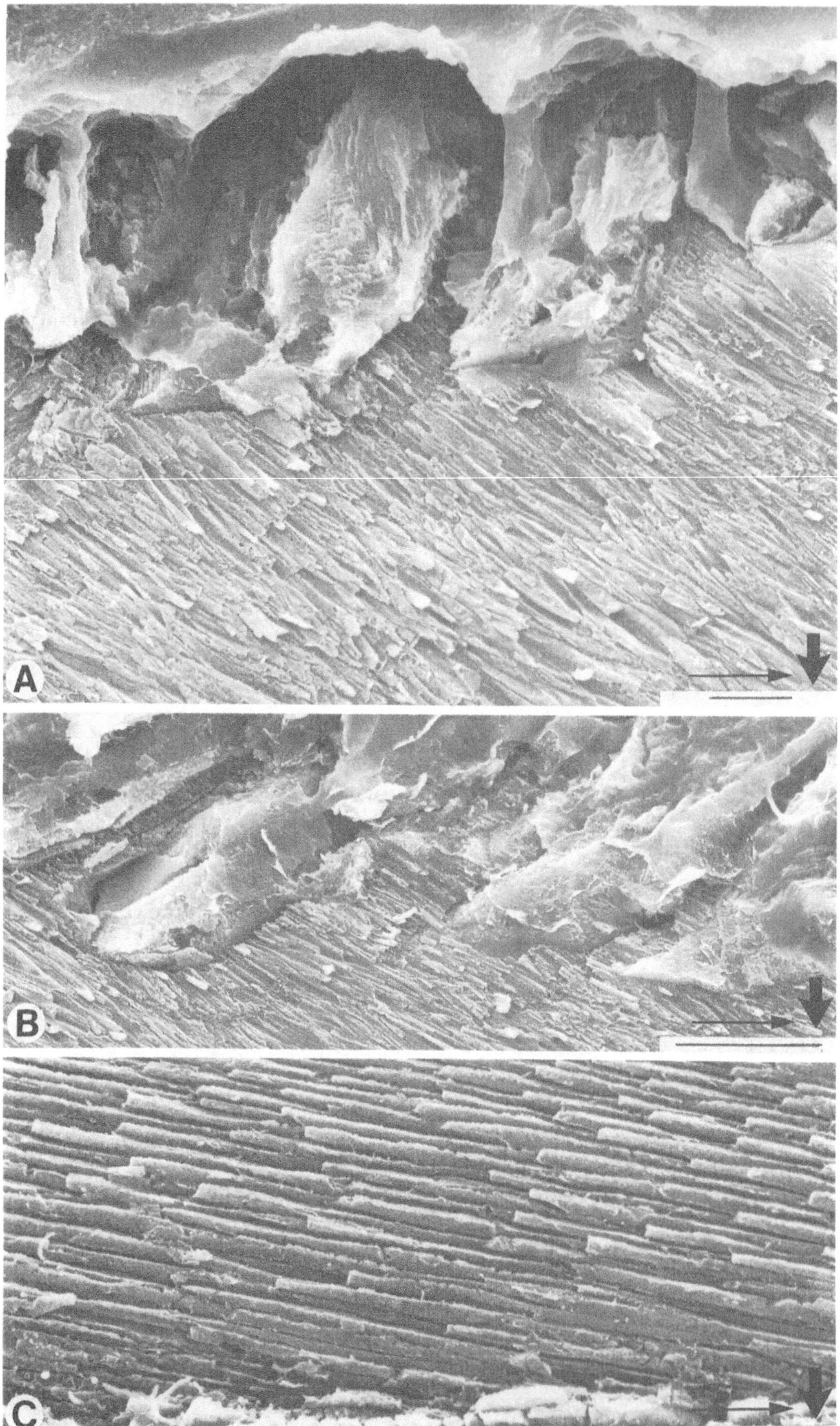
A
B
C

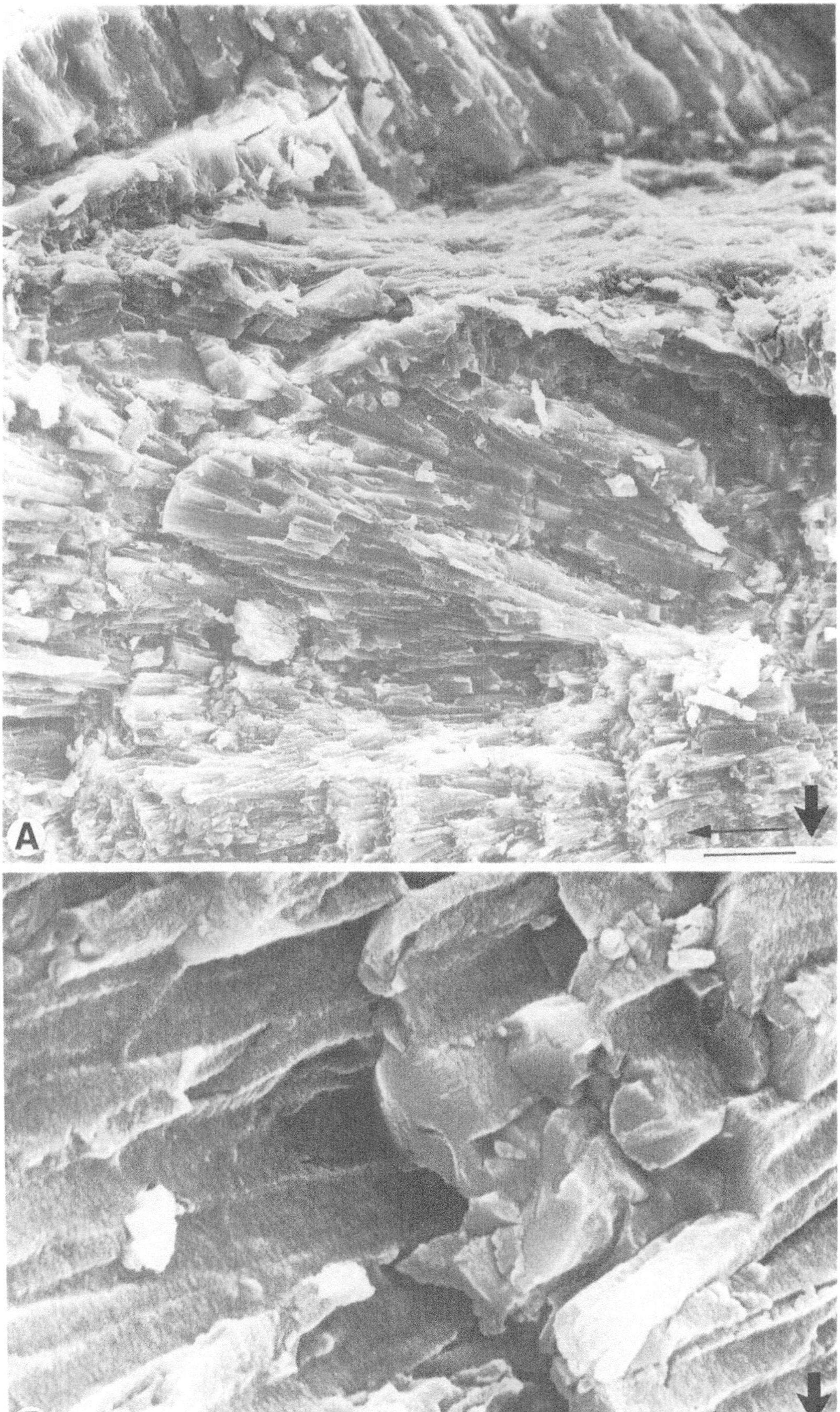
A
B

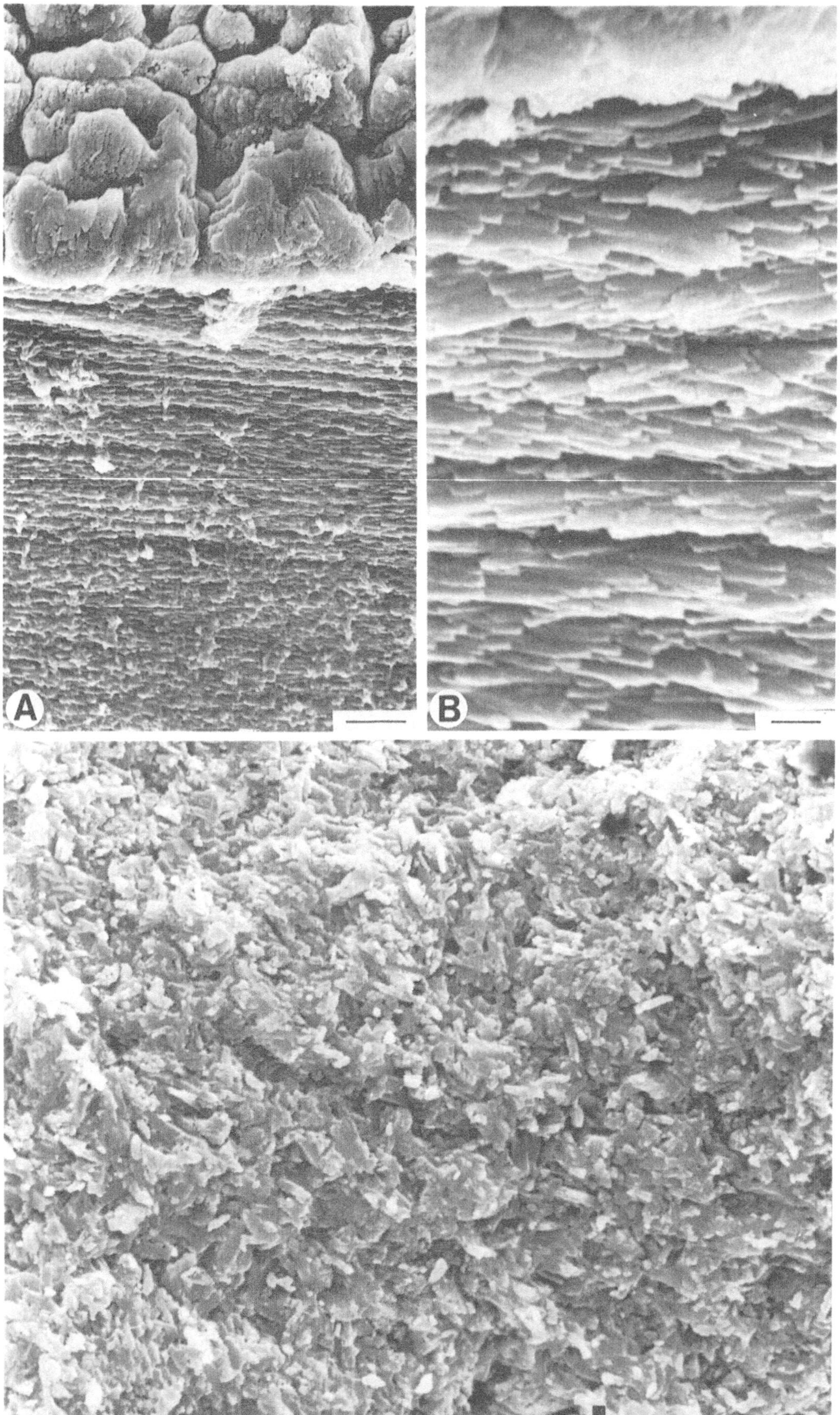
A
B
C

A

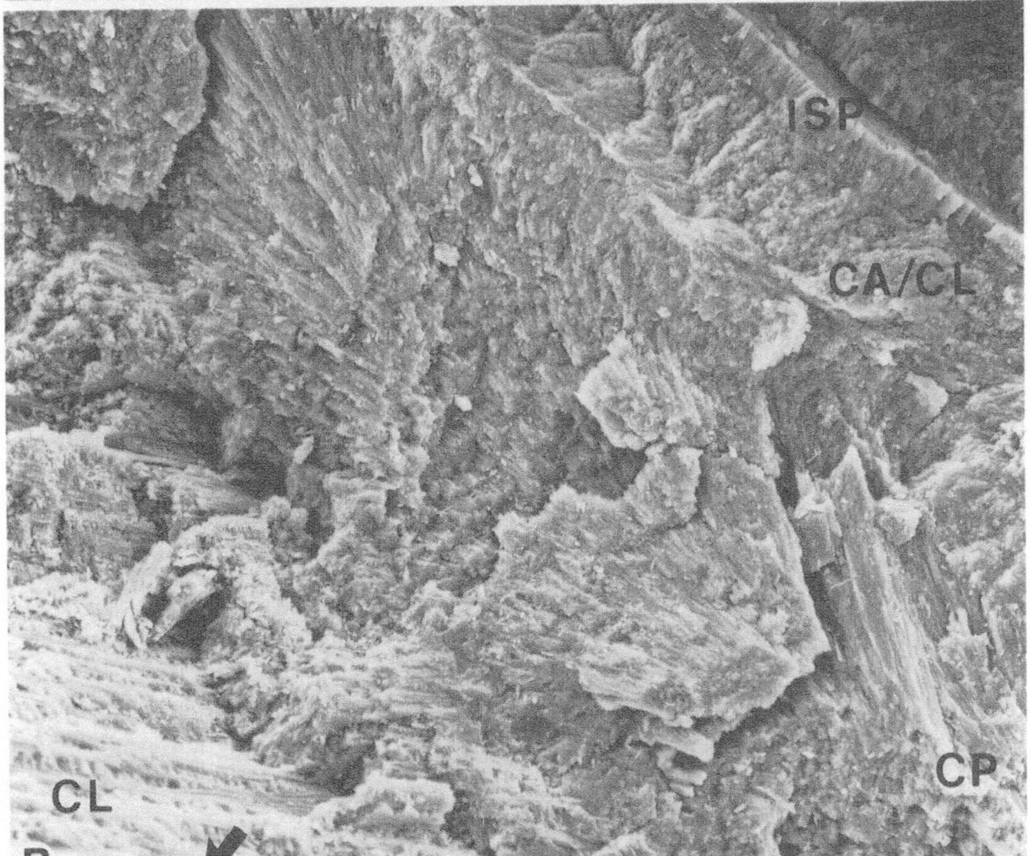
ISP
CA/CL
CL
CP
B

A

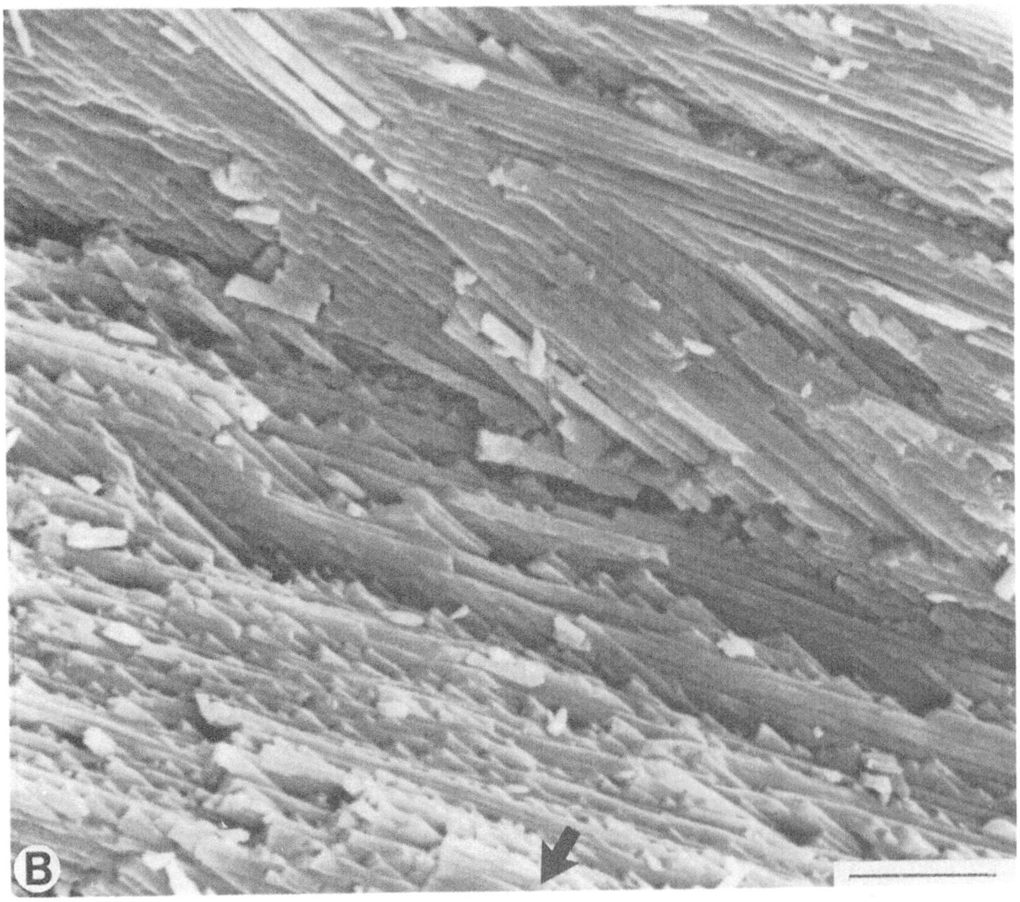
B

A

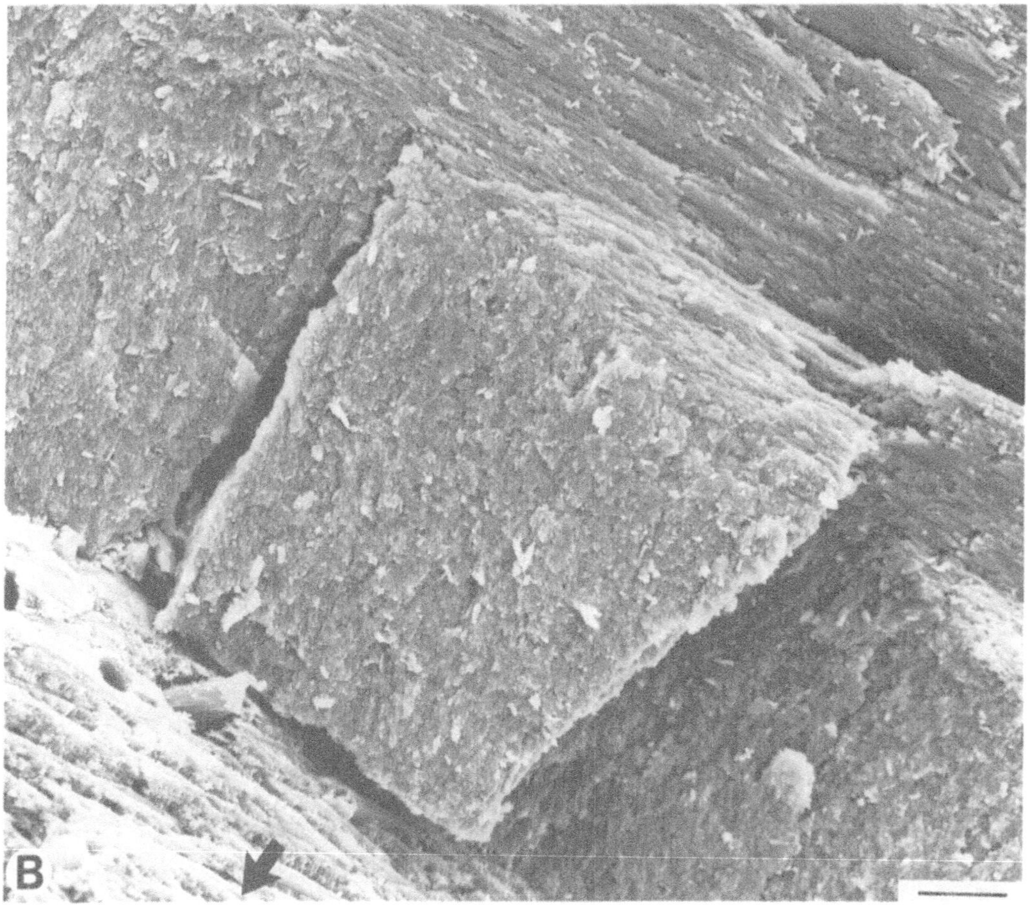
B

PLATE 126

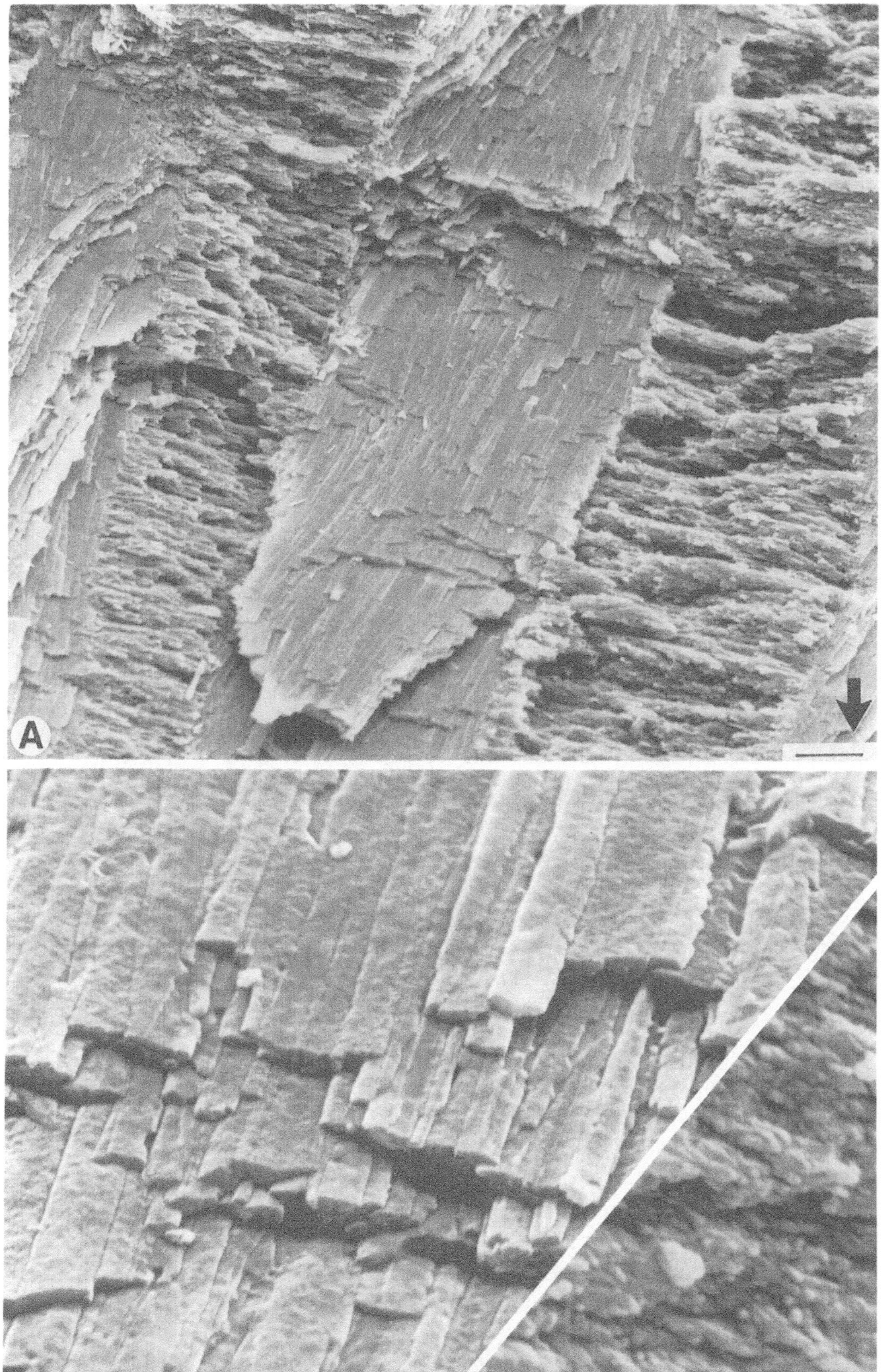
A
B
C

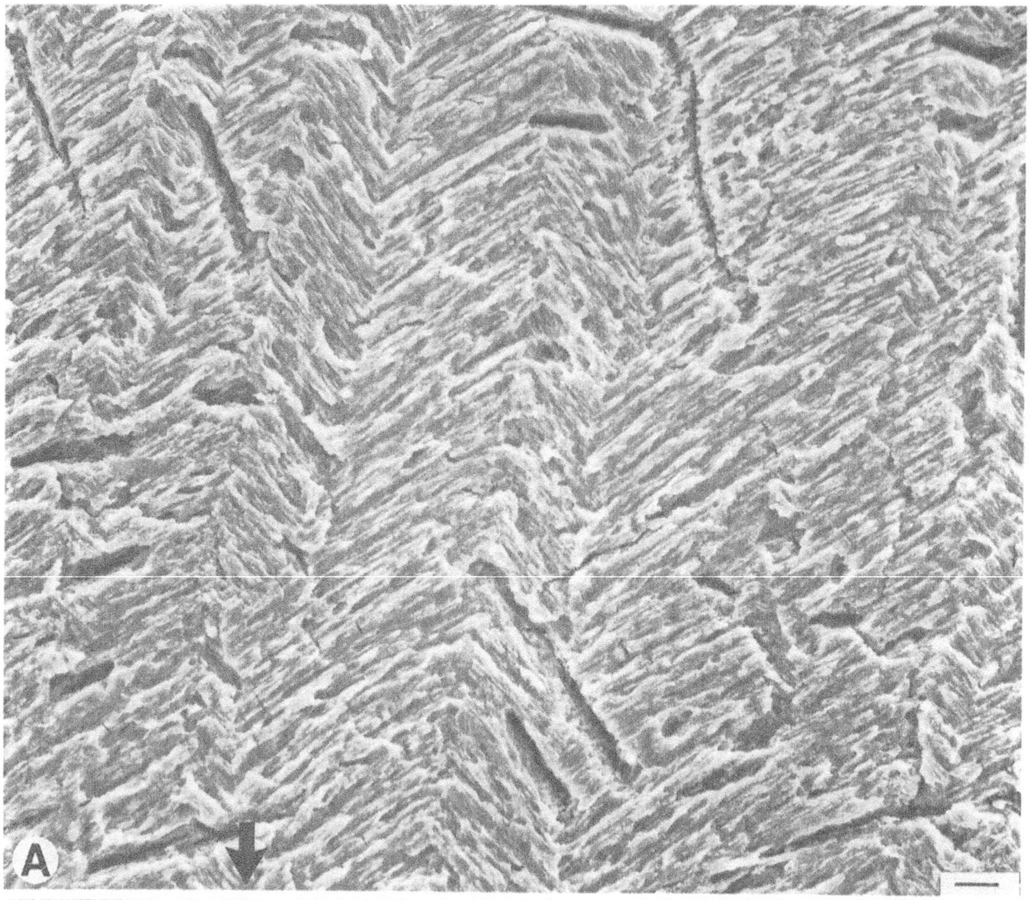
A

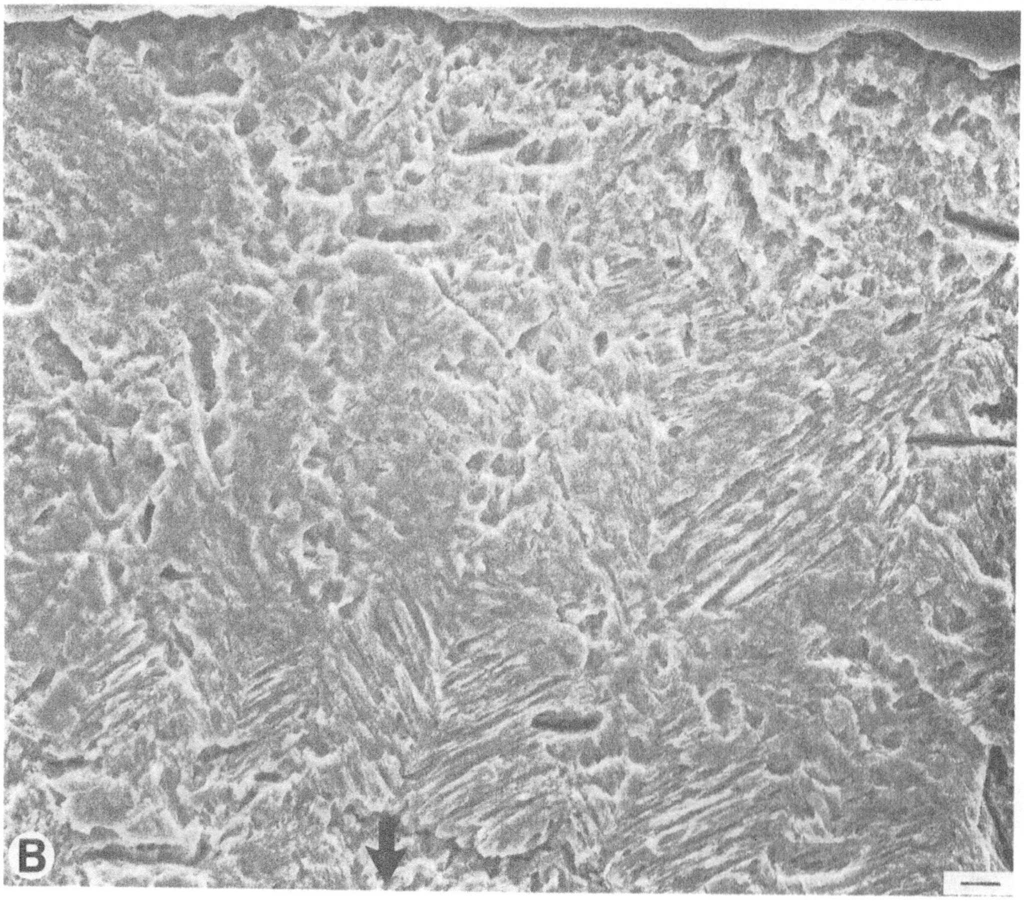
B

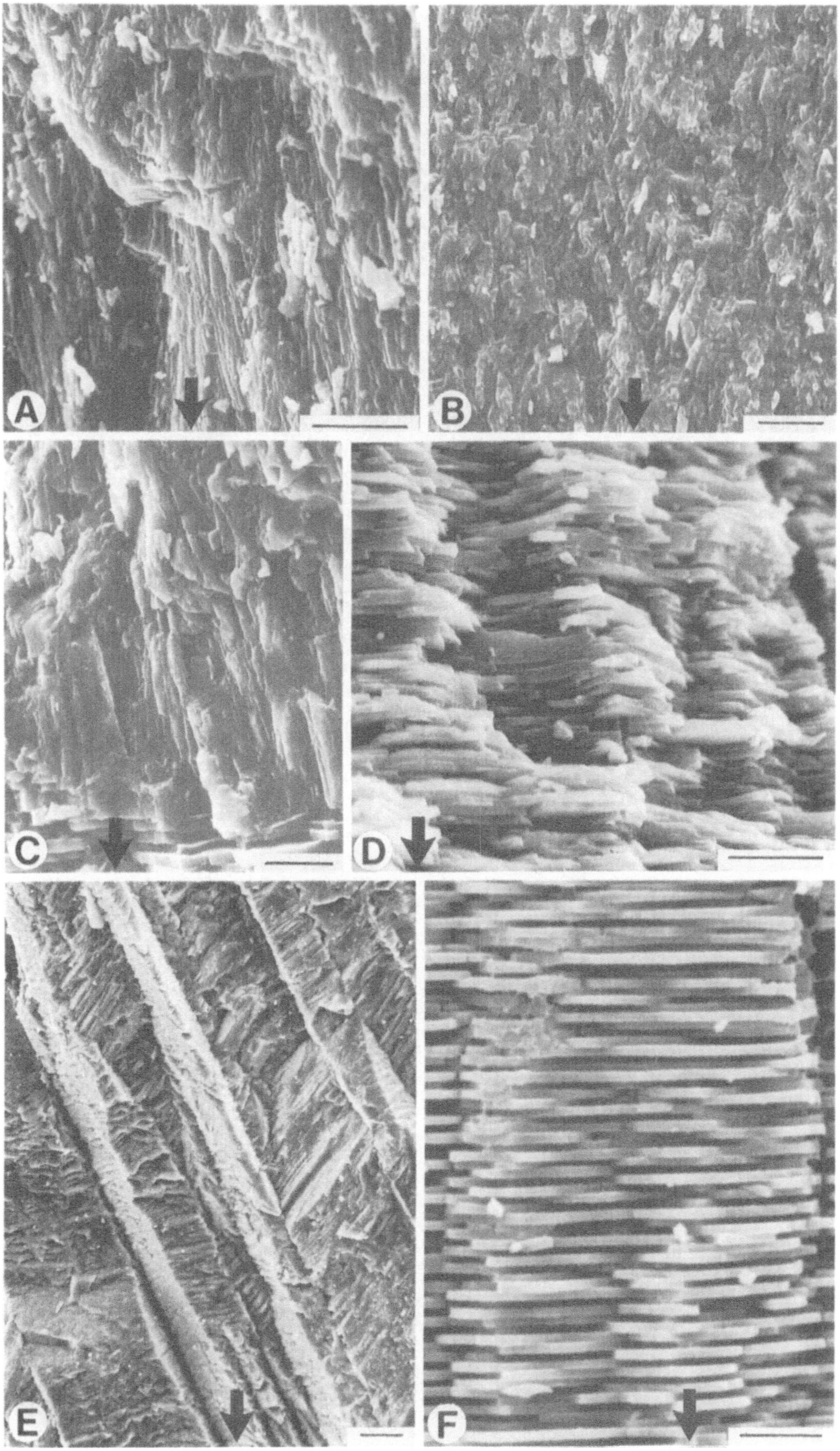
A
B
C
D
E
F

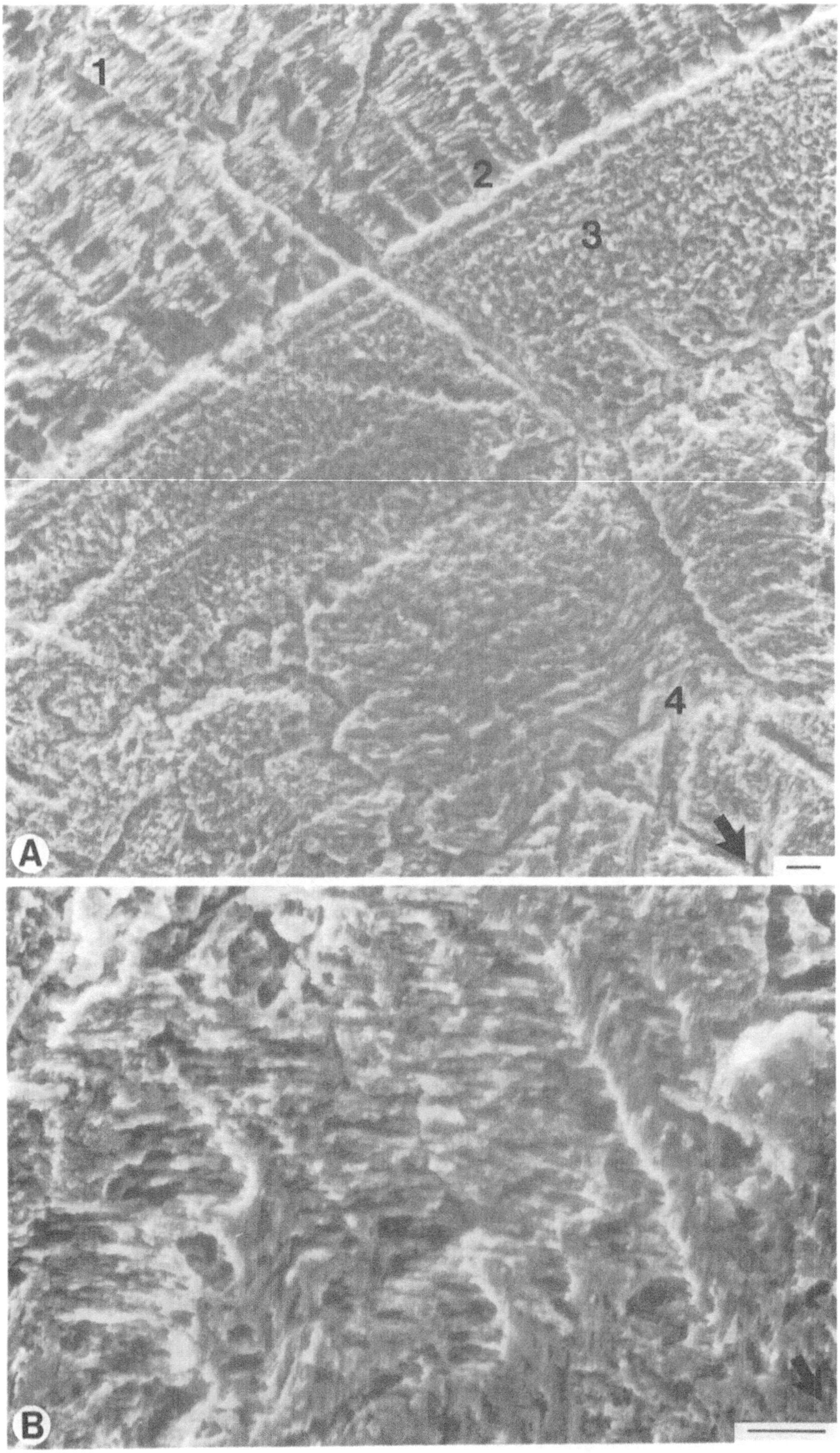
1
2
3
4
A
B

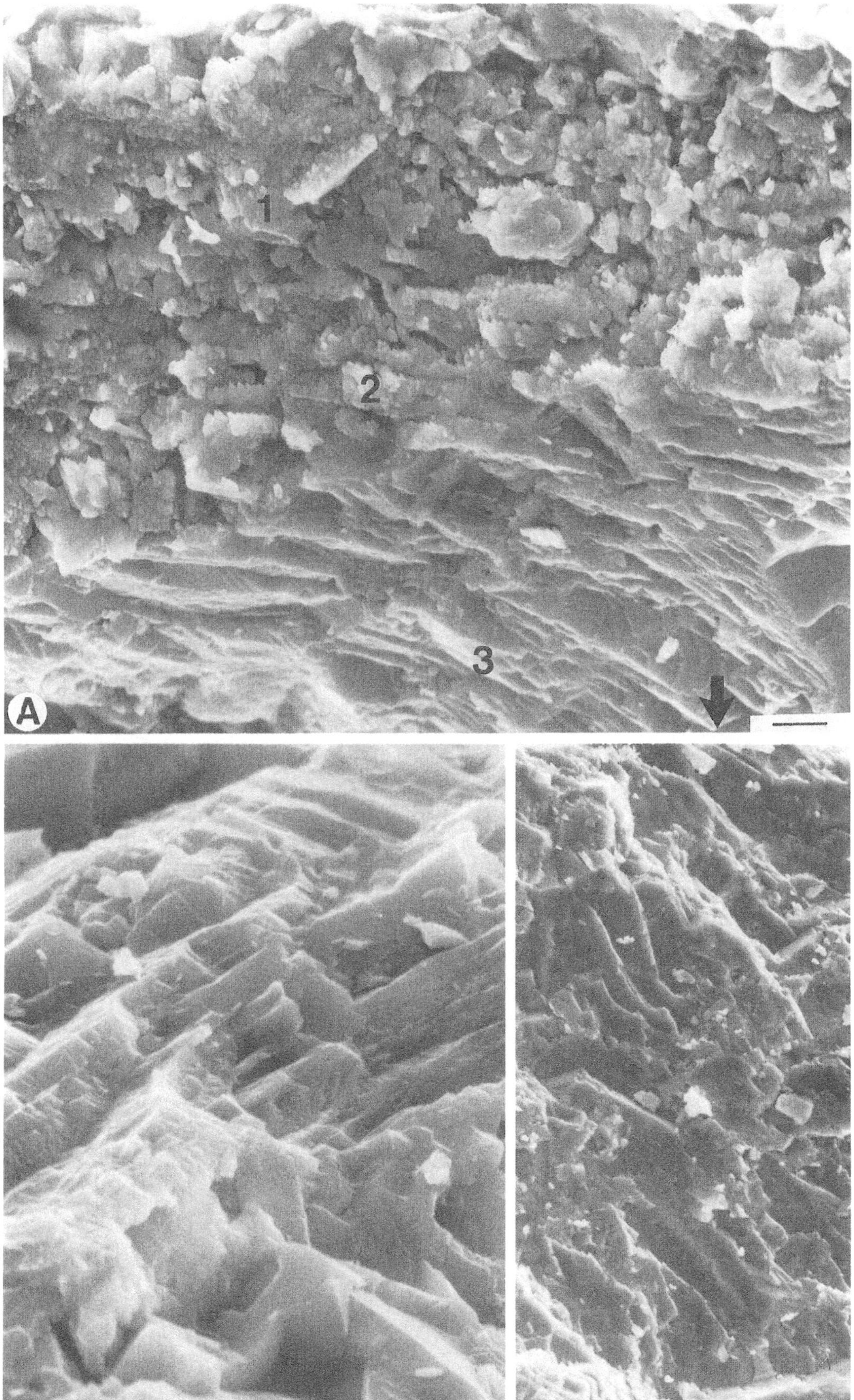
1
2
3
A
B
C

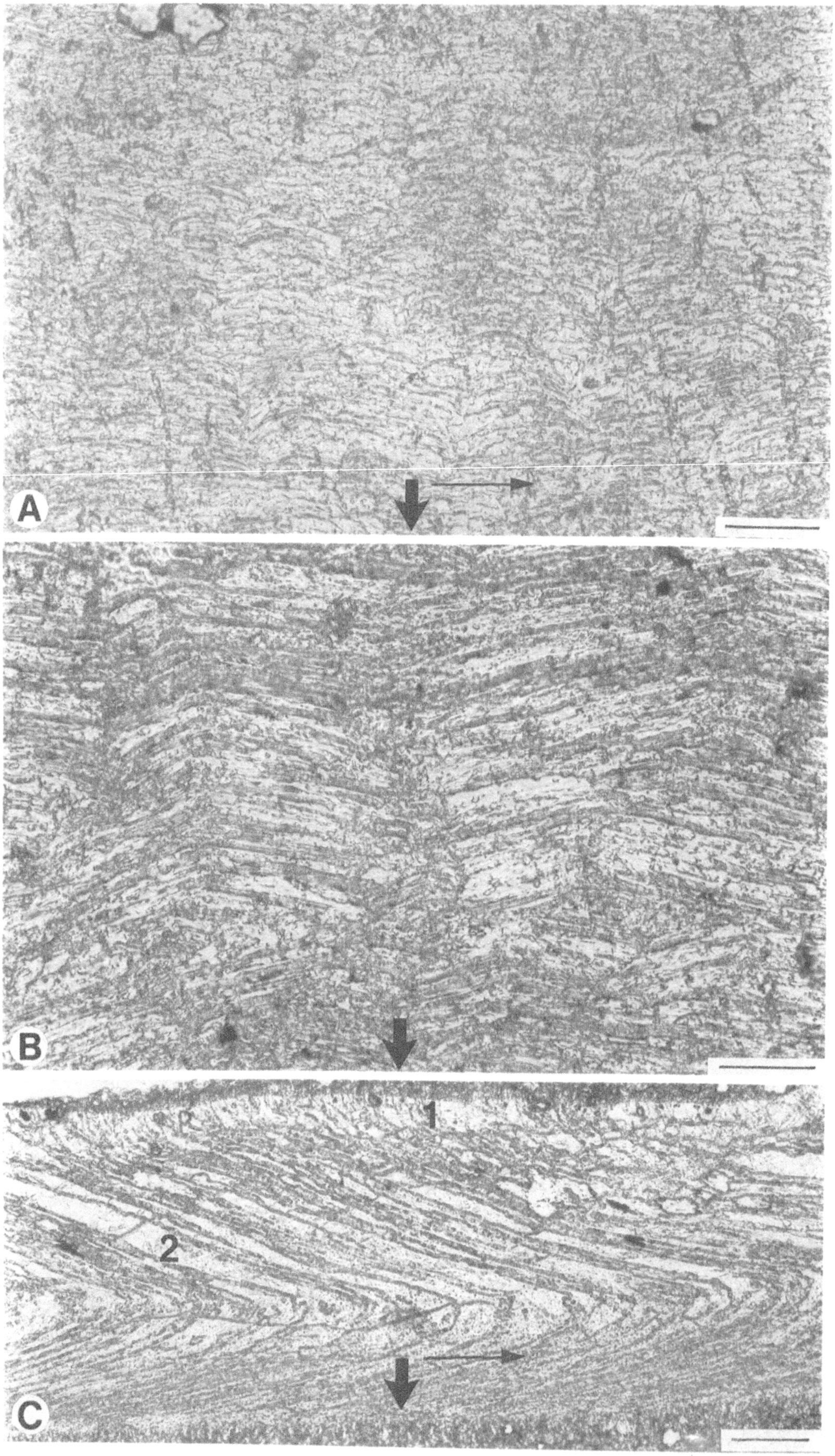
A
B
1
2
C

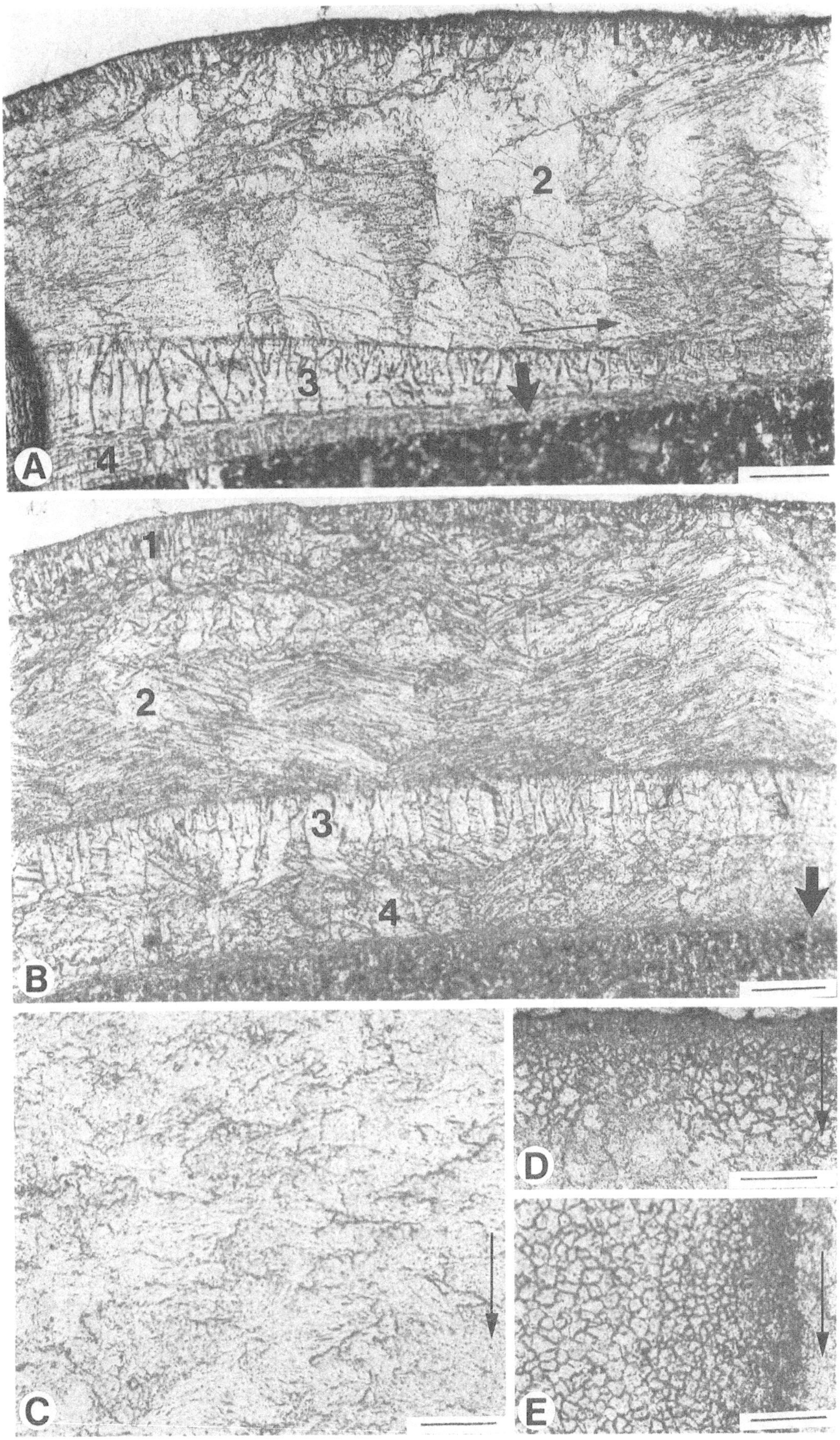
1
2
3
4
A
1
2
3
4
B
C
D
E

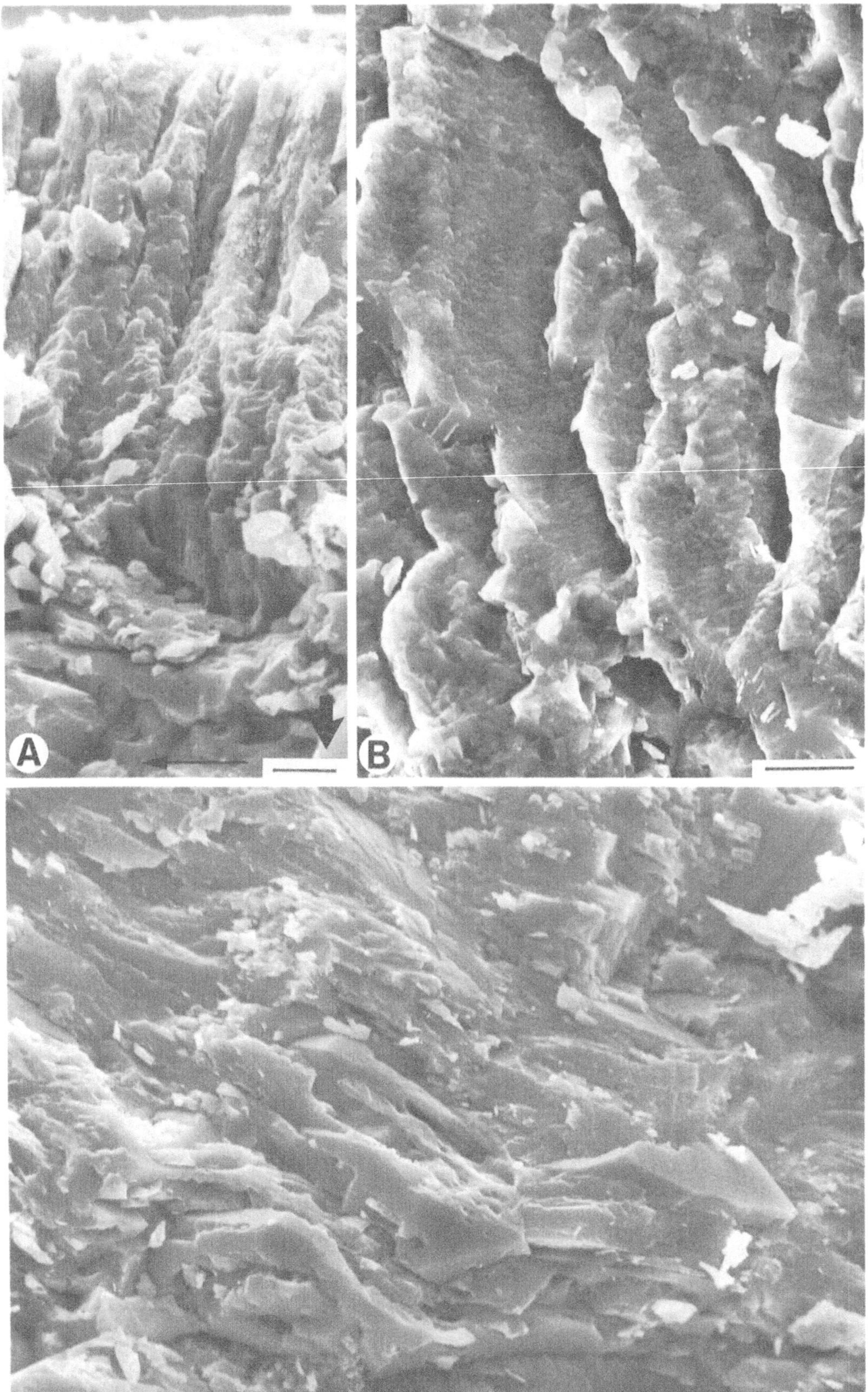
A
B
C

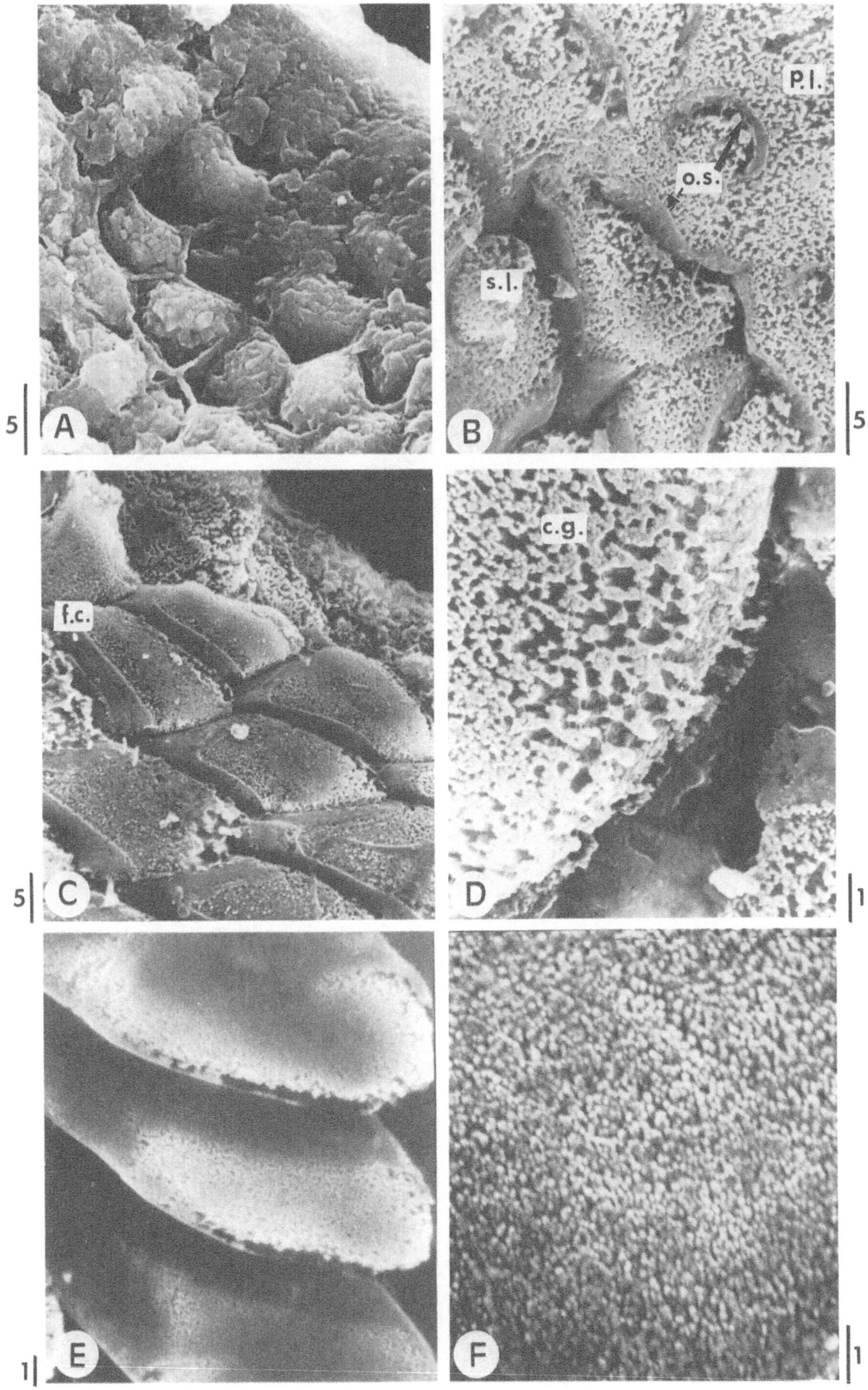
A
B
C
D
E
F
P.l.
o.s.
s.l.
f.c.
c.g.
5
5
5
1
1
1

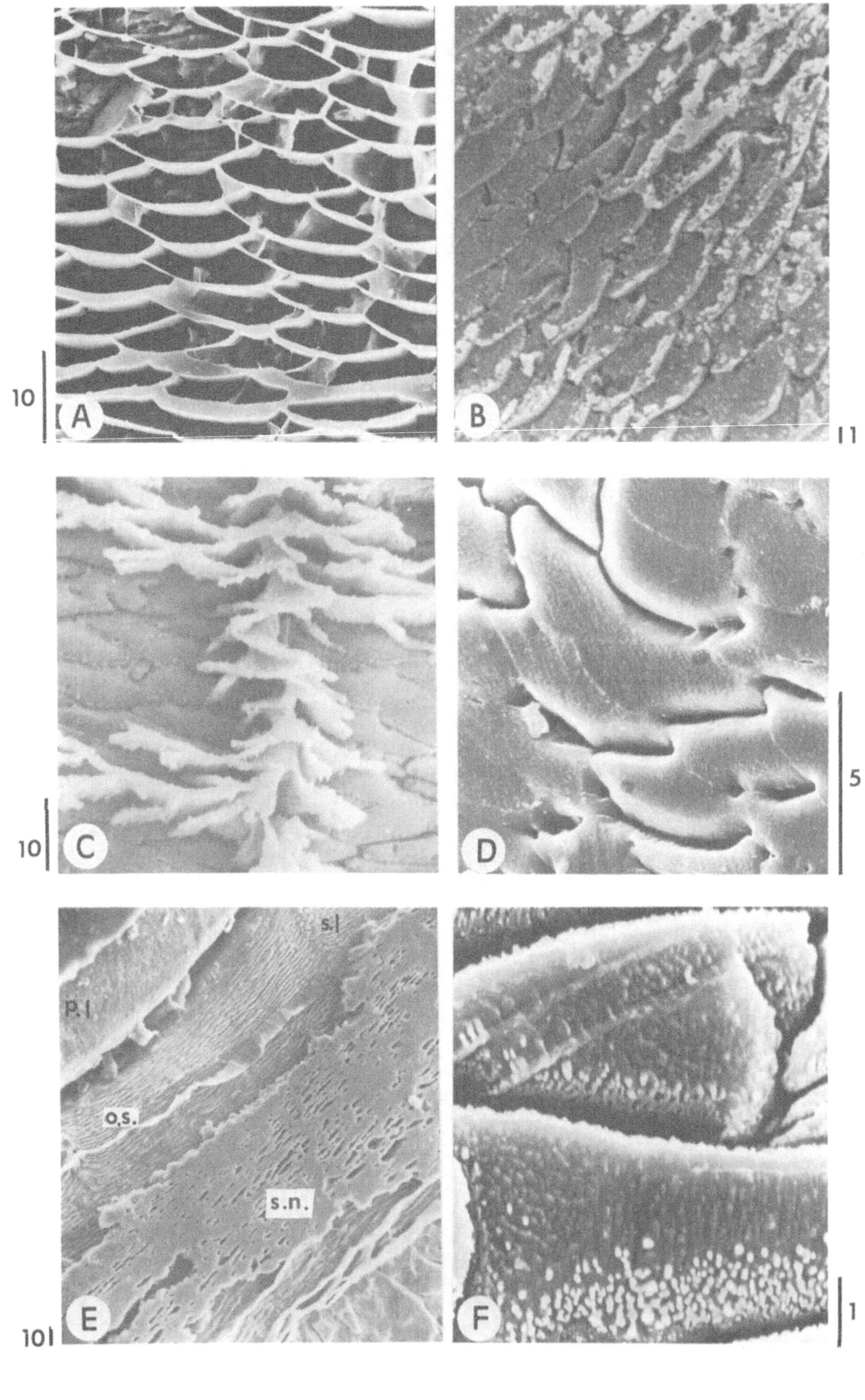
10
A
B
1
C
10
D
5
s.
p.
o.s.
s.n.
E
10
F
1

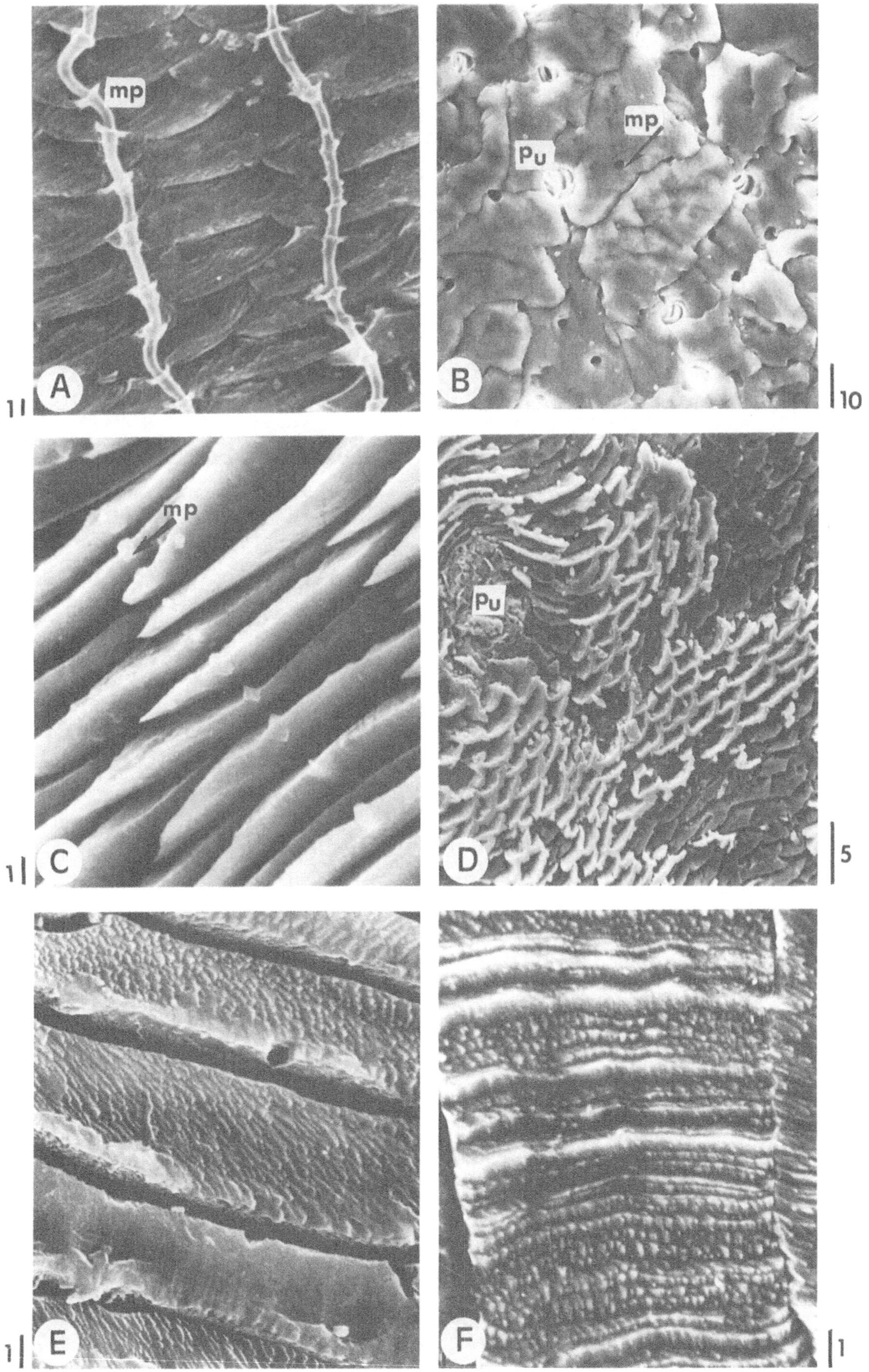
mp
A
1
Pu
mp
B
10
mp
C
1
Pu
D
5
E
1
F
1

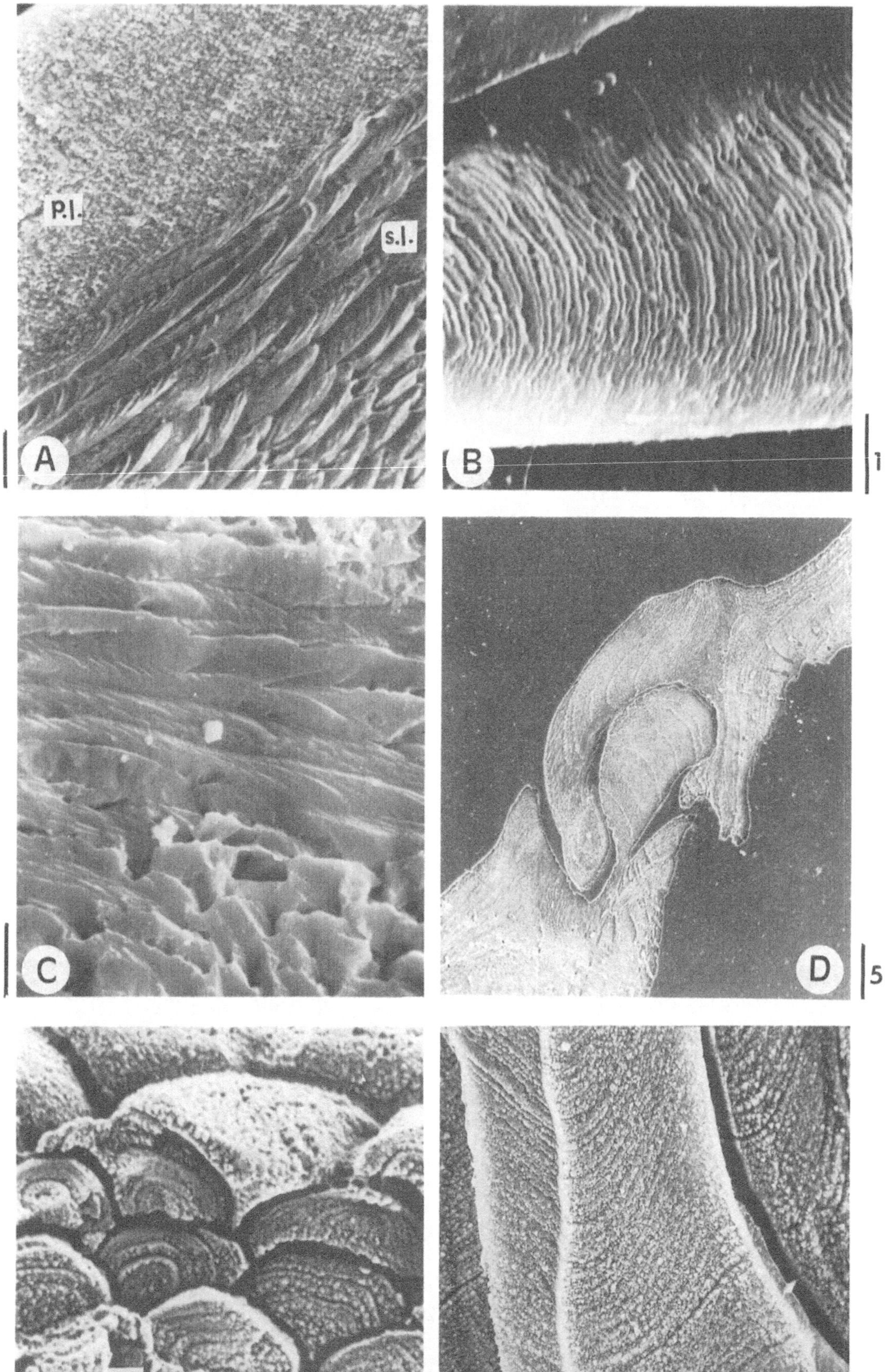
p.l.
s.l.
5
A
B
1
5
C
D
5
p.l.
1
E
F
1

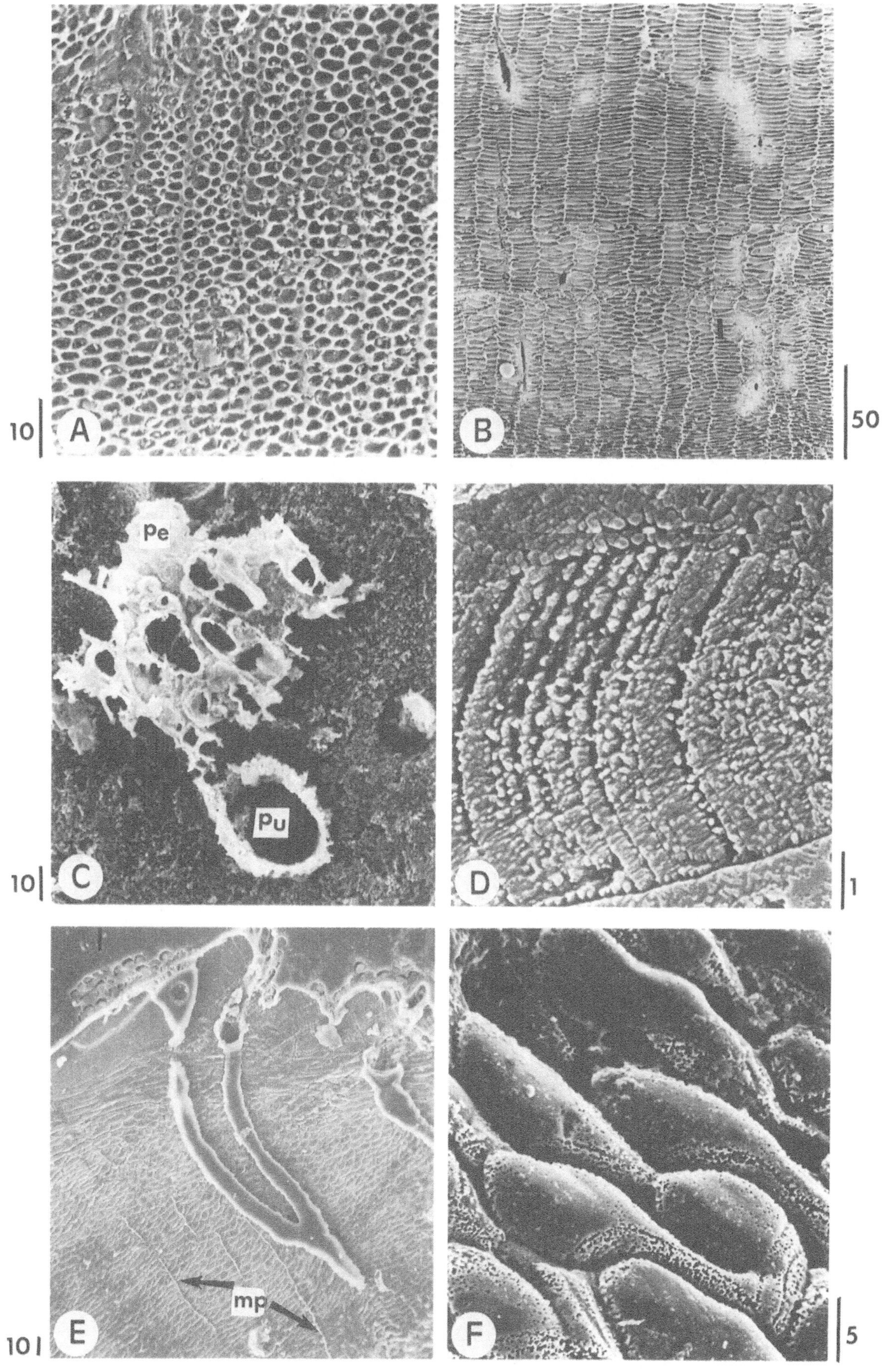
10
A
B
50
pe
Pu
10
C
D
1
mp
10
E
F
5

PLATE 140

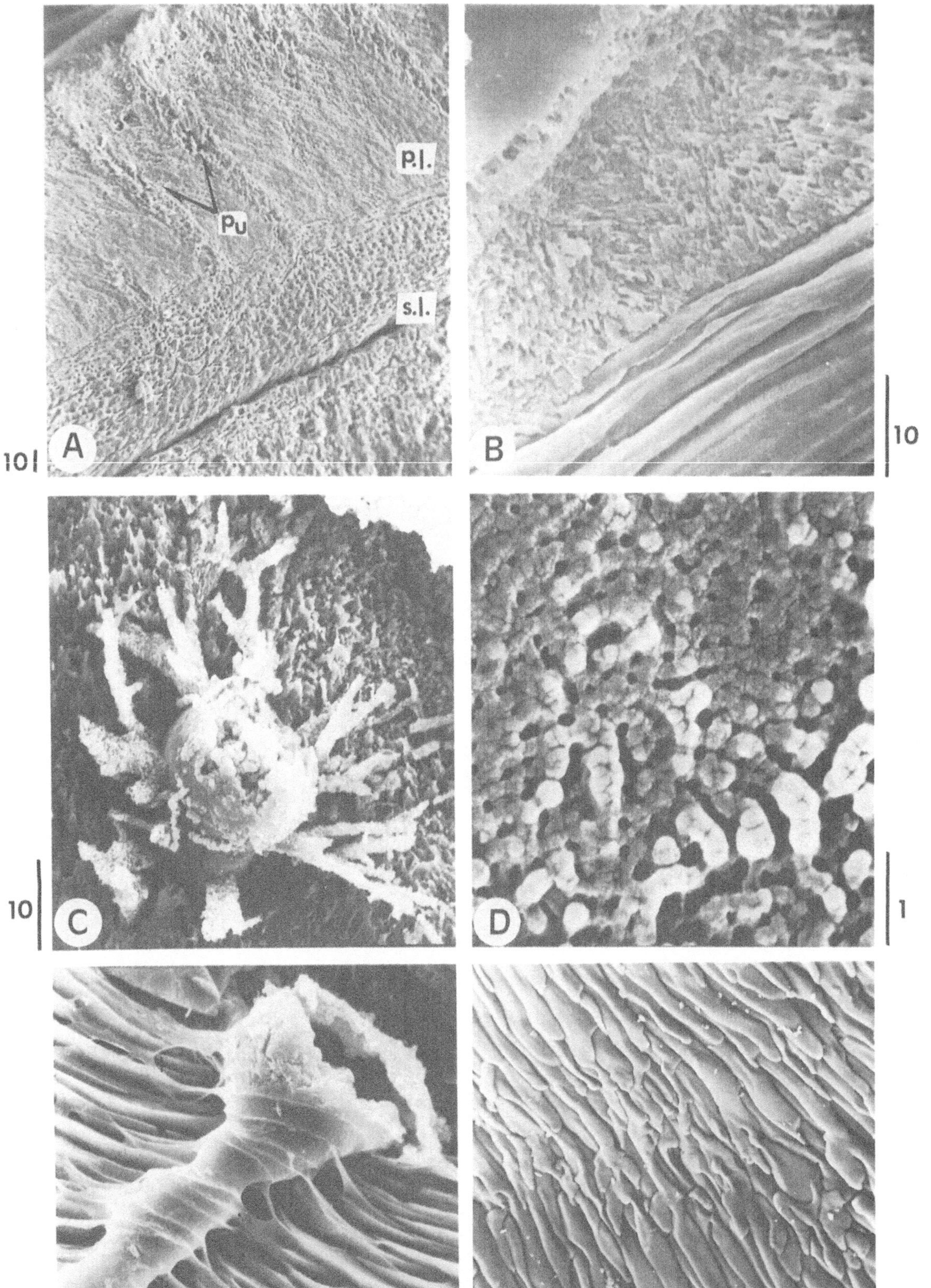

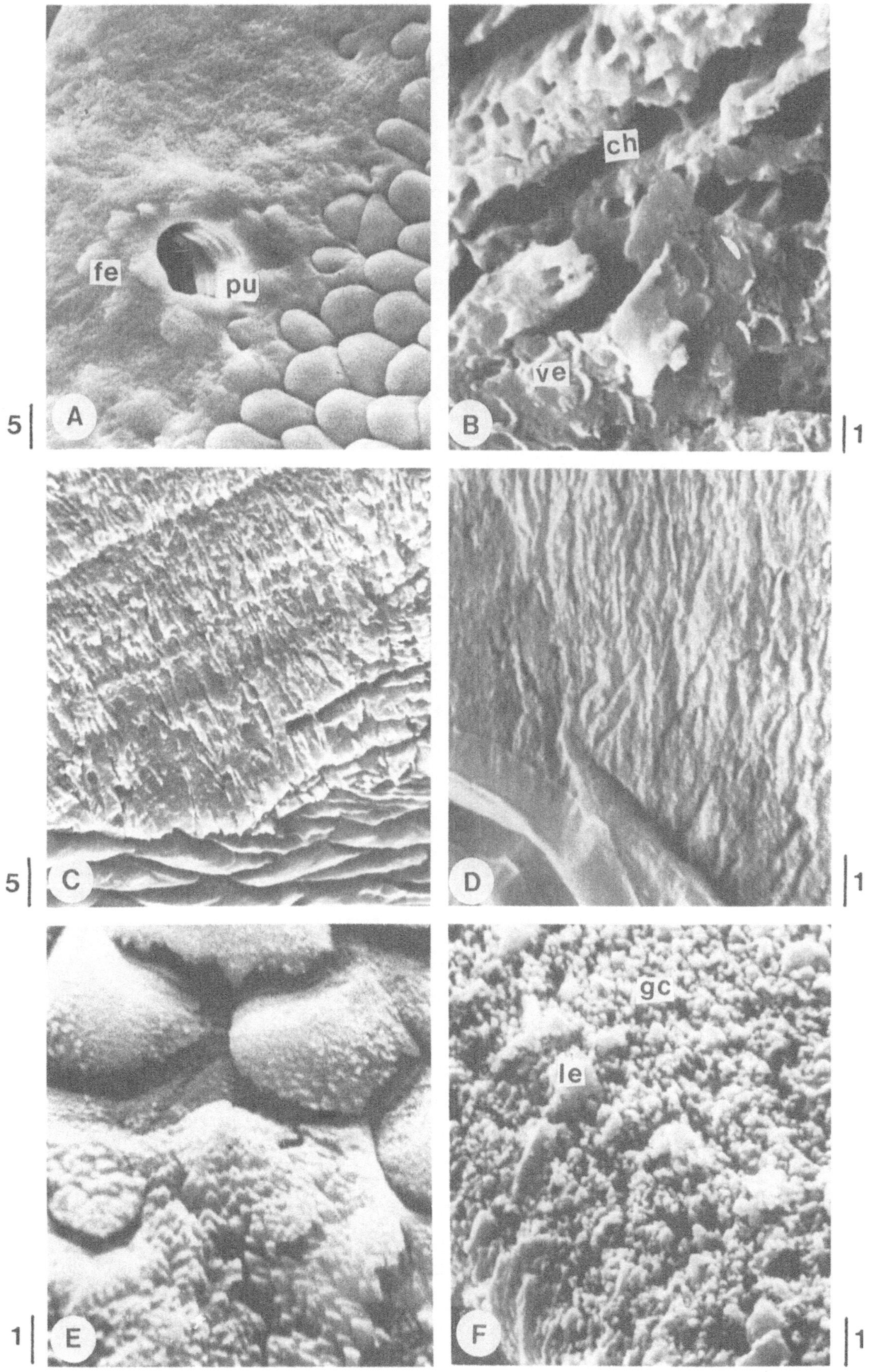
fe
pu
5
A
ch
ve
B
1
5
C
D
1
1
E
gc
le
F
1

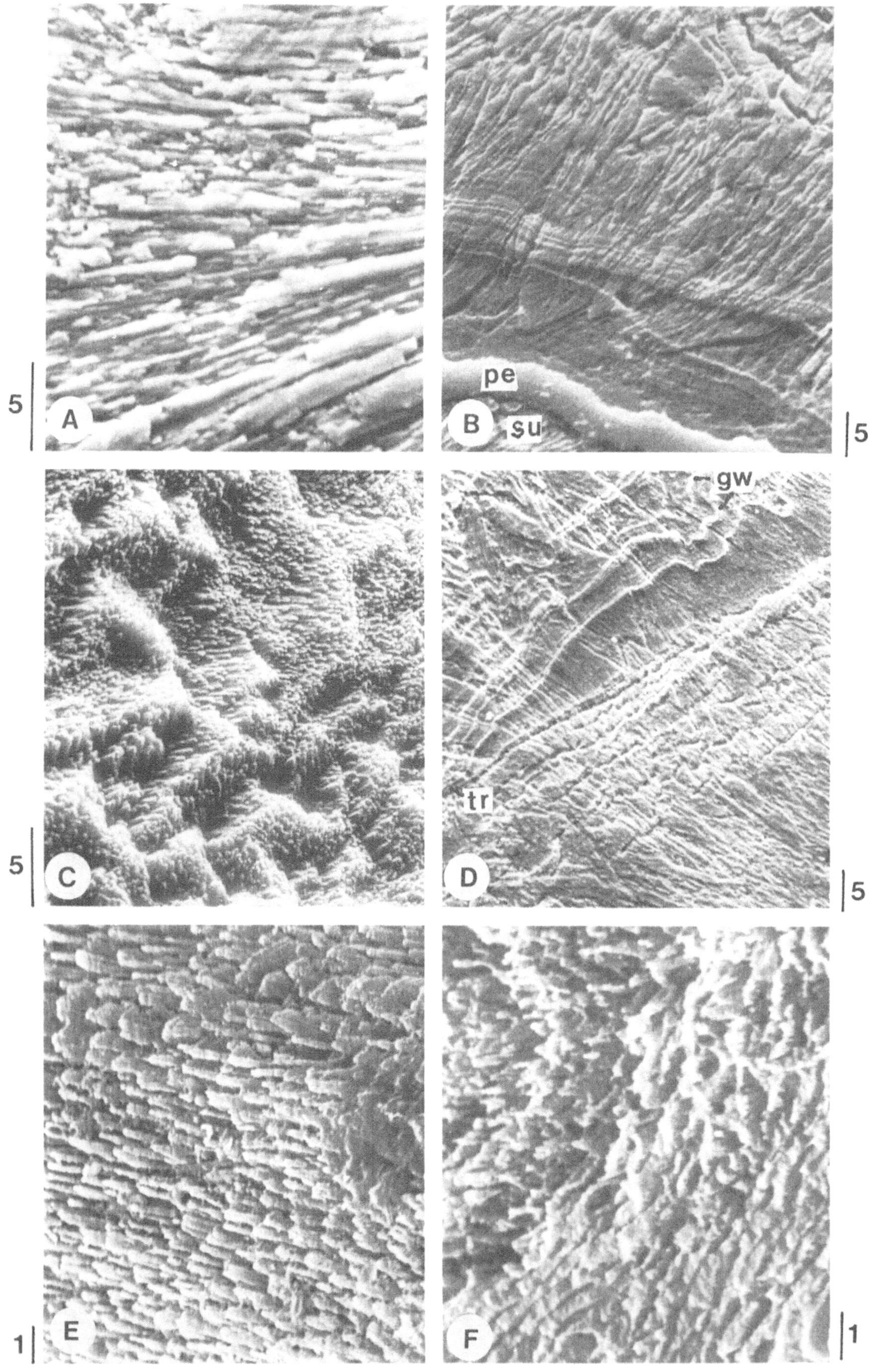
5
A
pe
B
su
5
C
5
gw
tr
D
5
1
E
F
1

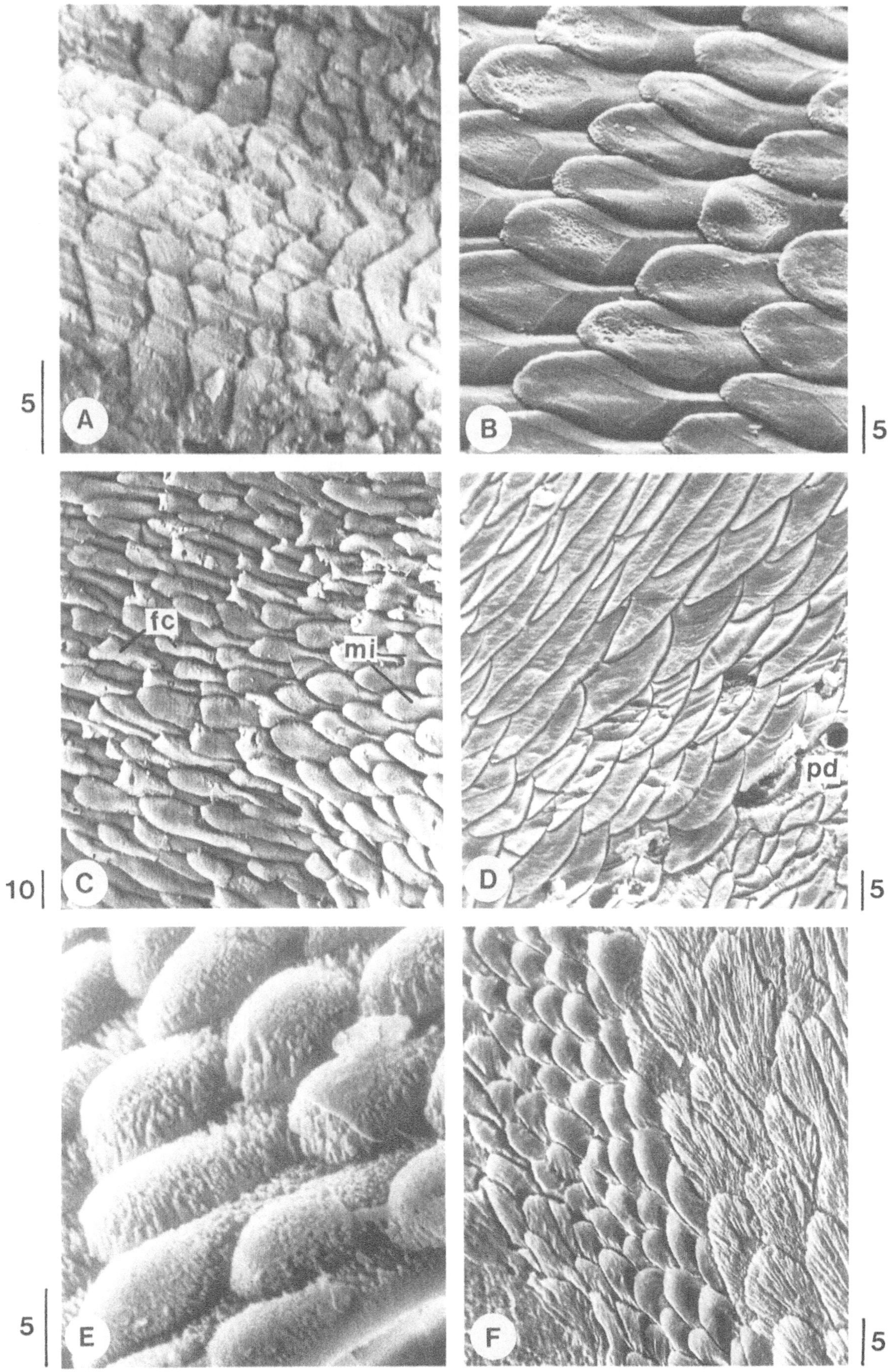
5
A
B
5
fc
mi
pd
10
C
D
5
5
E
F
5

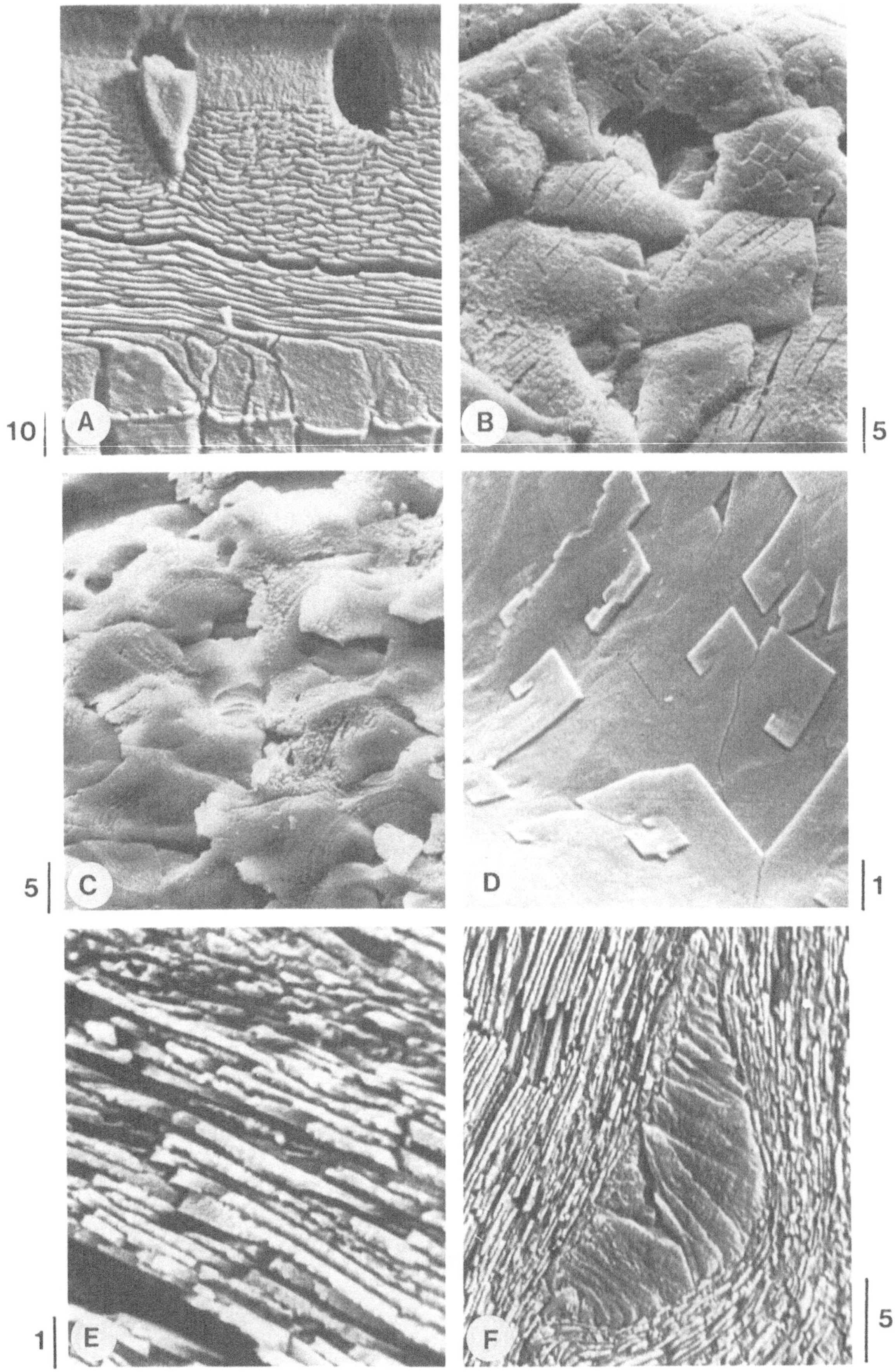
10
A
B
5
5
C
D
1
1
E
F
5

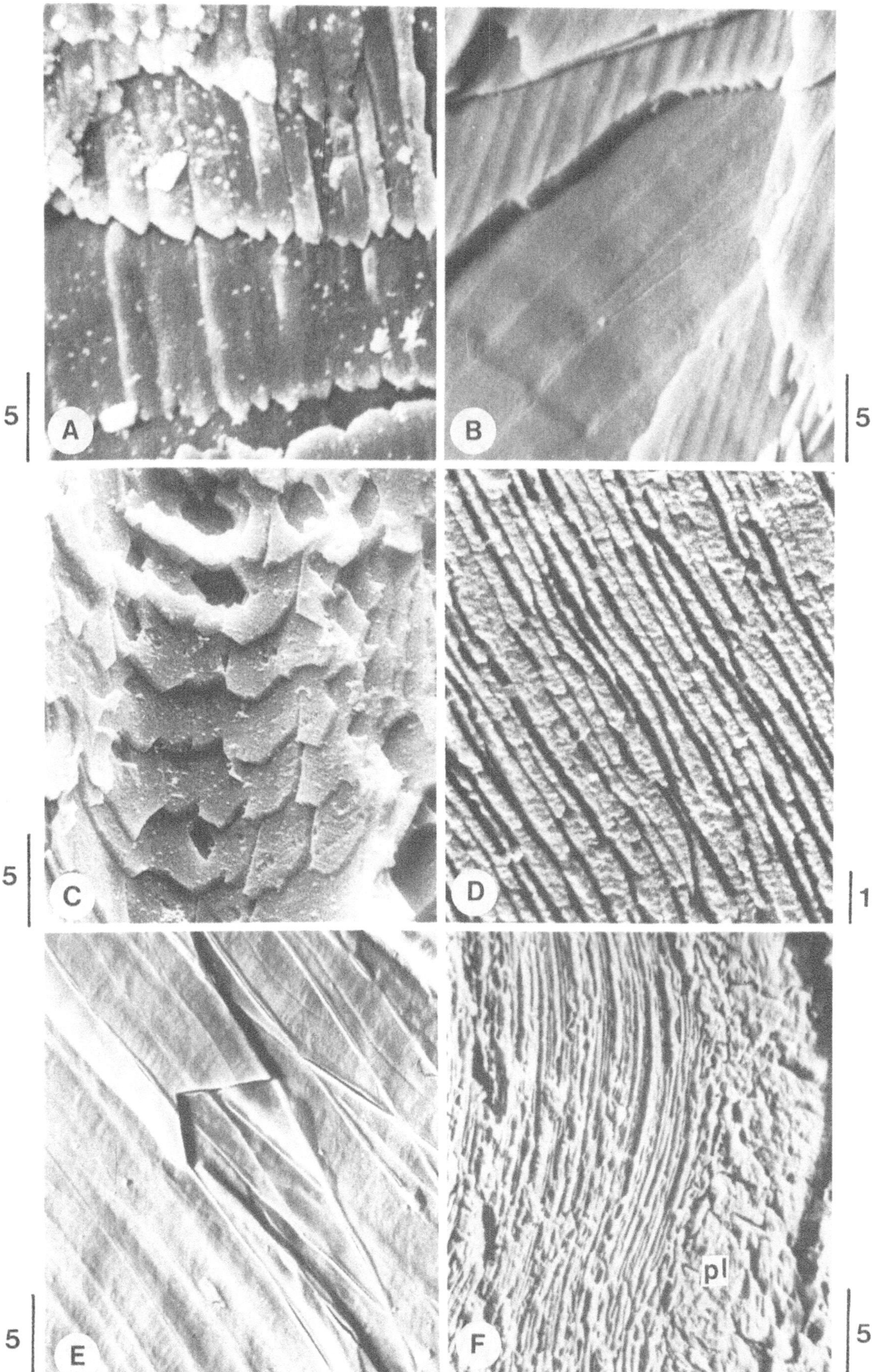
5
A
B
5
5
C
D
1
pl
5
E
F
5

PLATE 146

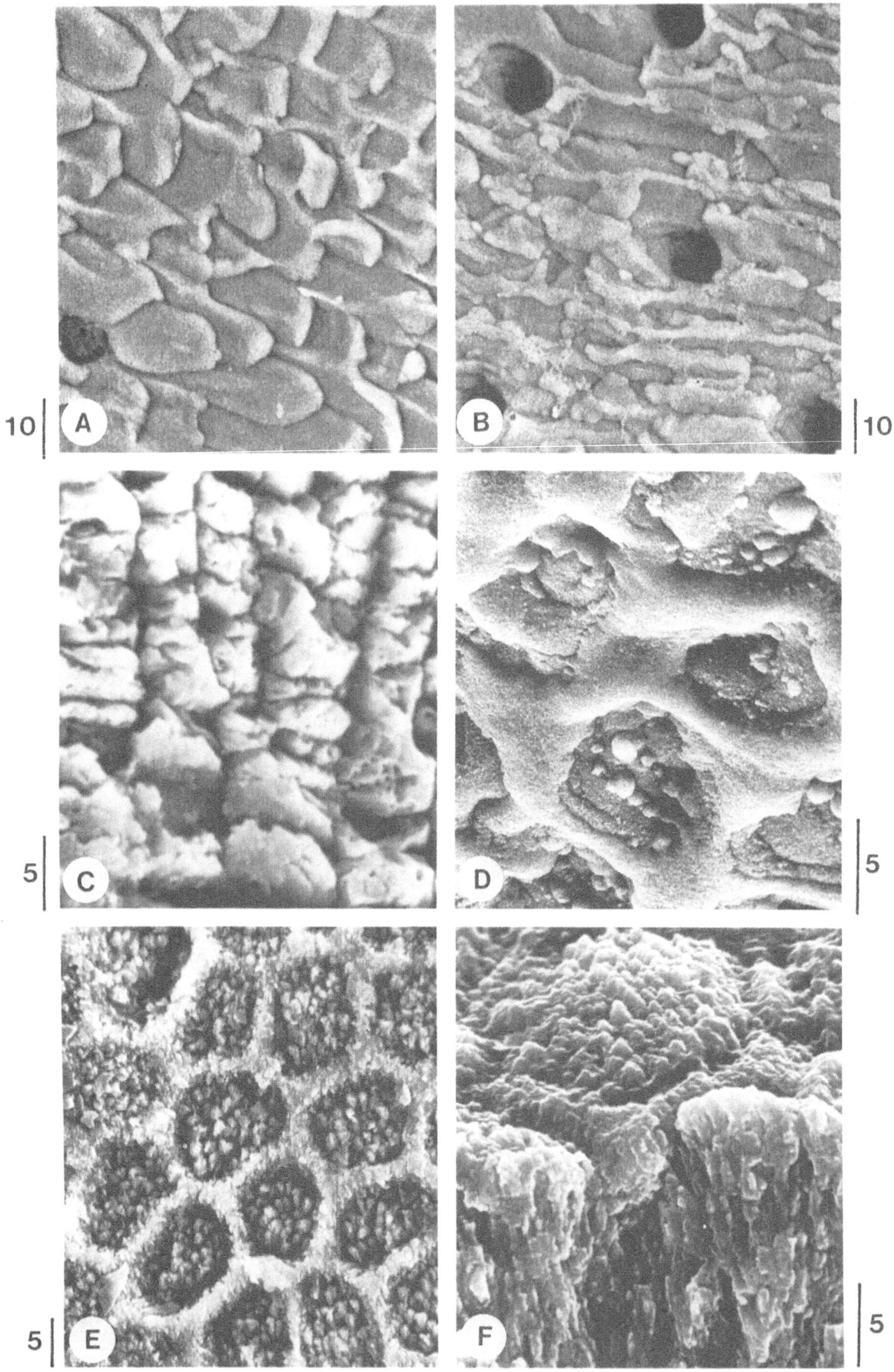

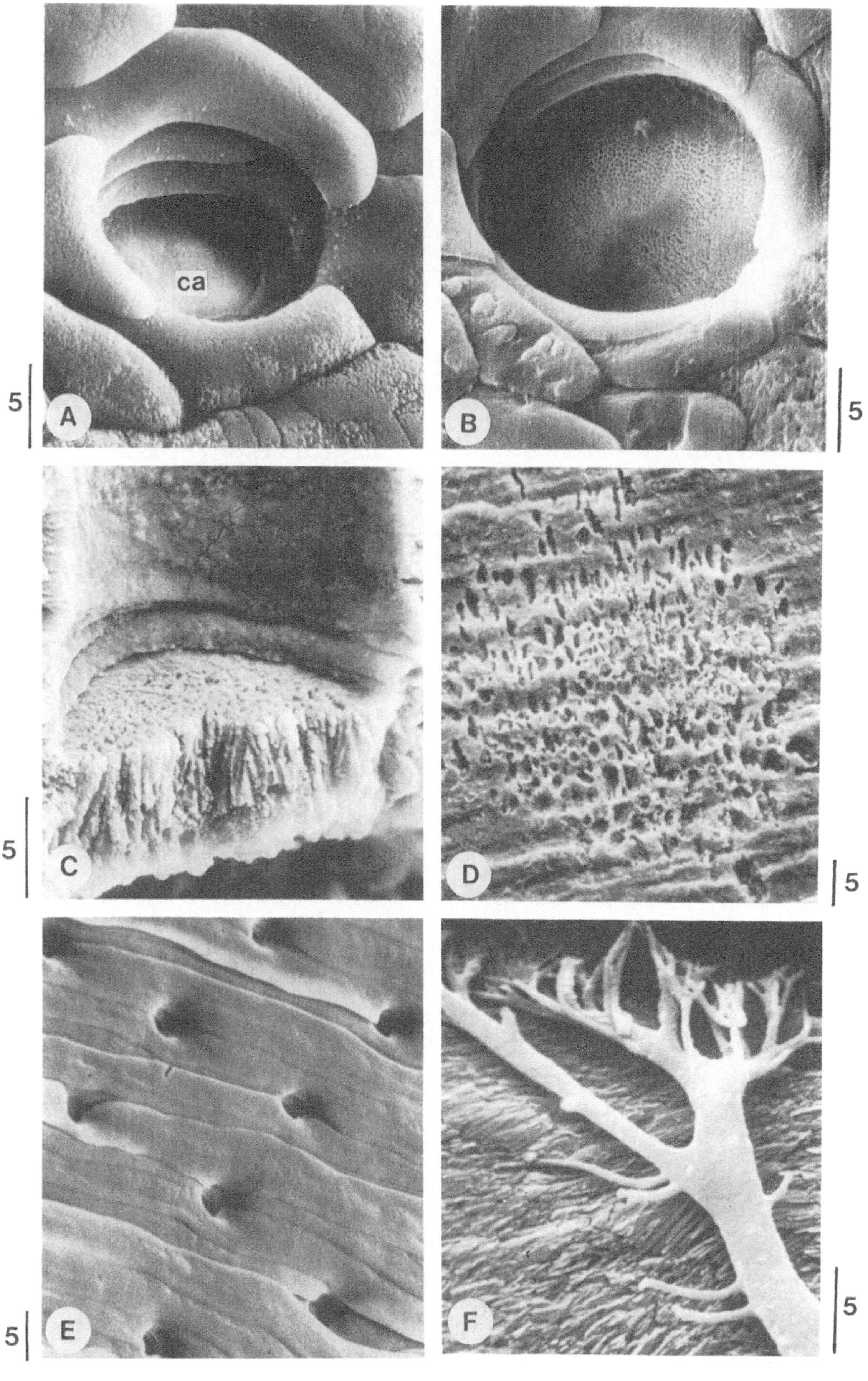
ca
5
A
B
5
5
C
D
5
5
E
F
5

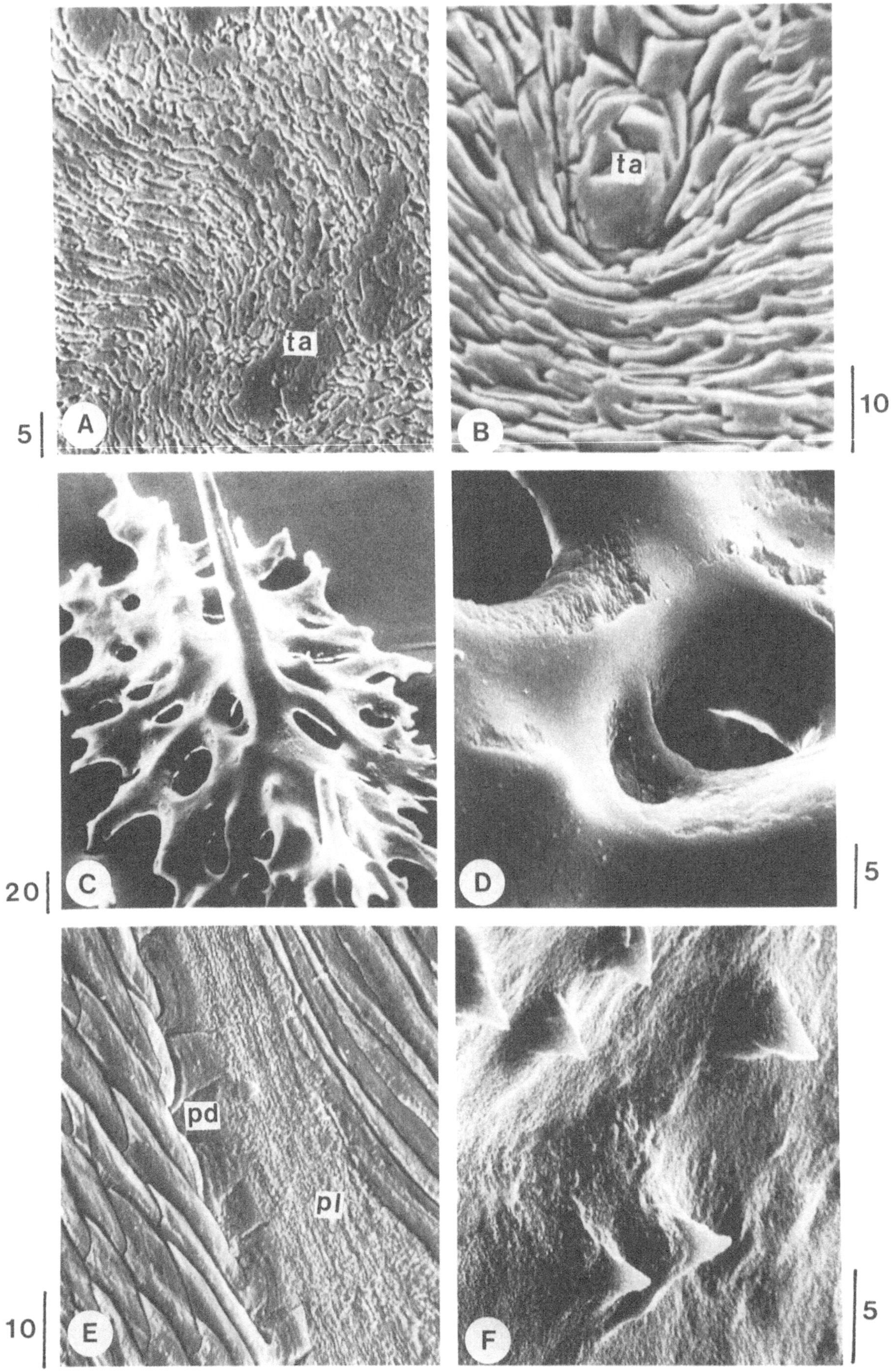
ta
A
5
ta
B
10
C
20
D
5
pd
pl
E
10
F
5

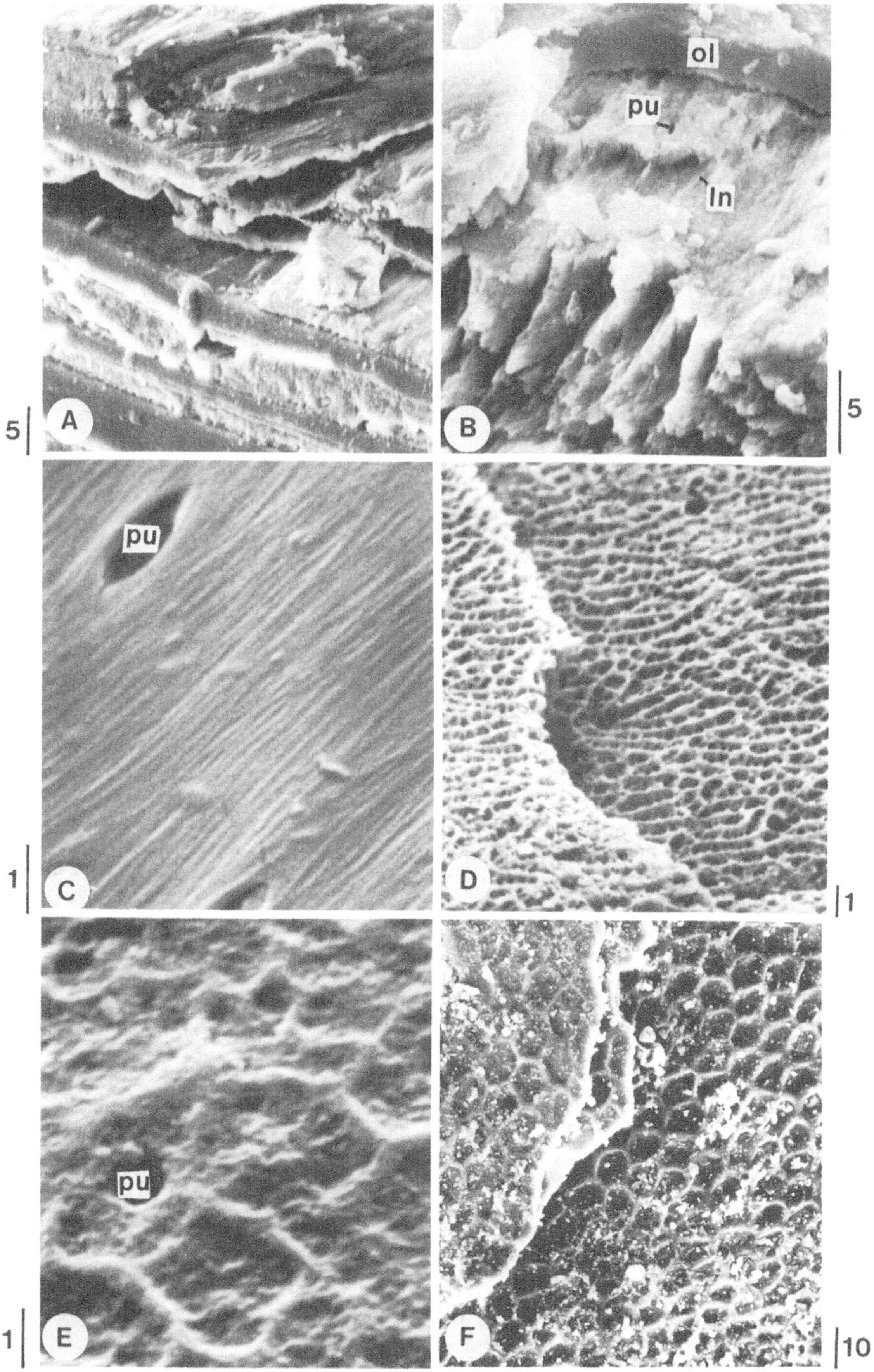
A
B
C
D
E
F
ol
pu
ln
pu
pu
5
5
1
1
1
10

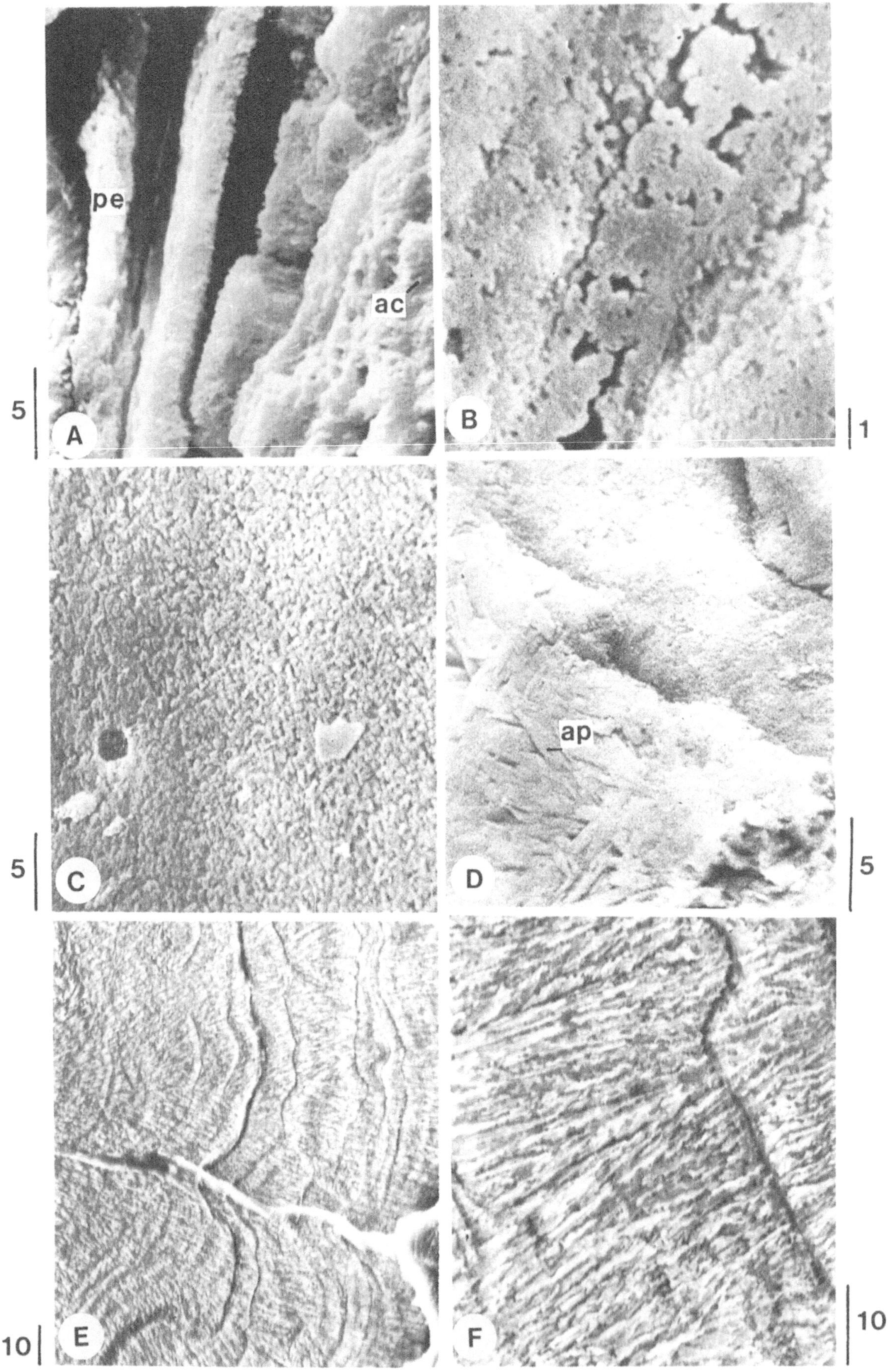
pe
ac
5
A
B
1
5
C
ap
D
5
10
E
F
10

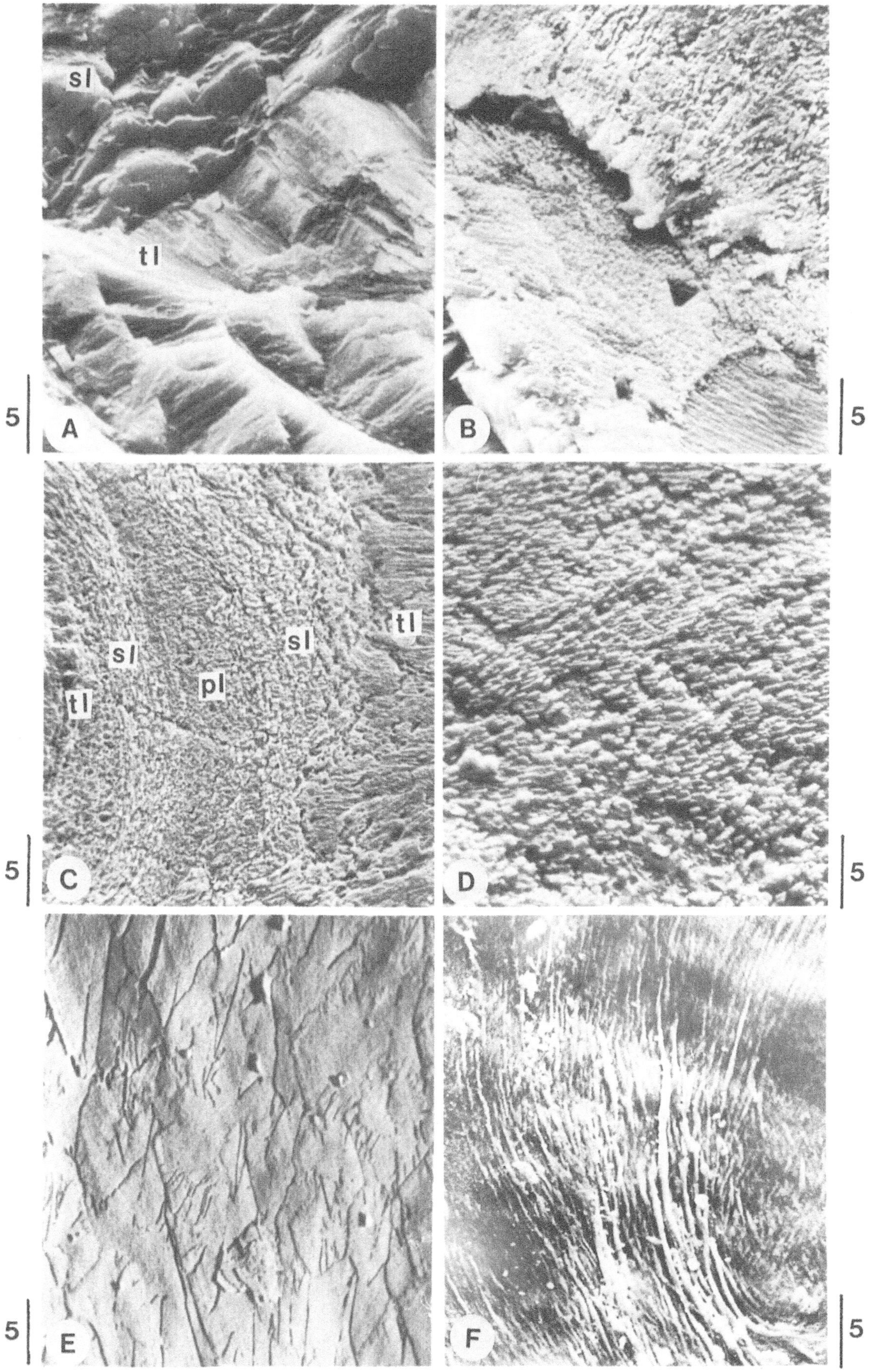
sl
tl
5
A
B
5
tl
sl
sl
pl
tl
5
C
D
5
5
E
F
5

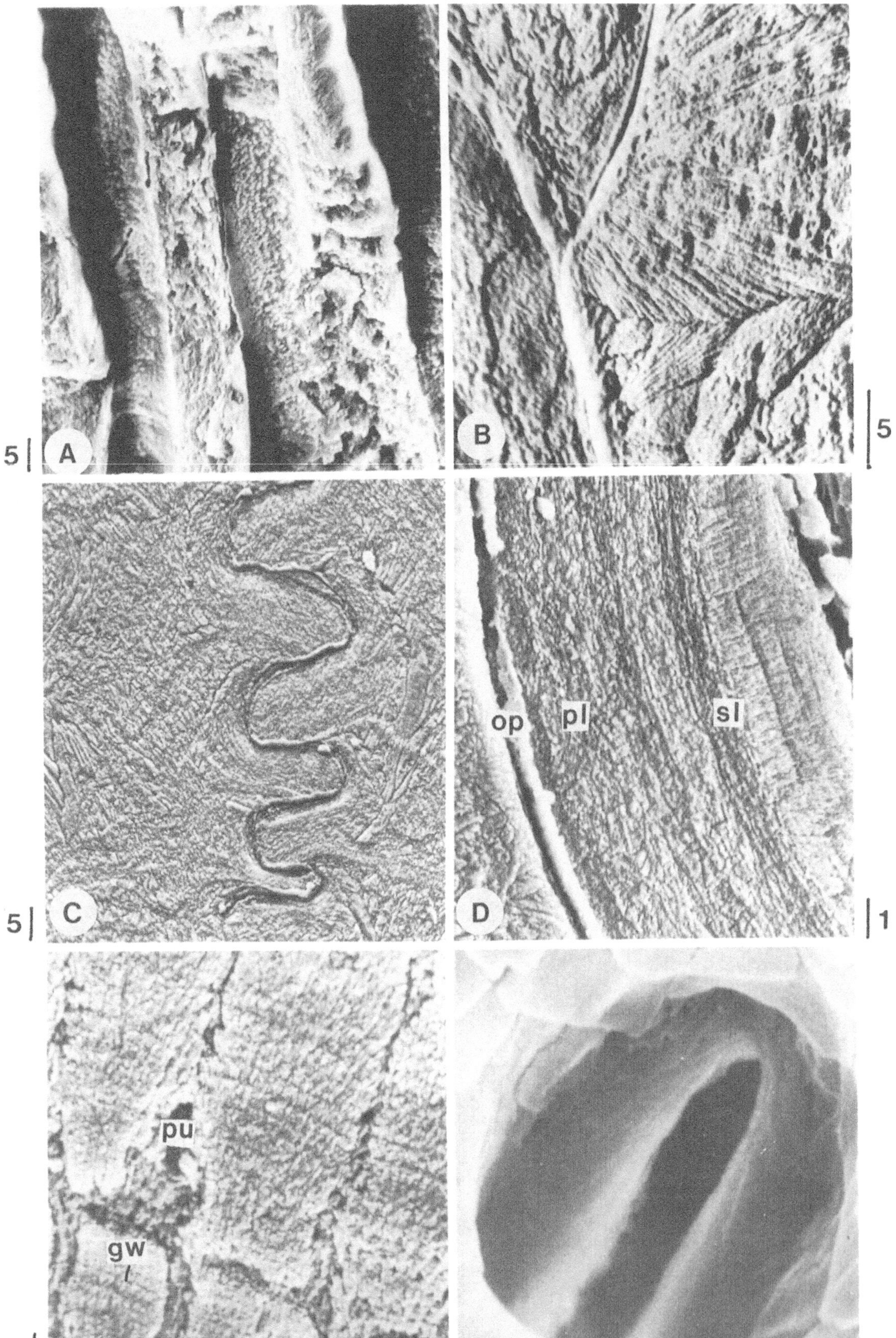
5
A
B
5
C
op
pl
sl
D
5
1
pu
gw
5
E
F
1

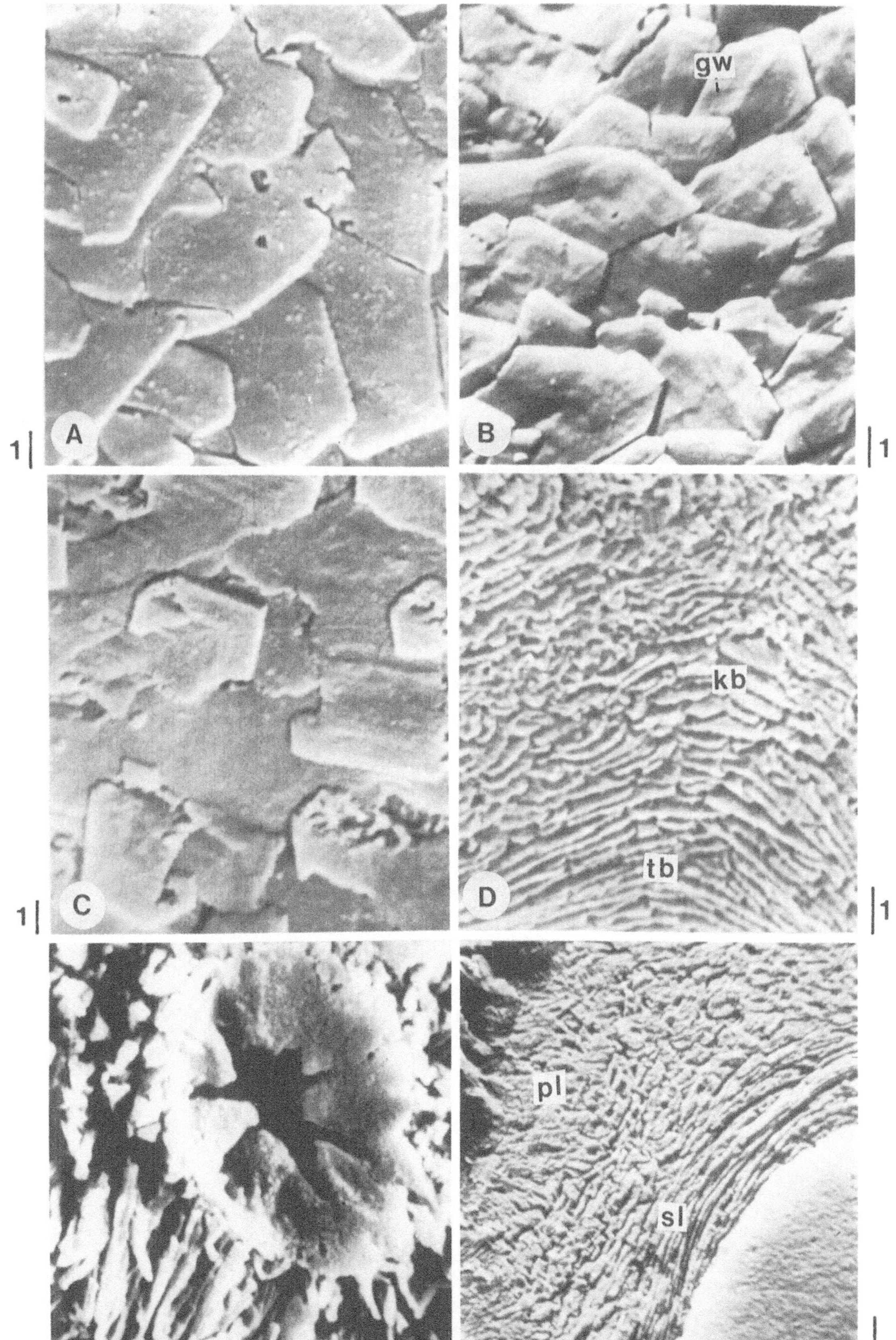
A
1
B
gw
1
C
1
D
kb
tb
1
E
1
F
pl
sl
5

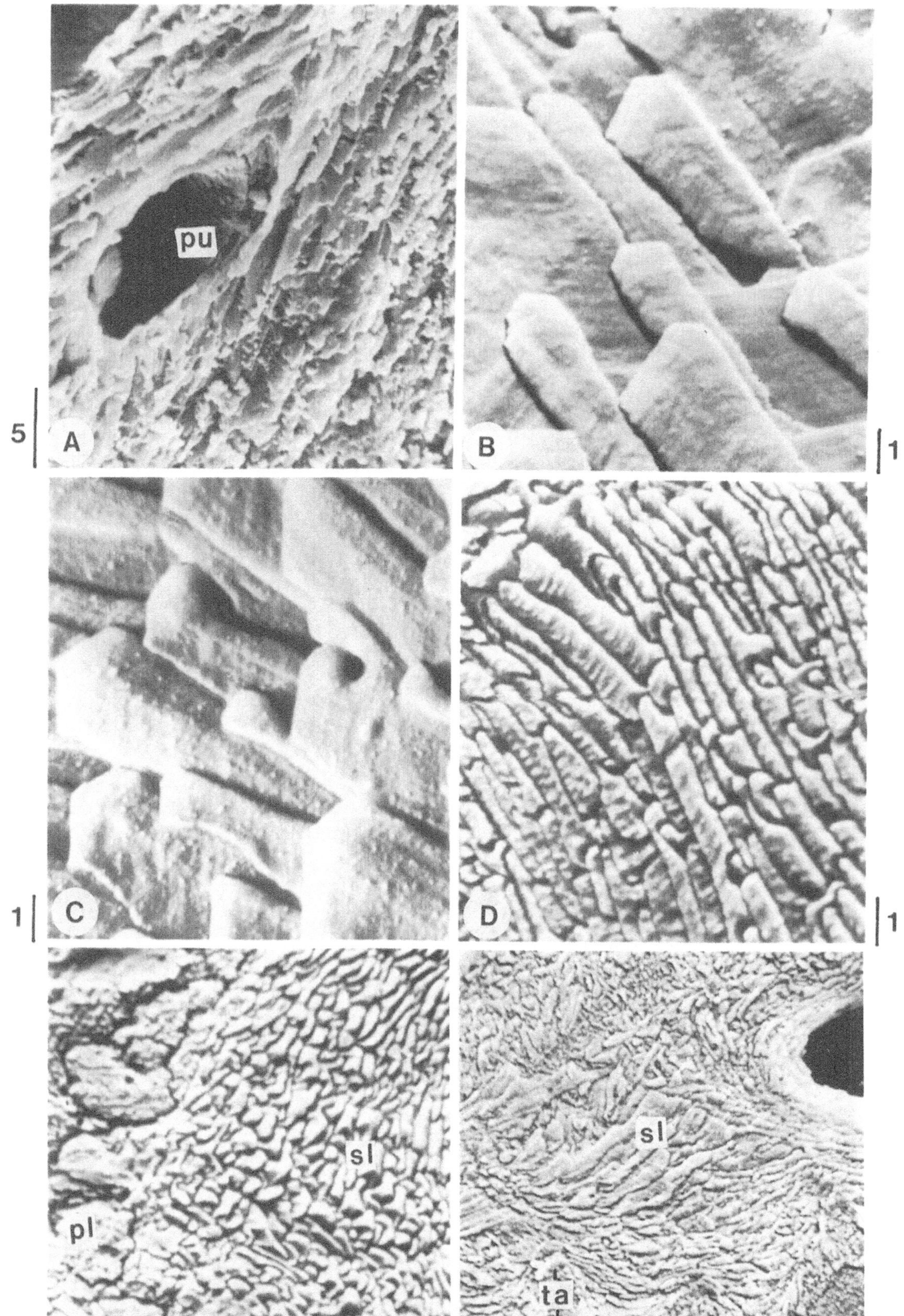
pu
5
A
B
1
1
C
D
1
sl
pl
5
E
sl
ta
F
pl
5

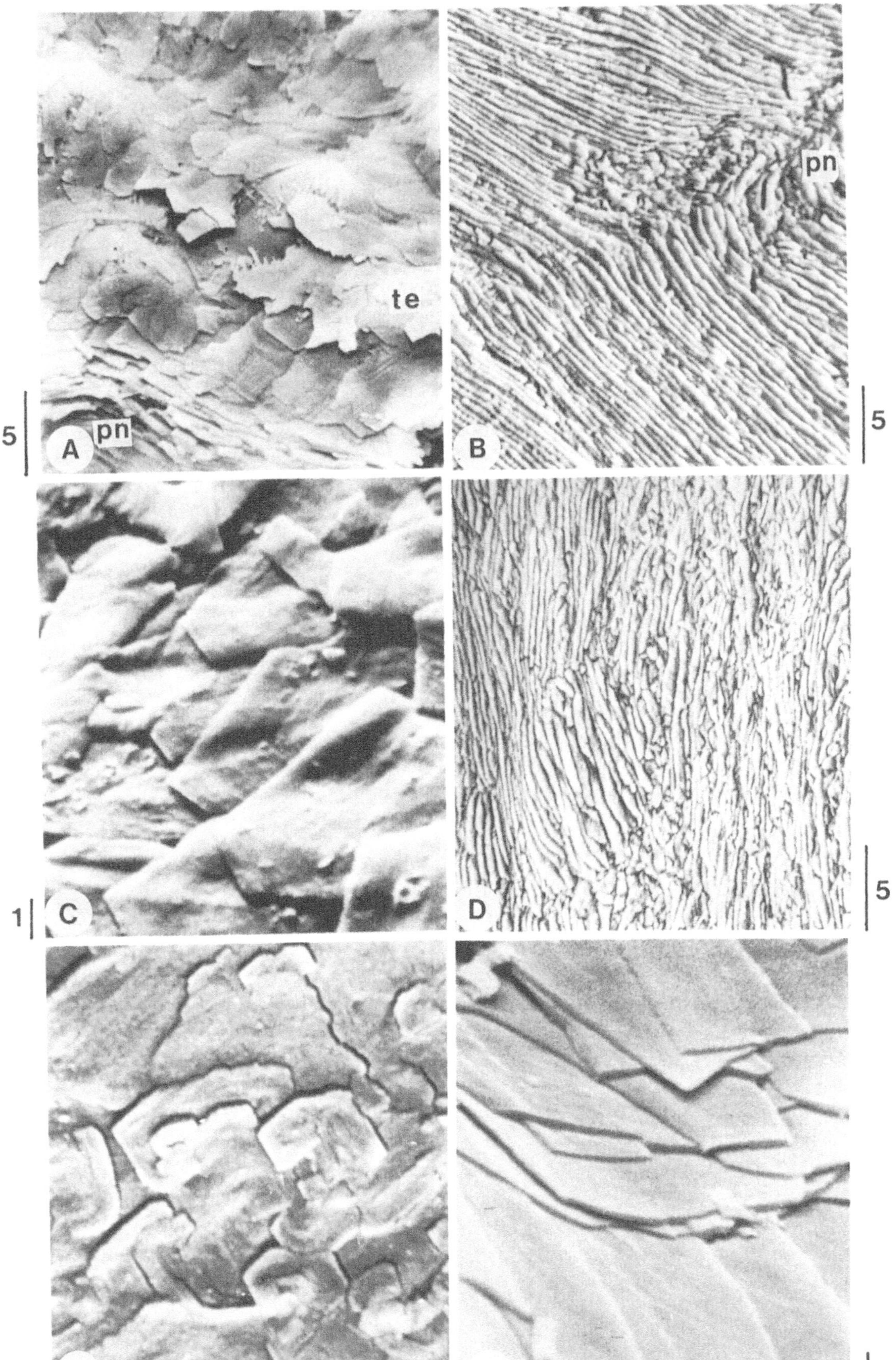
te
pn
A
5
pn
B
5
1
C
D
5
1
E
F
1

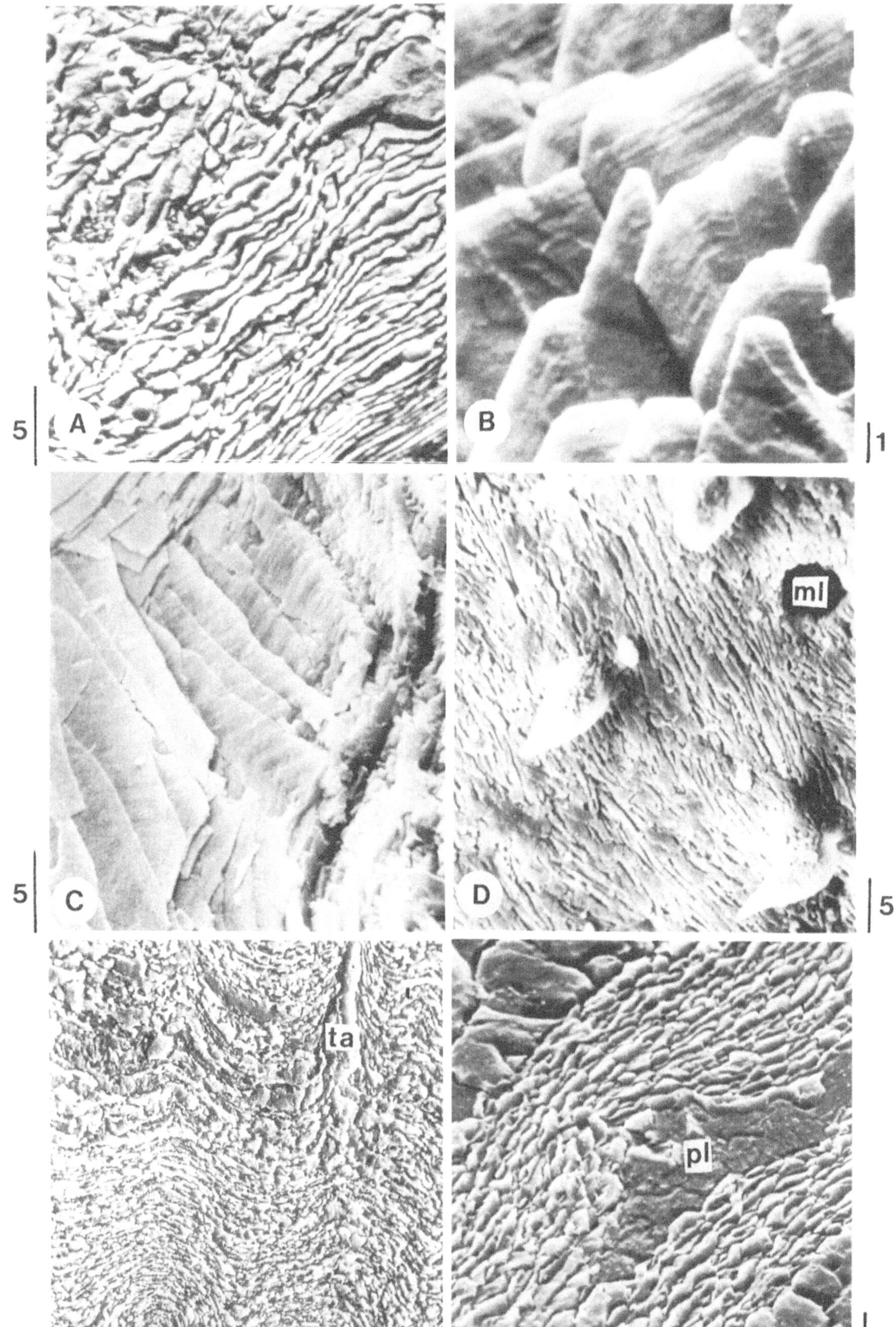
5
A
B
1
ml
5
C
D
5
ta
pl
10
E
F
5

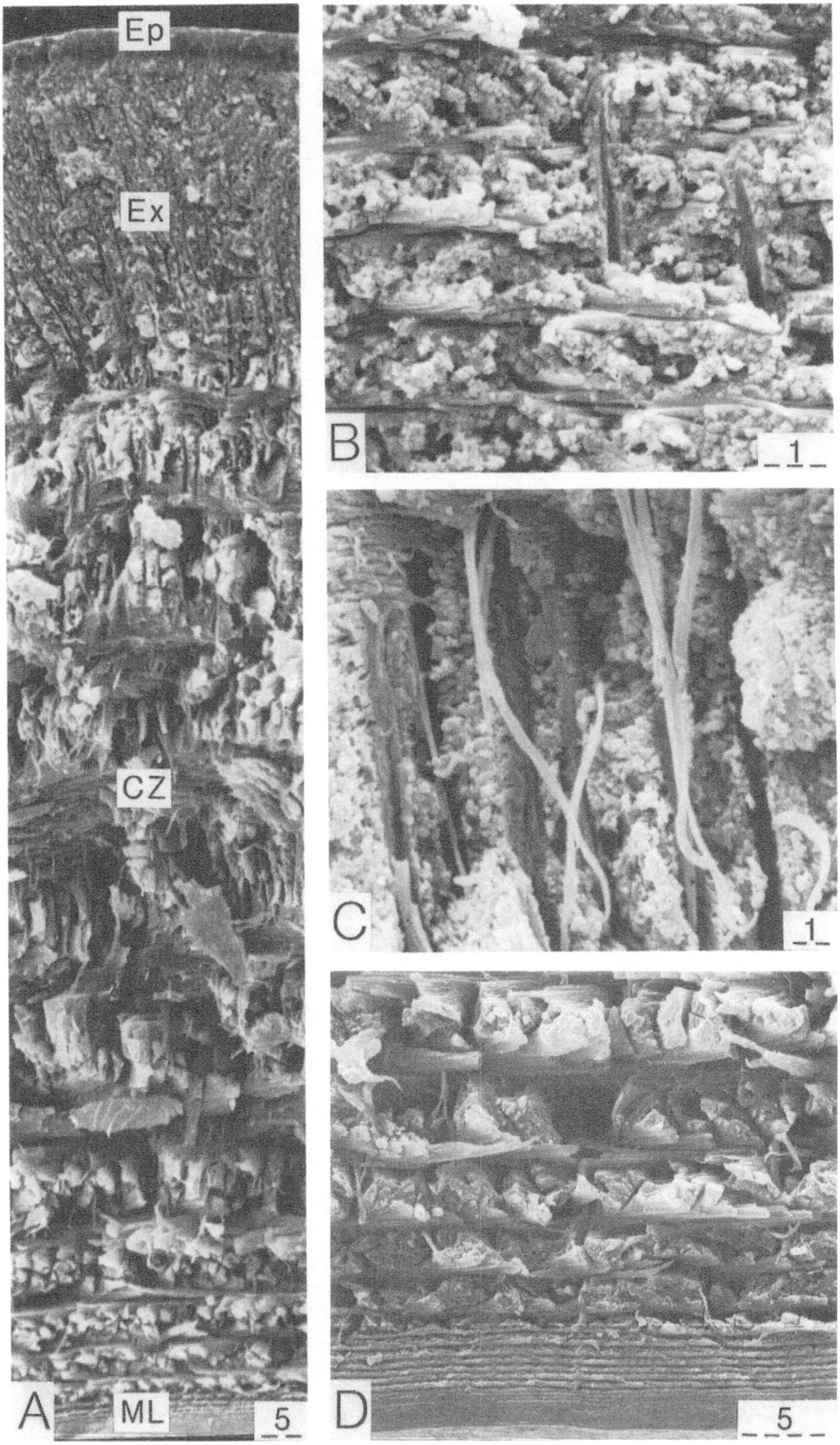
Ep
Ex
CZ
A
ML
5
B
1
C
1
D
5

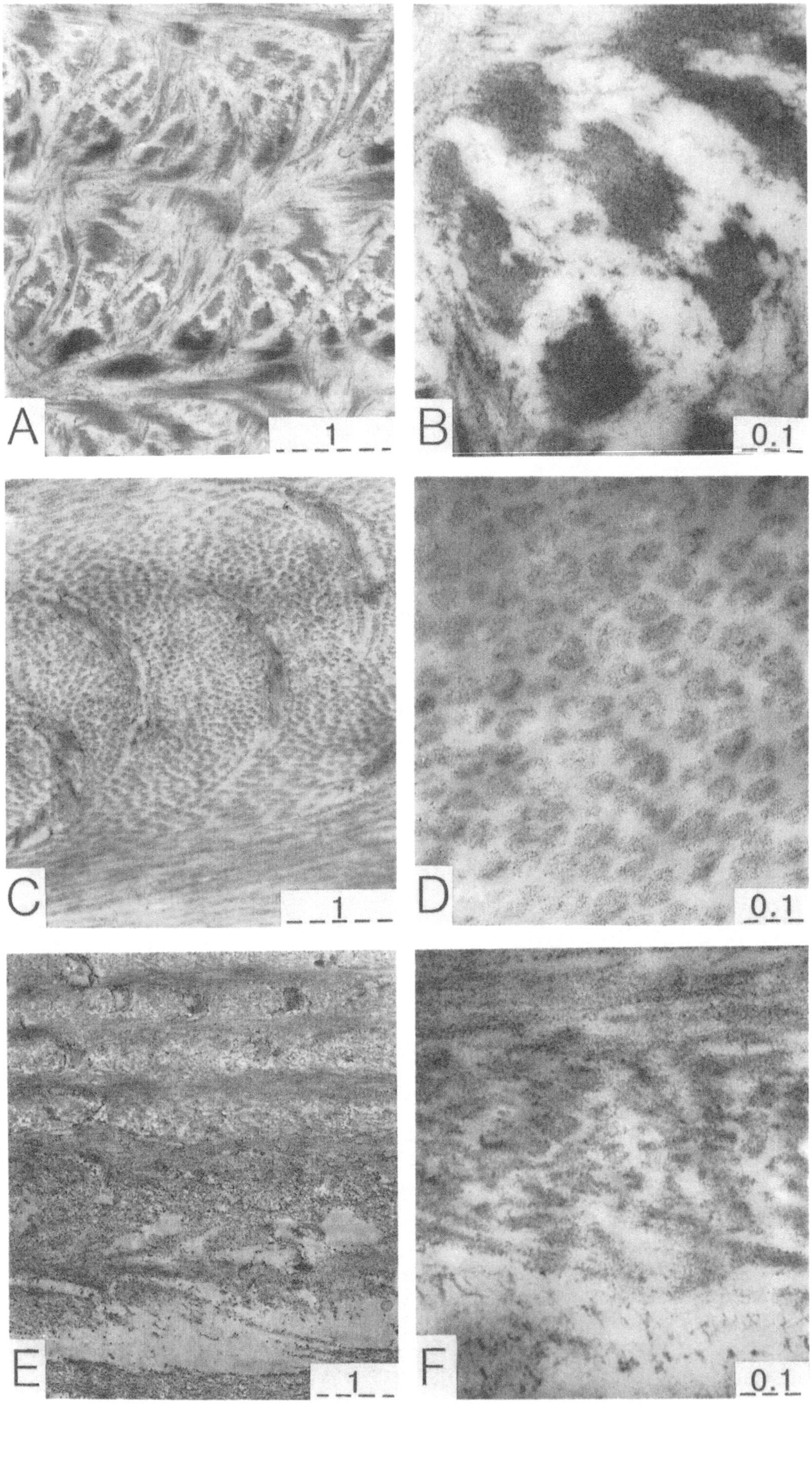
A
1
B
0.1
C
1
D
0.1
E
1
F
0.1

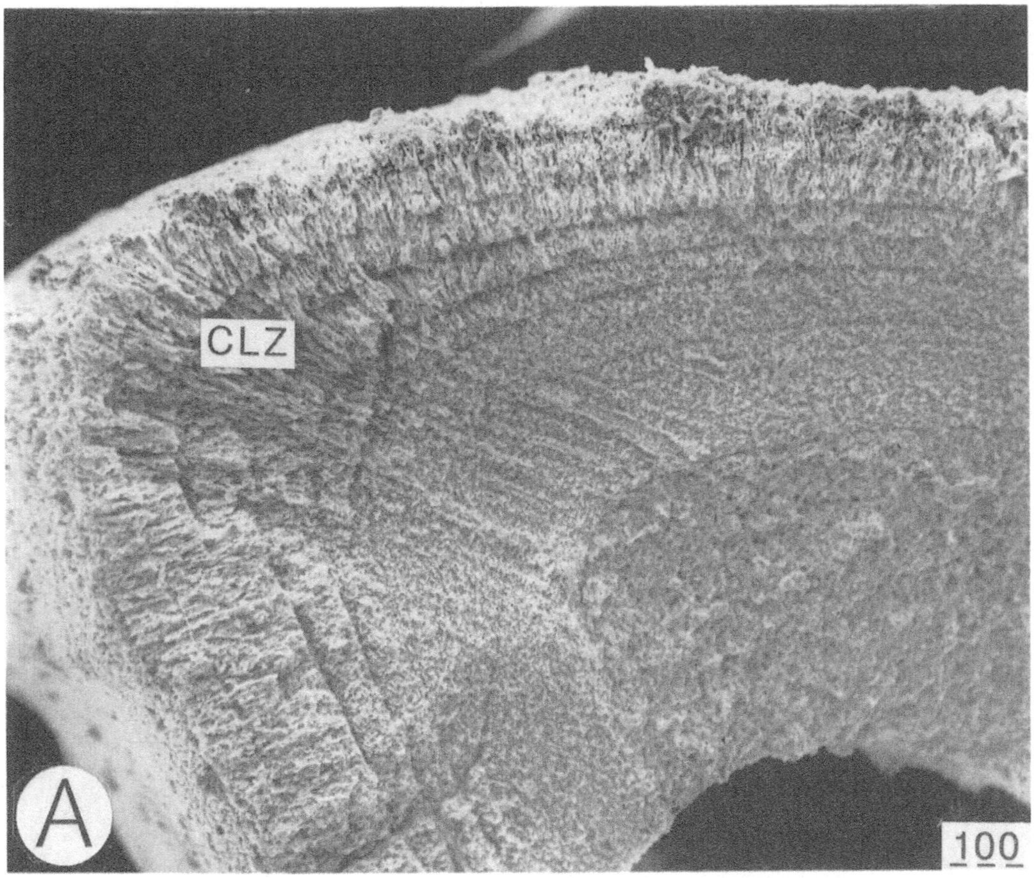
CLZ
A
100

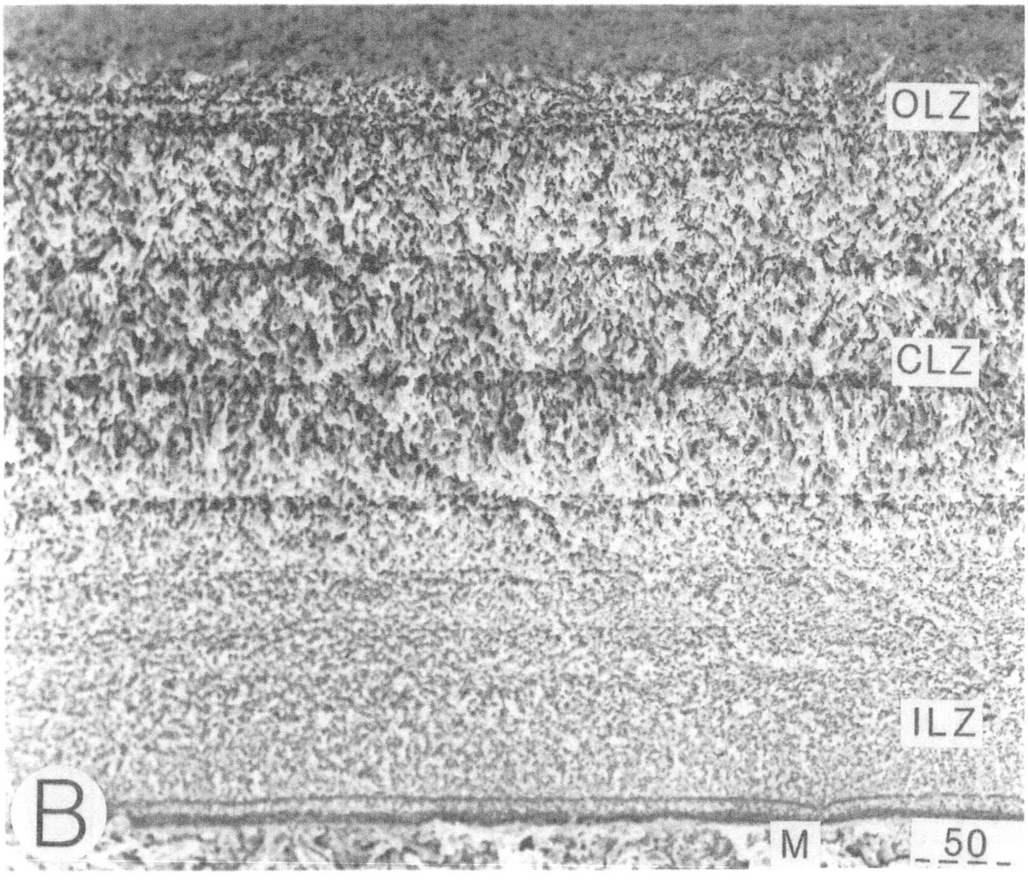
OLZ
CLZ
ILZ
B
M
50

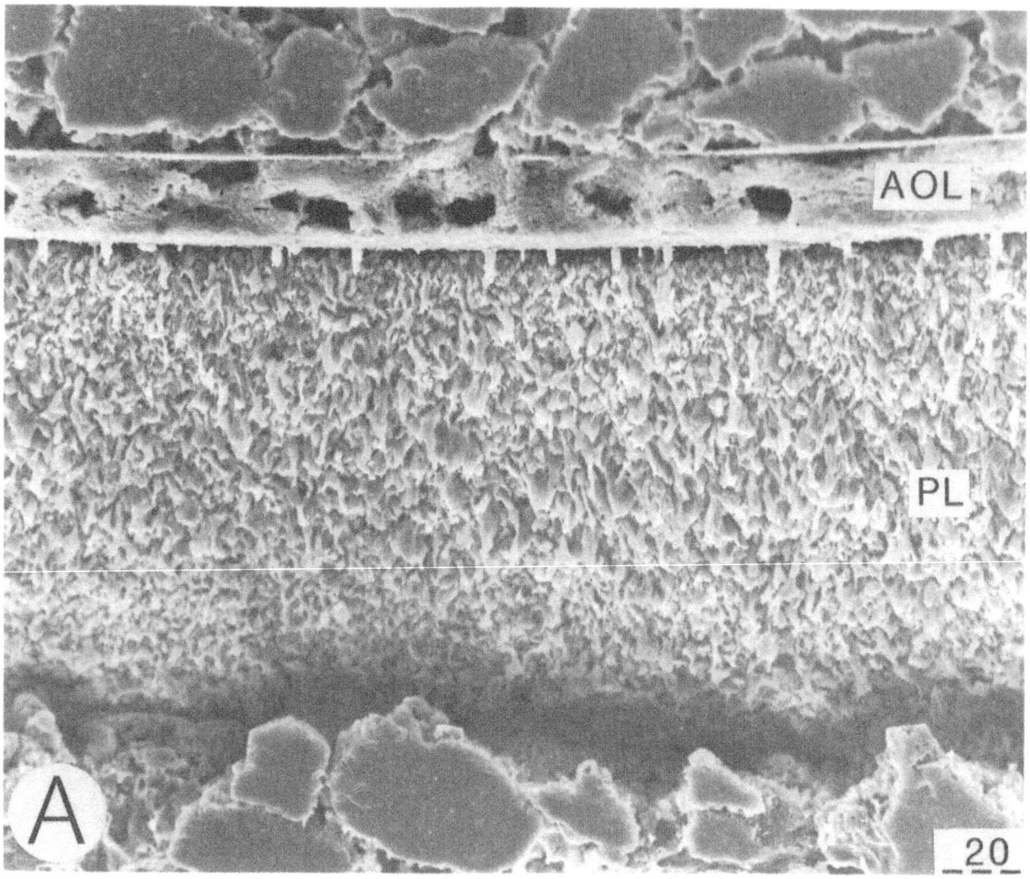
AOL
PL
A
20

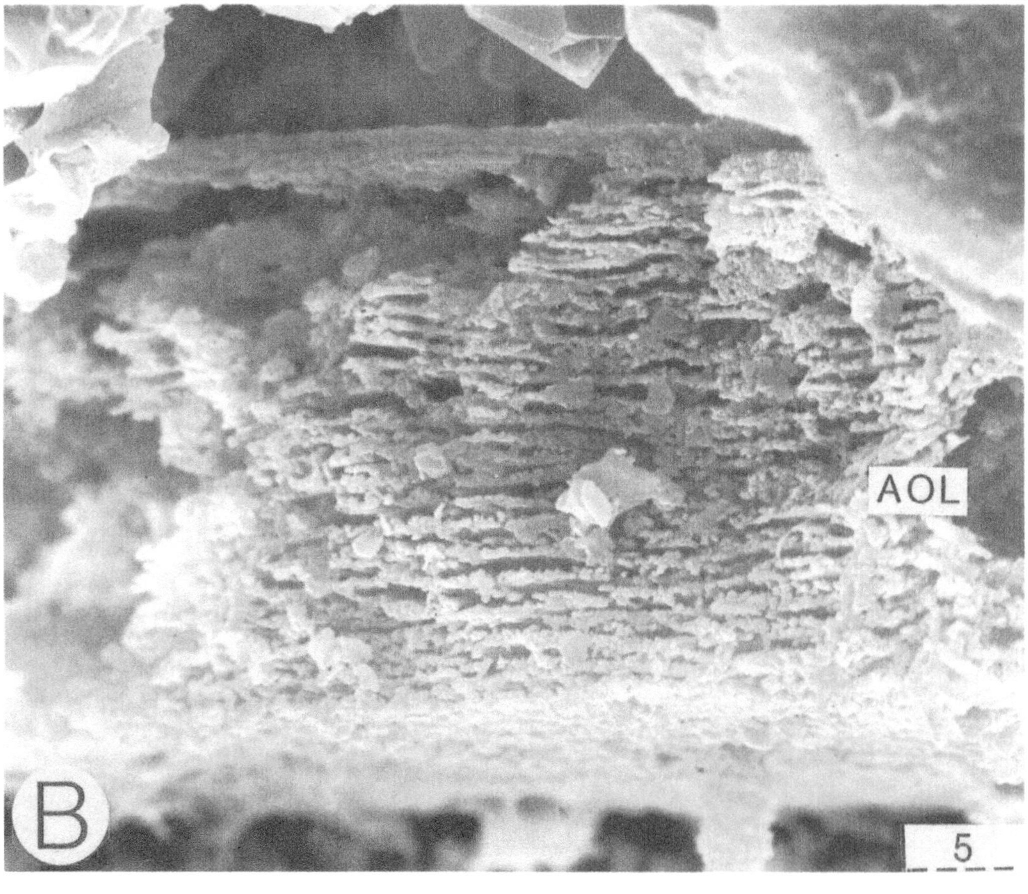
AOL
B
5

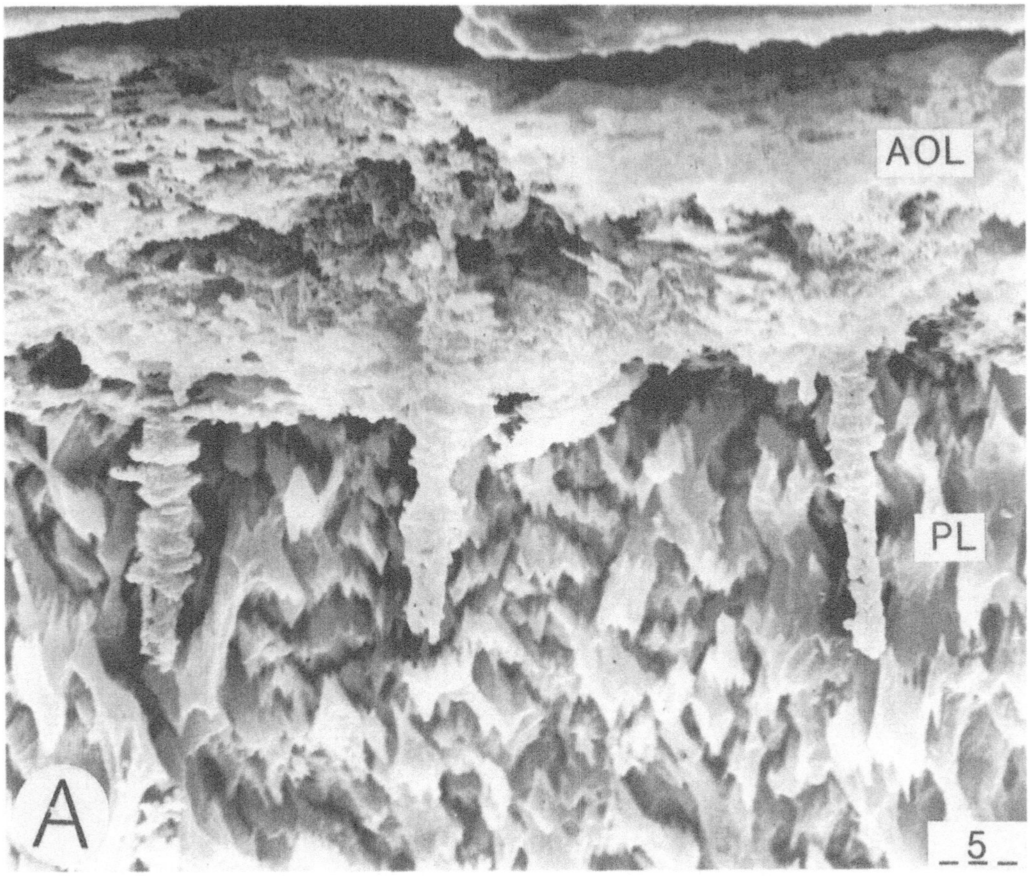
AOL
PL
A
5

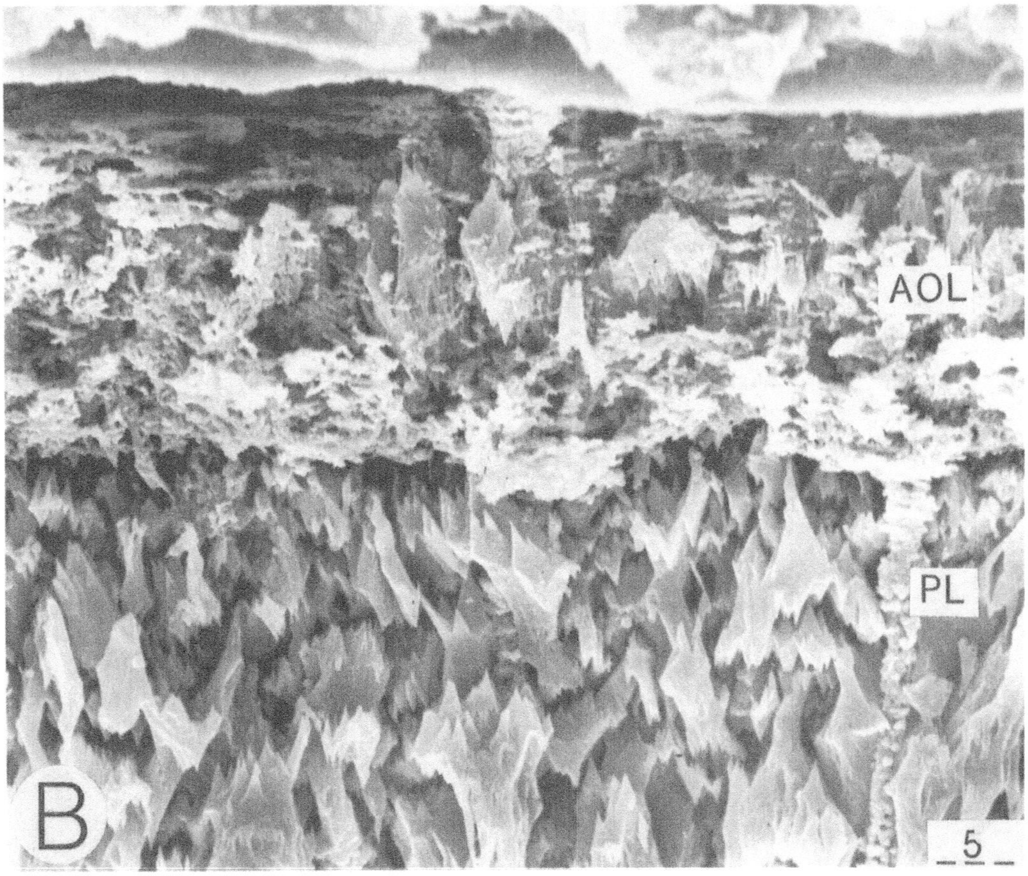
AOL
PL
B
5

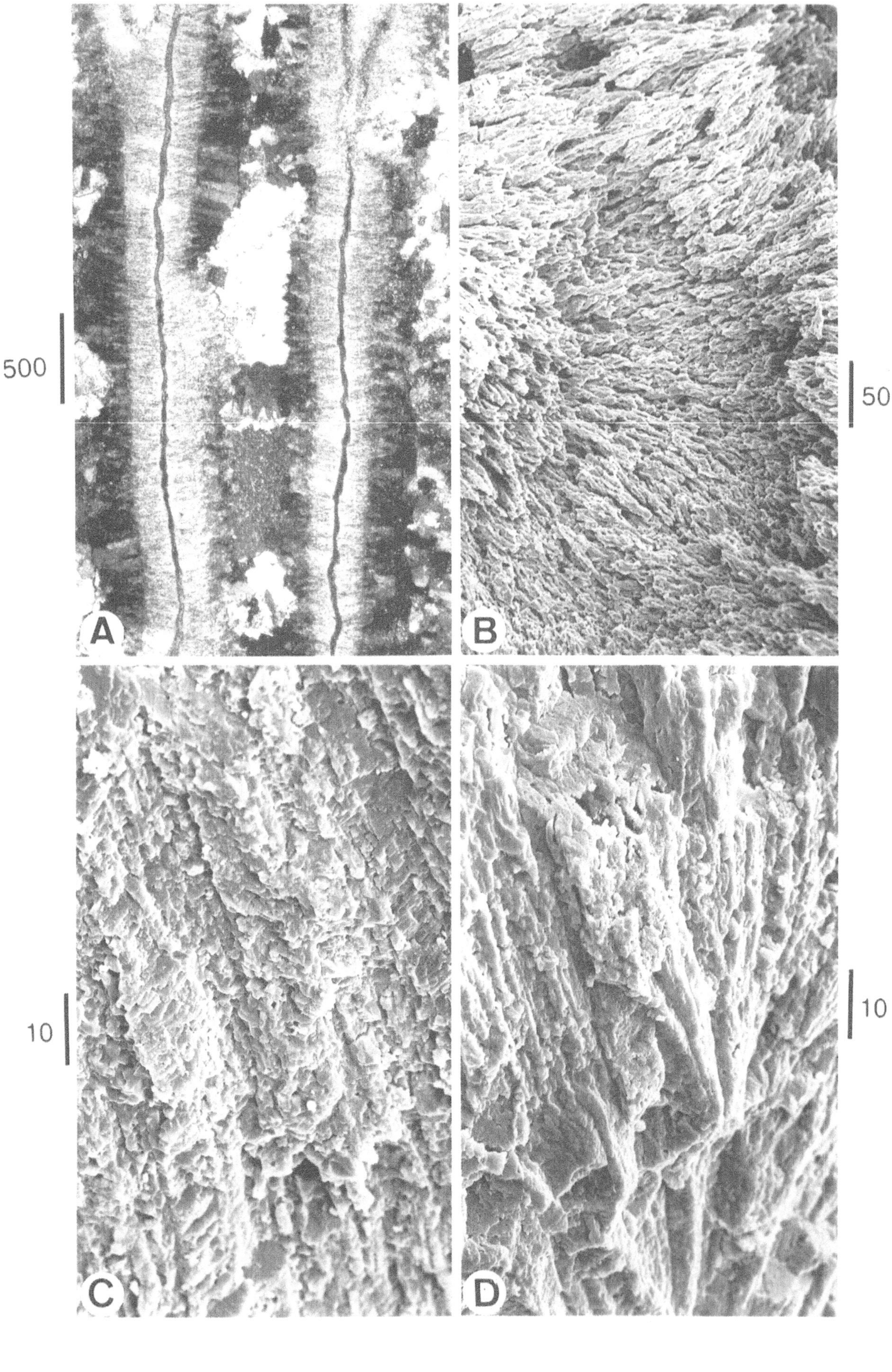
500
50
10
10
A
B
C
D

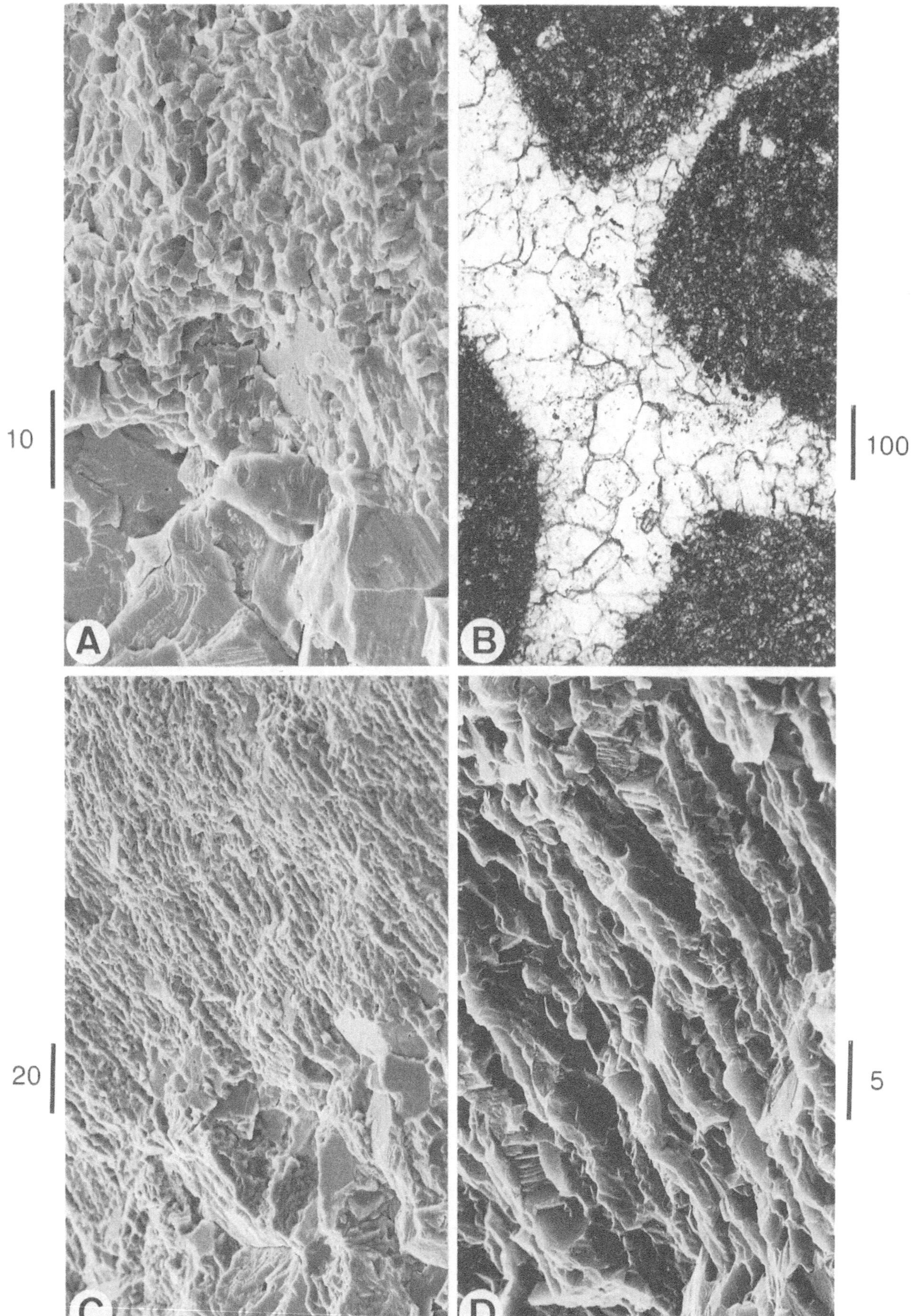
10
100
20
5
A
B
C
D

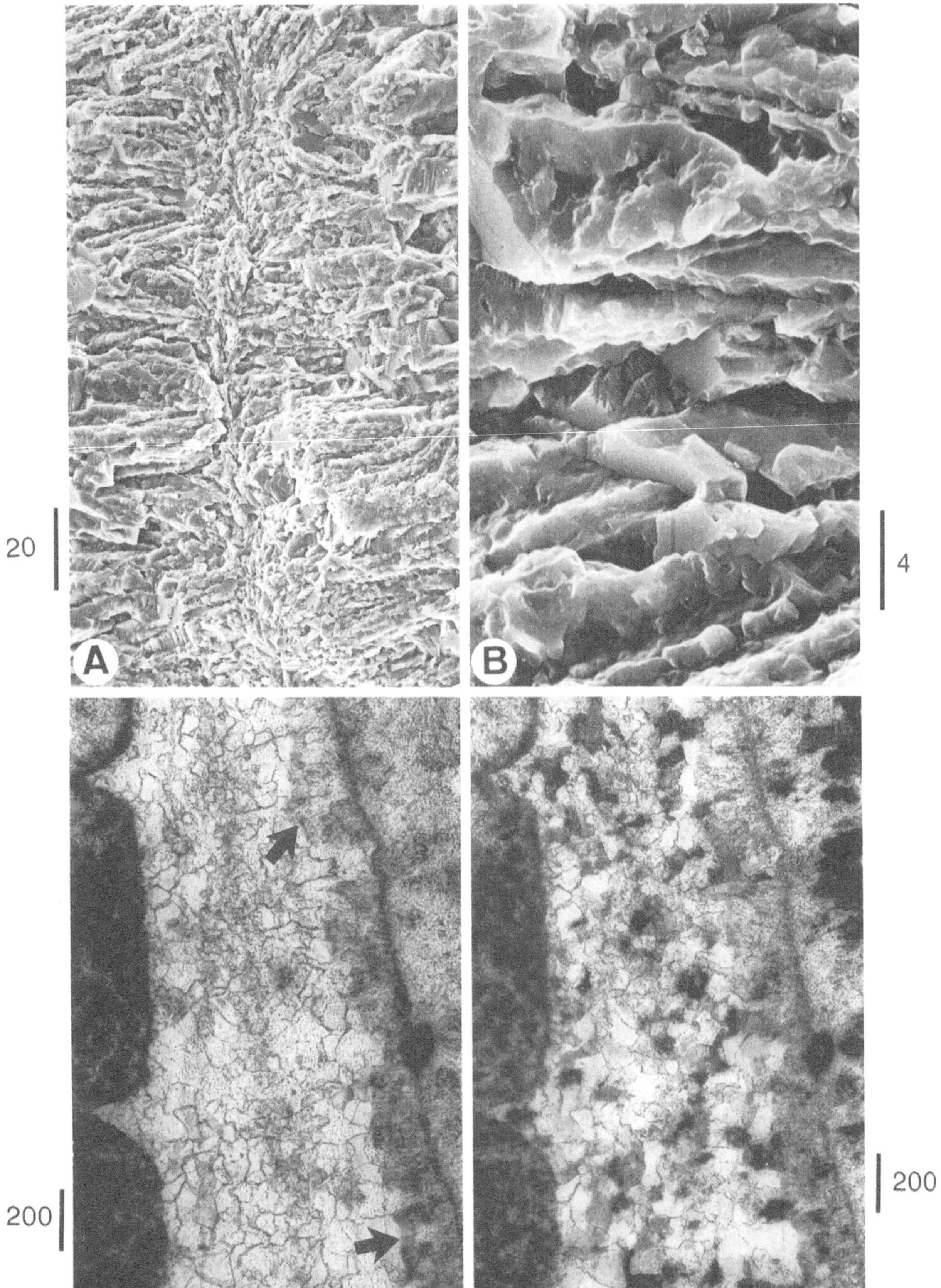
20
4
200
200
A
B
C
D

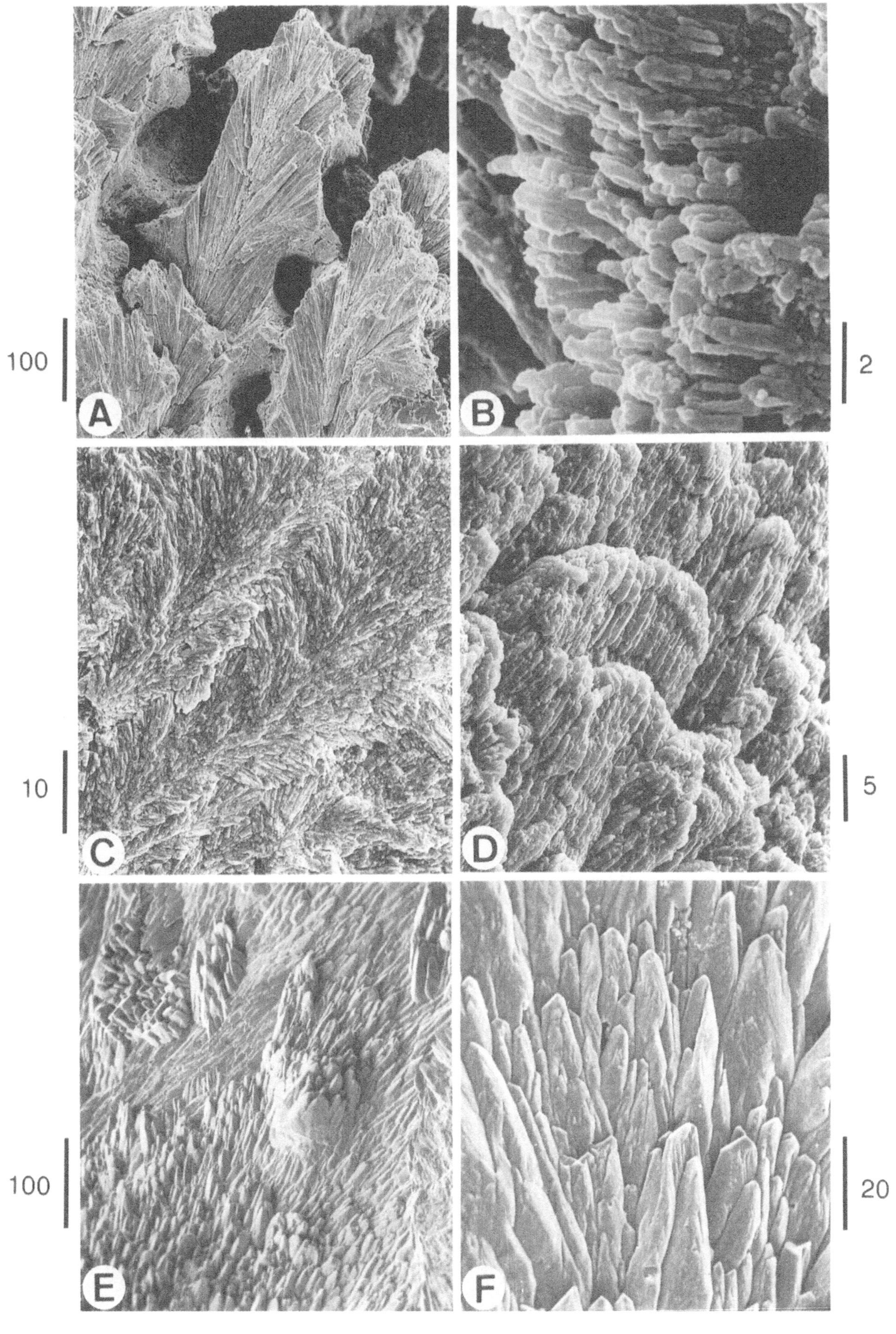
A
100
B
2
C
10
D
5
E
100
F
20

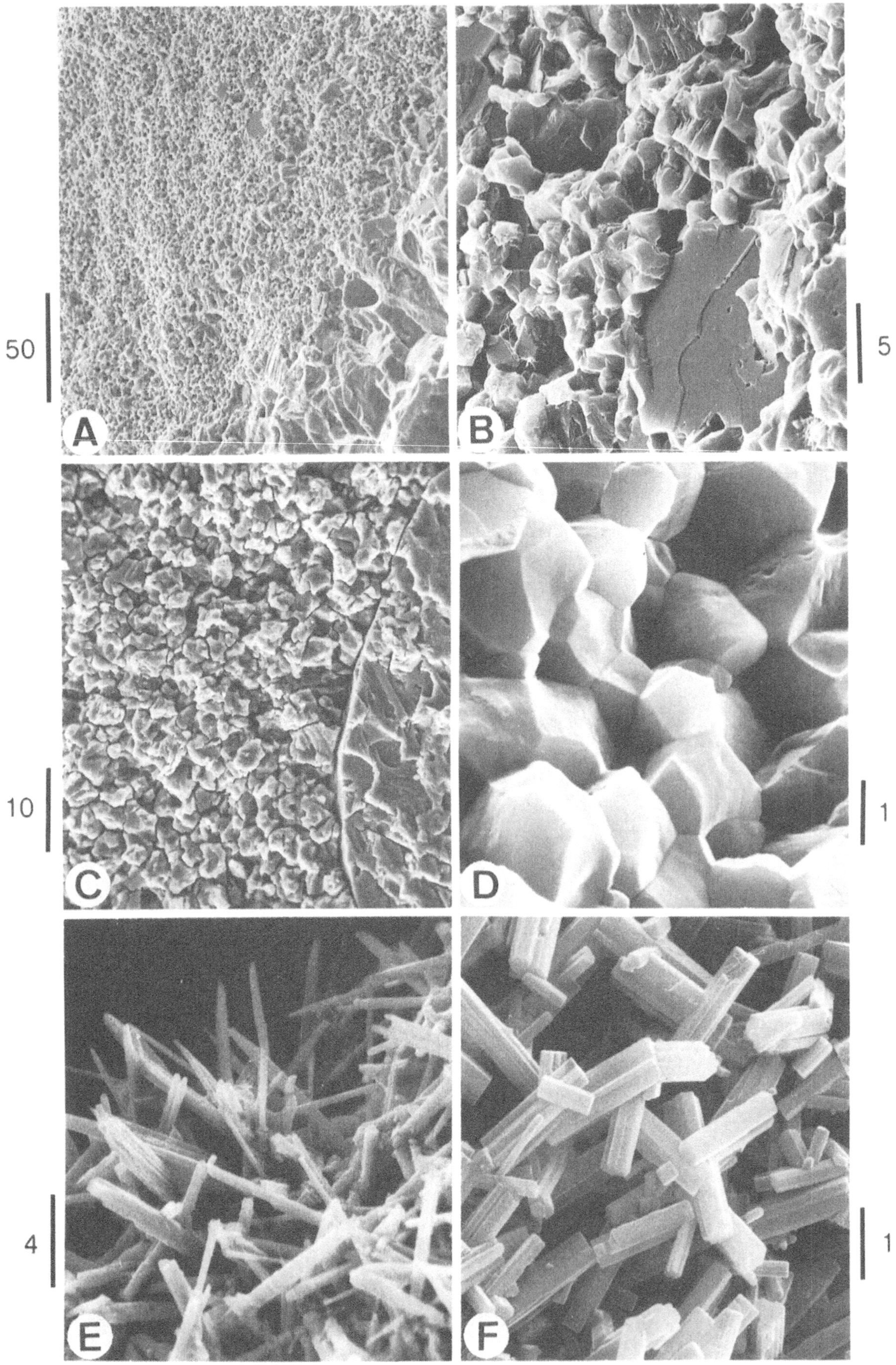
50
A
B
5
10
C
D
1
4
E
F
1

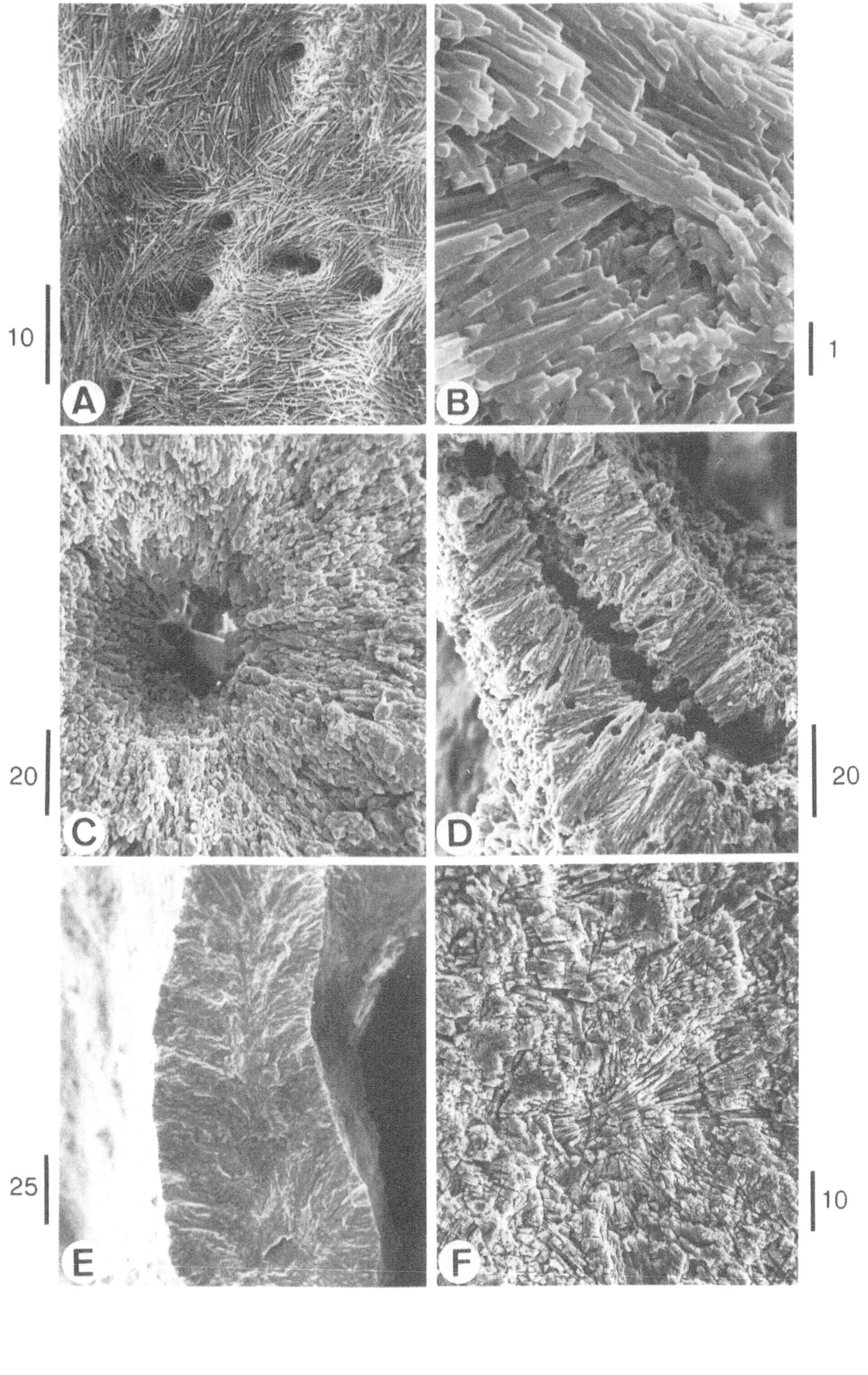
10
A
1
B
20
C
20
D
25
E
10
F

50
A
B
10
10
C
D
5
50
E
F
50

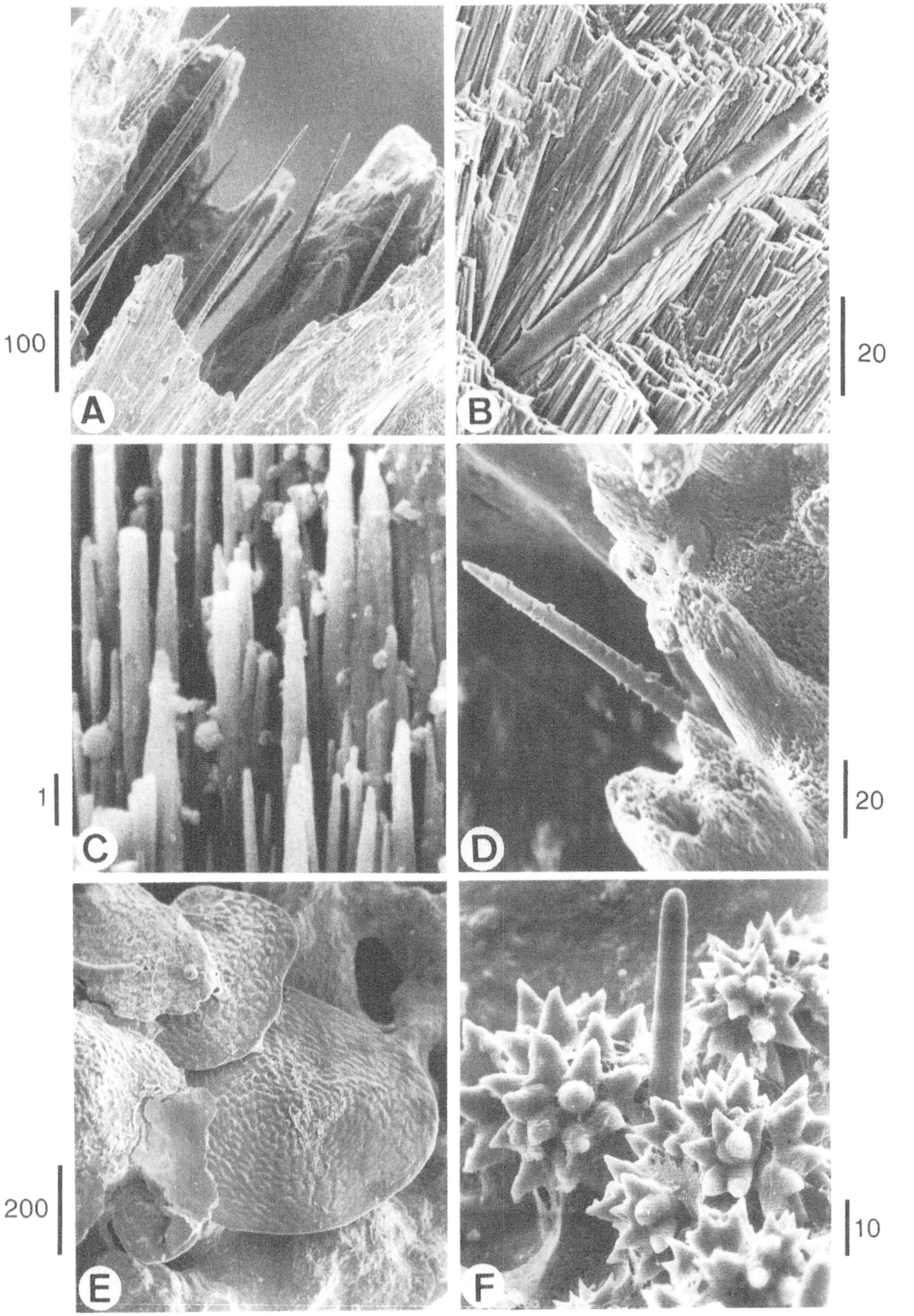
100
A
20
B
1
C
20
D
200
E
10
F

40
A
40
B
40
C

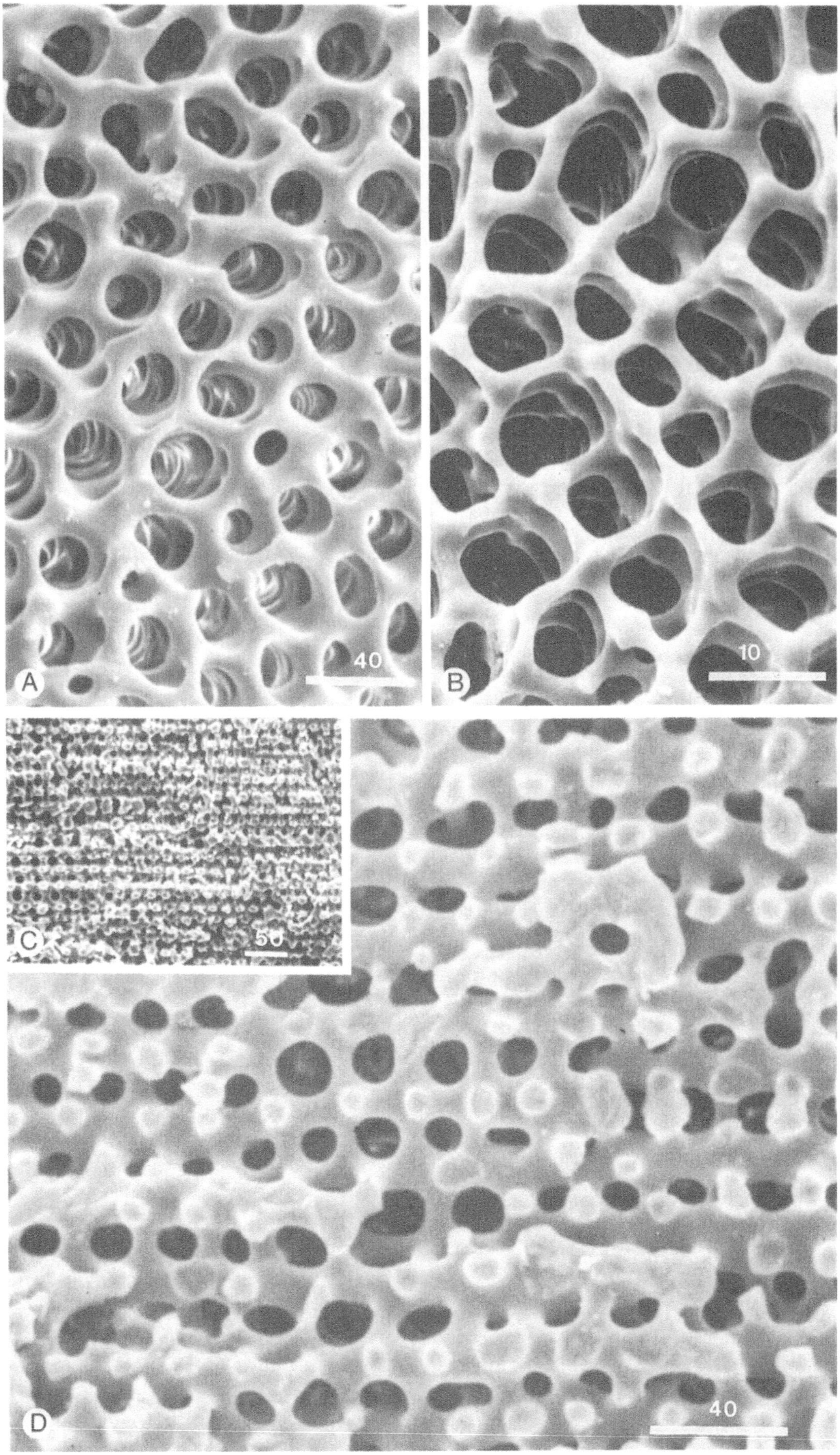
A
40
B
10
C
50
D
40

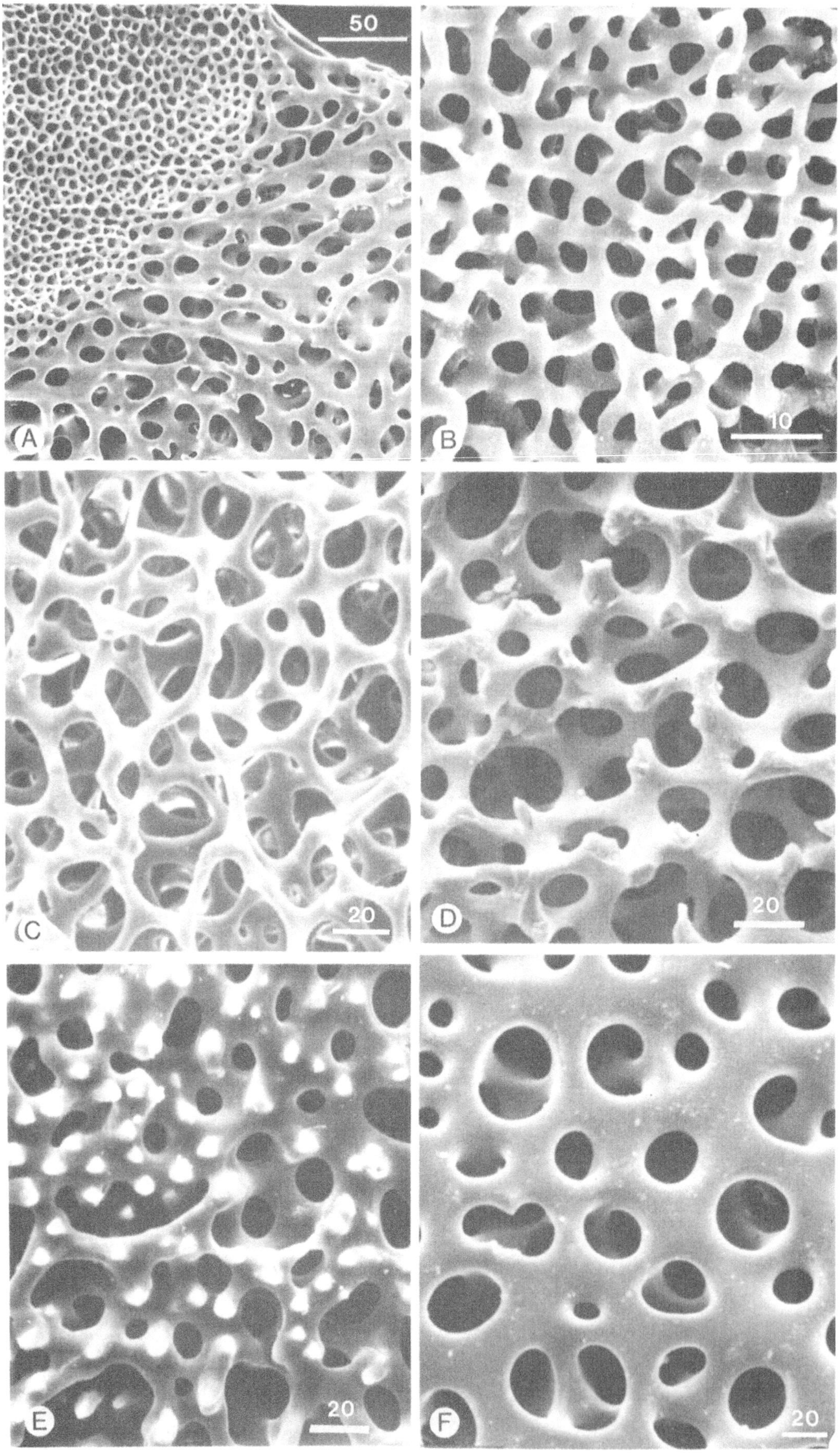
50
A
B
10
C
20
D
20
E
20
F
20

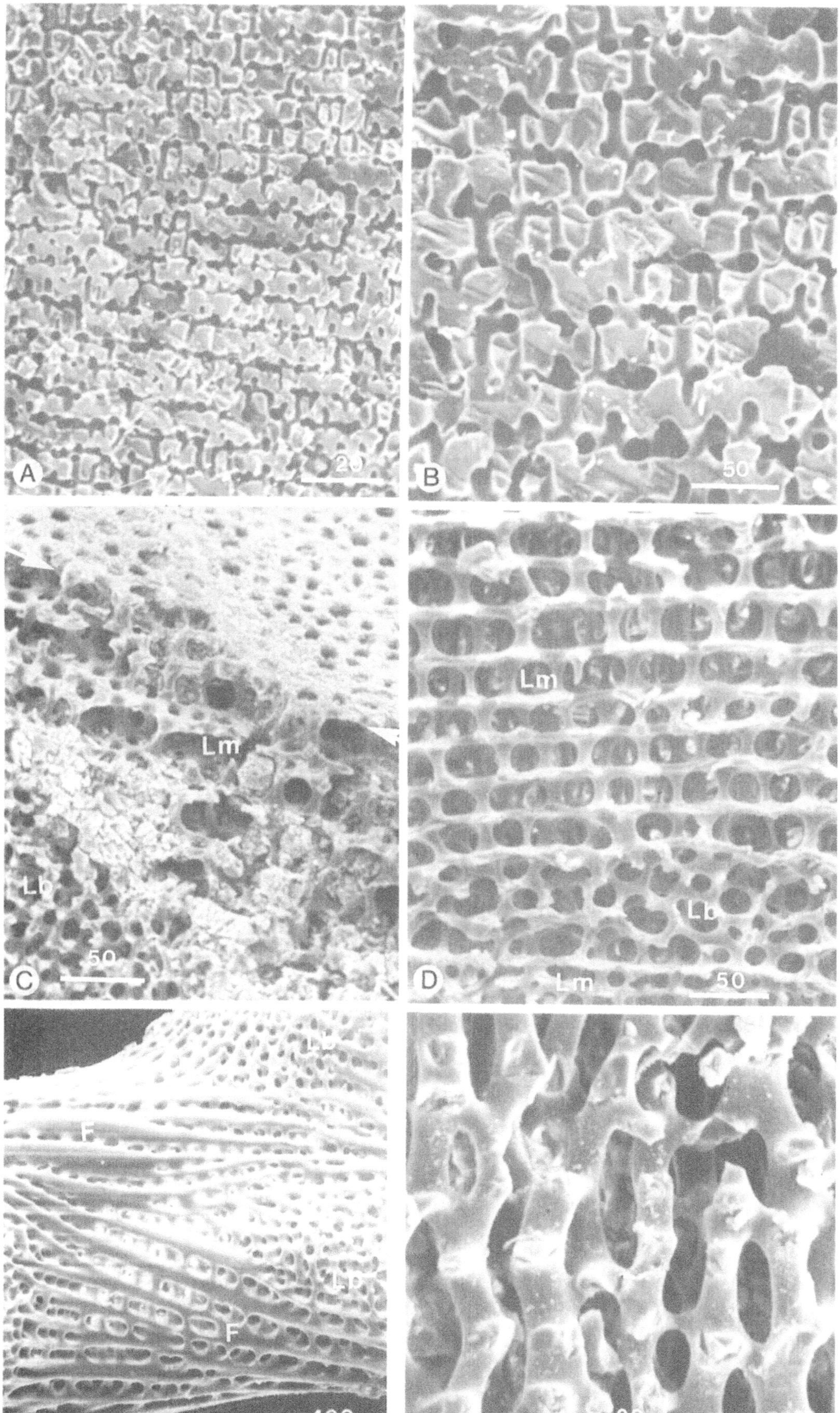
A
20
B
50
Lm
Lb
C
50
Lm
Lb
D
Lm
50
F
Lp
F
E
400
F
20

PLATE 174

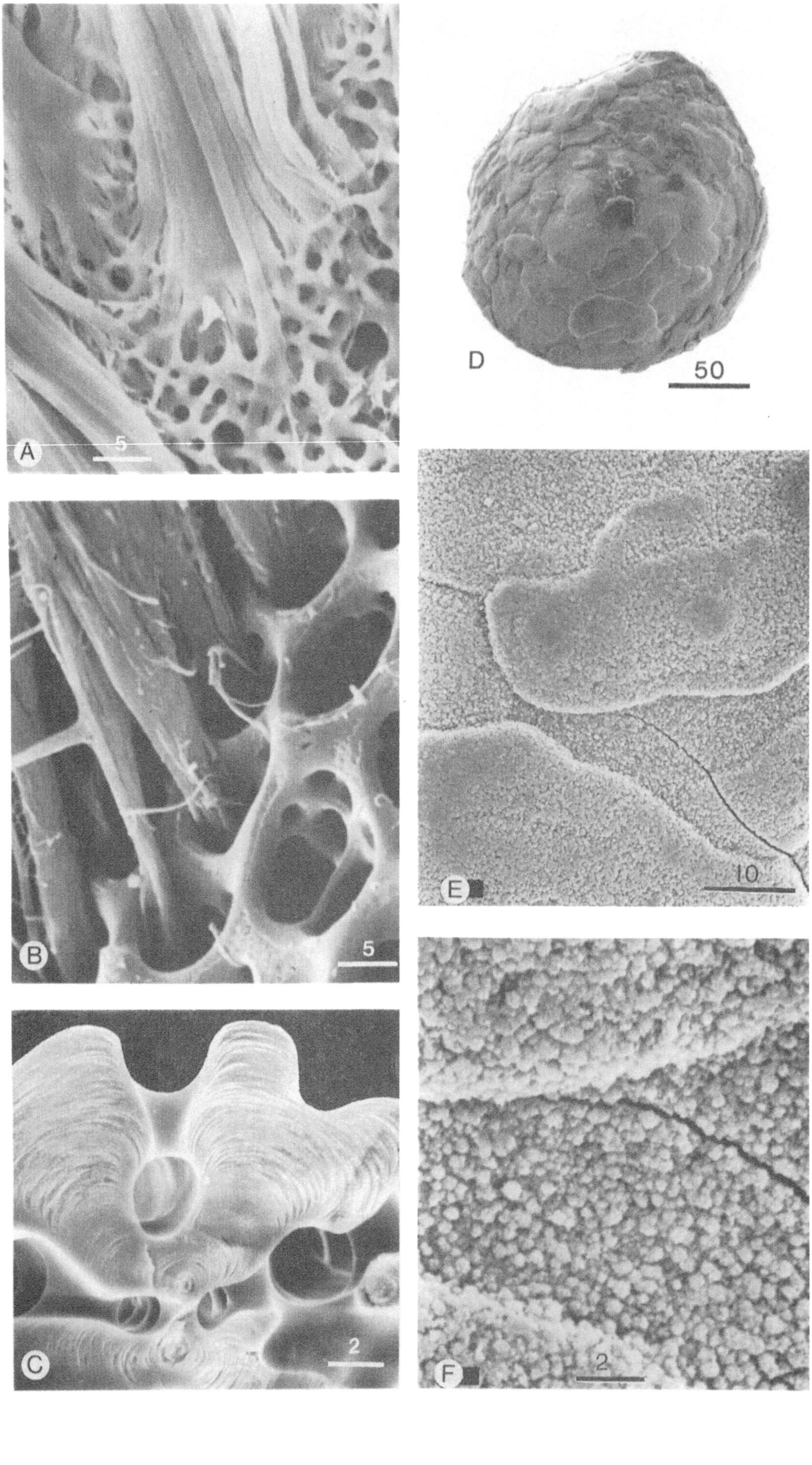

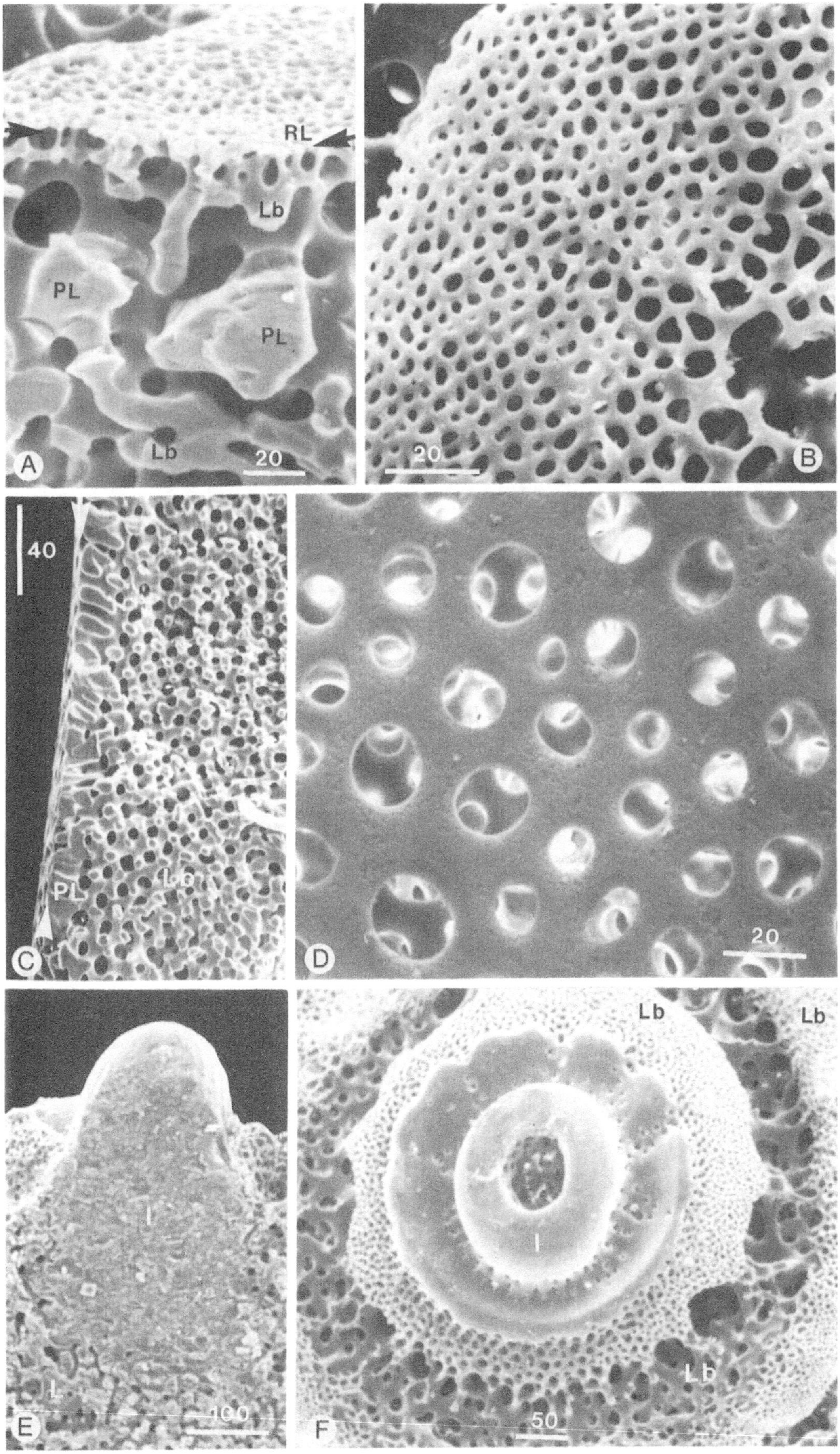
RL
Lb
PL
PL
Lb
20
A
20
B
40
PL
Lb
C
20
D
Lb
Lb
I
L
100
E
I
Lb
50
F

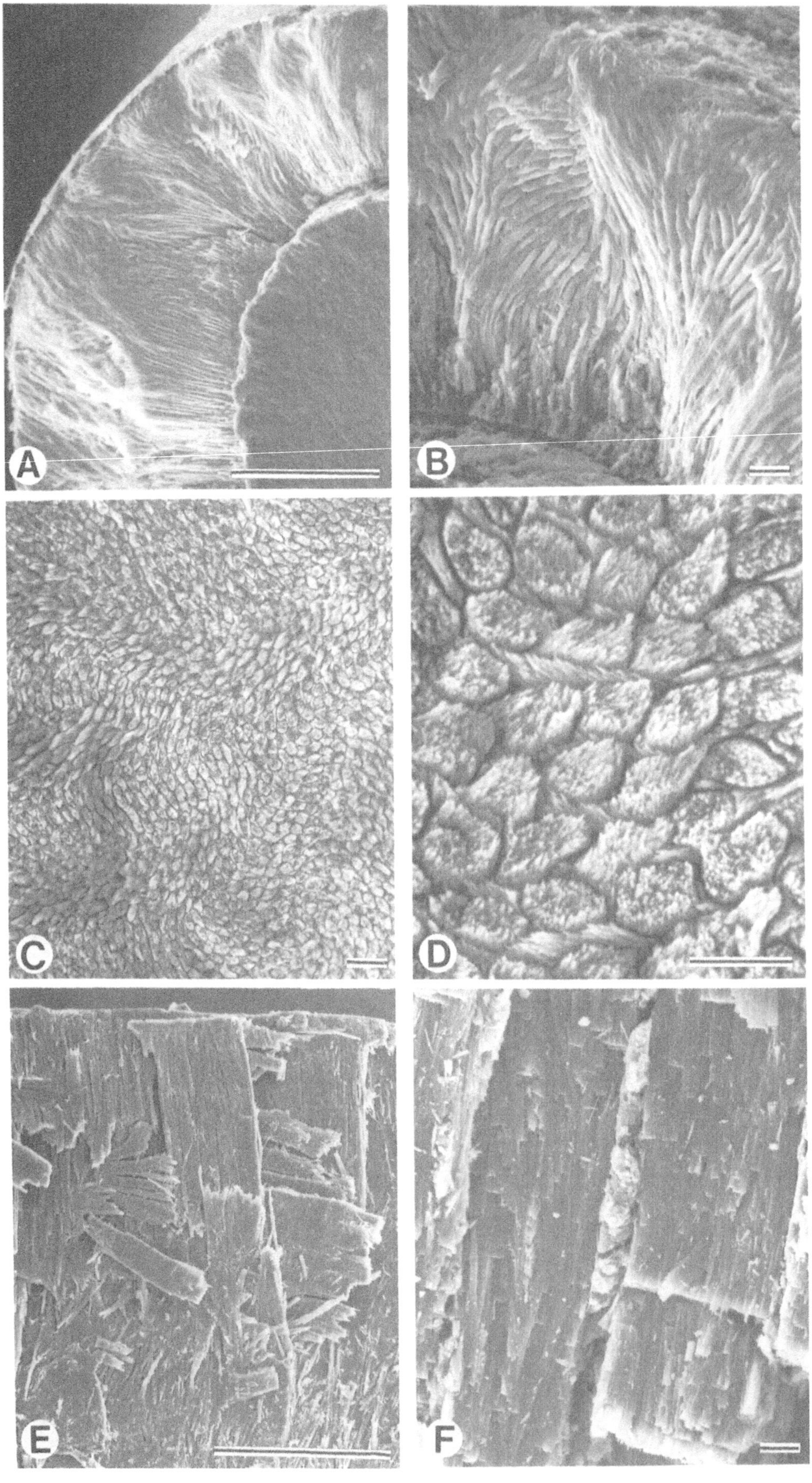
A
B
C
D
E
F

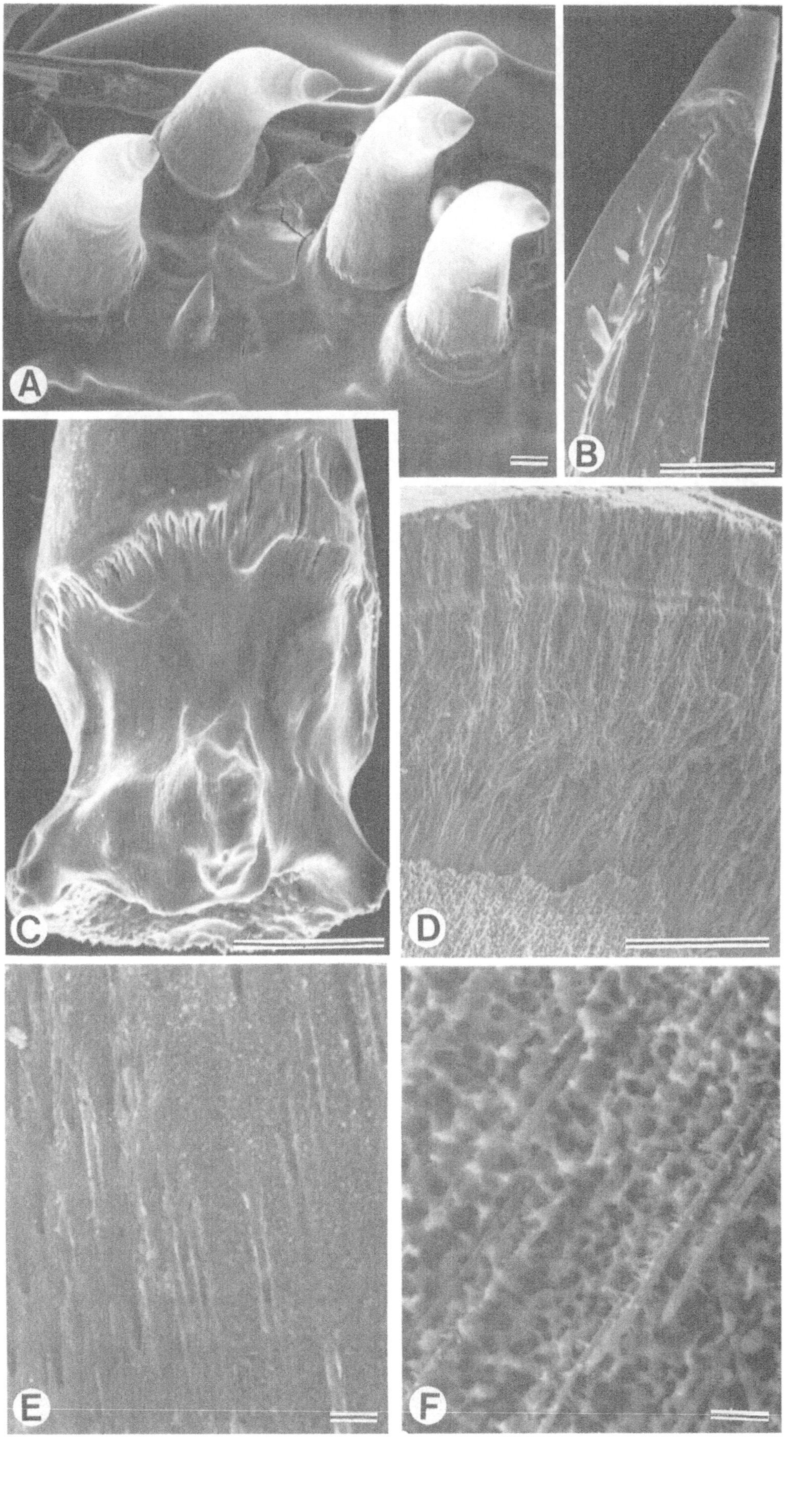
A
B
C
D
E
F

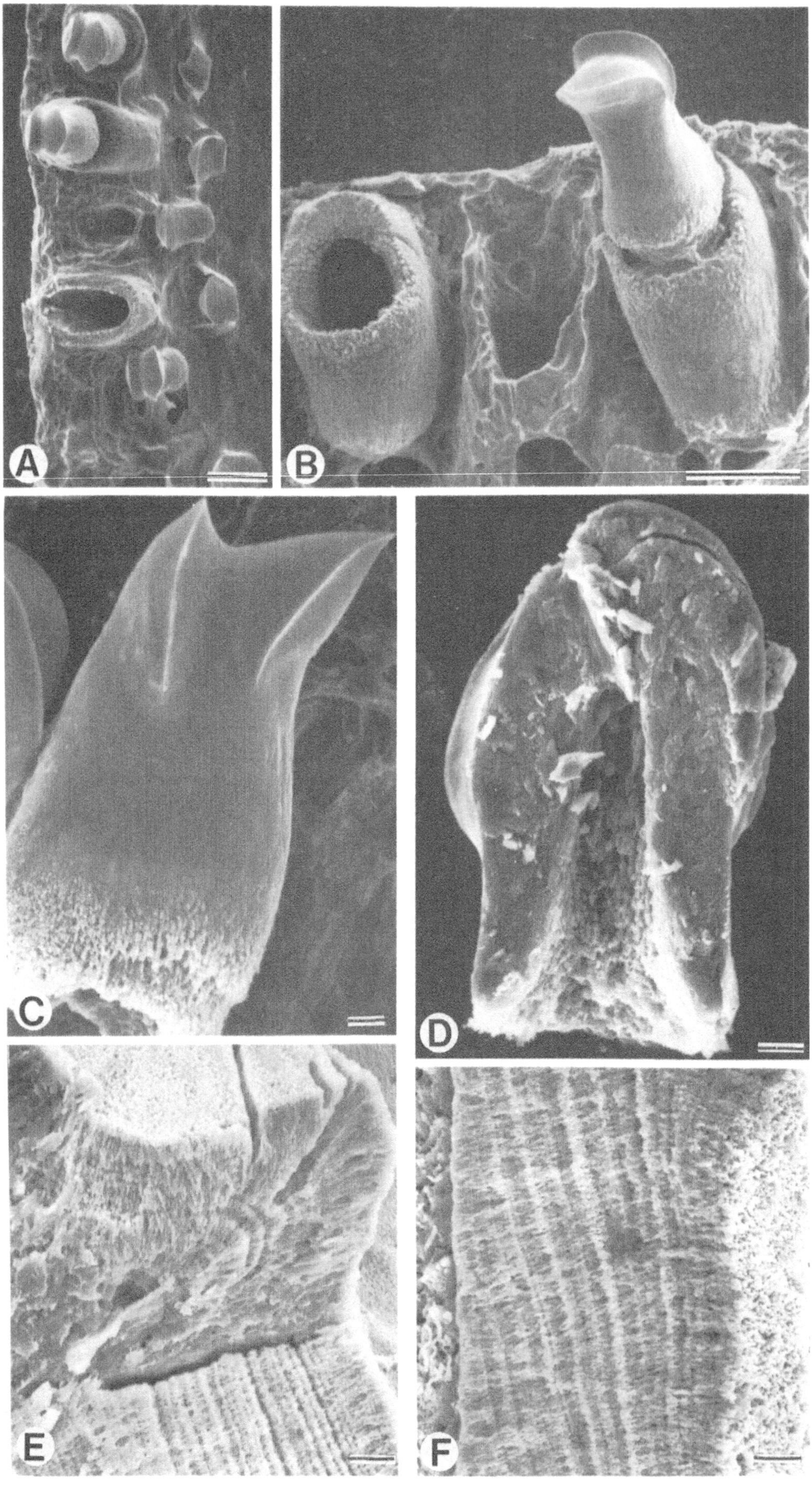
A
B
C
D
E
F

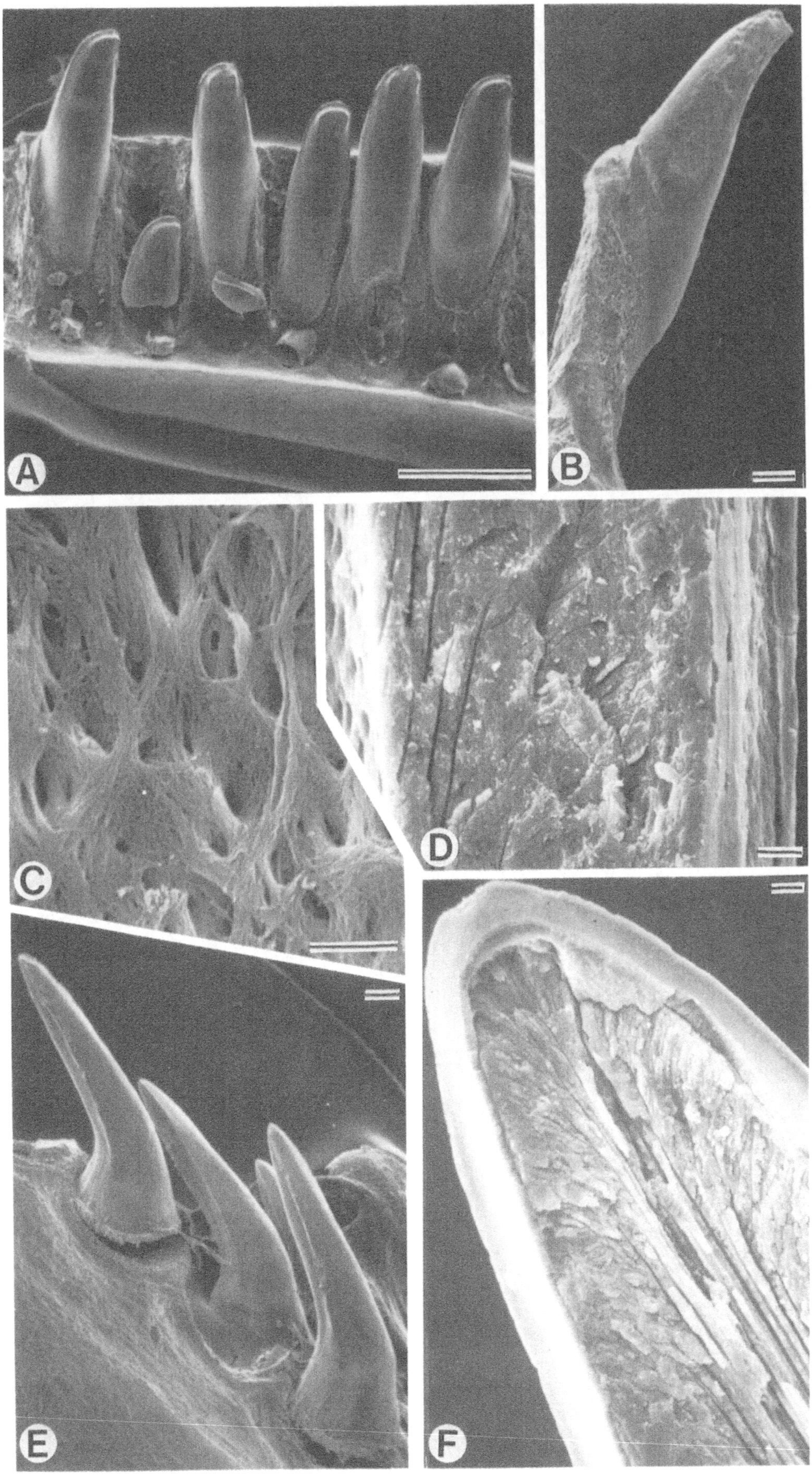
A
B
C
D
E
F

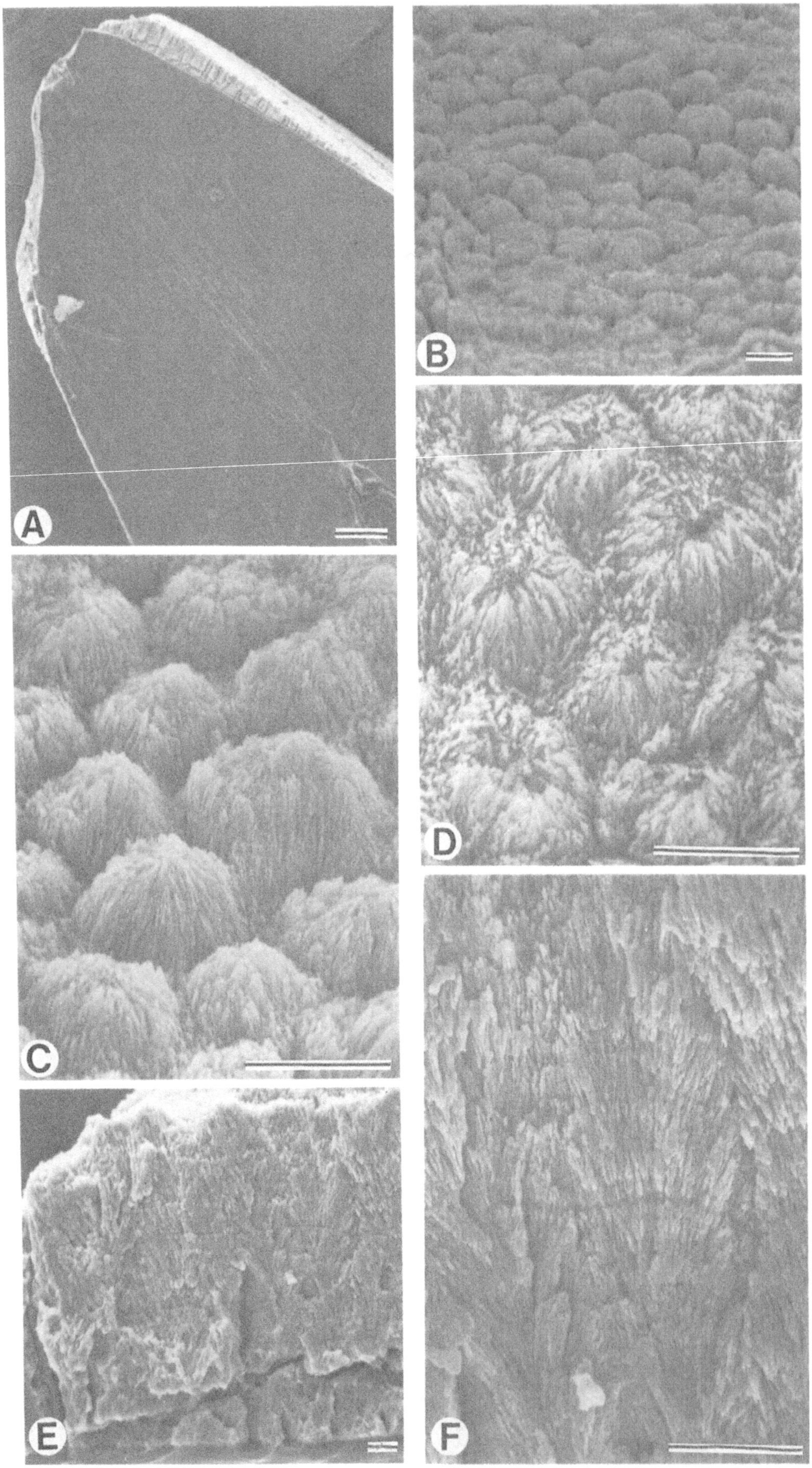
A
B
C
D
E
F

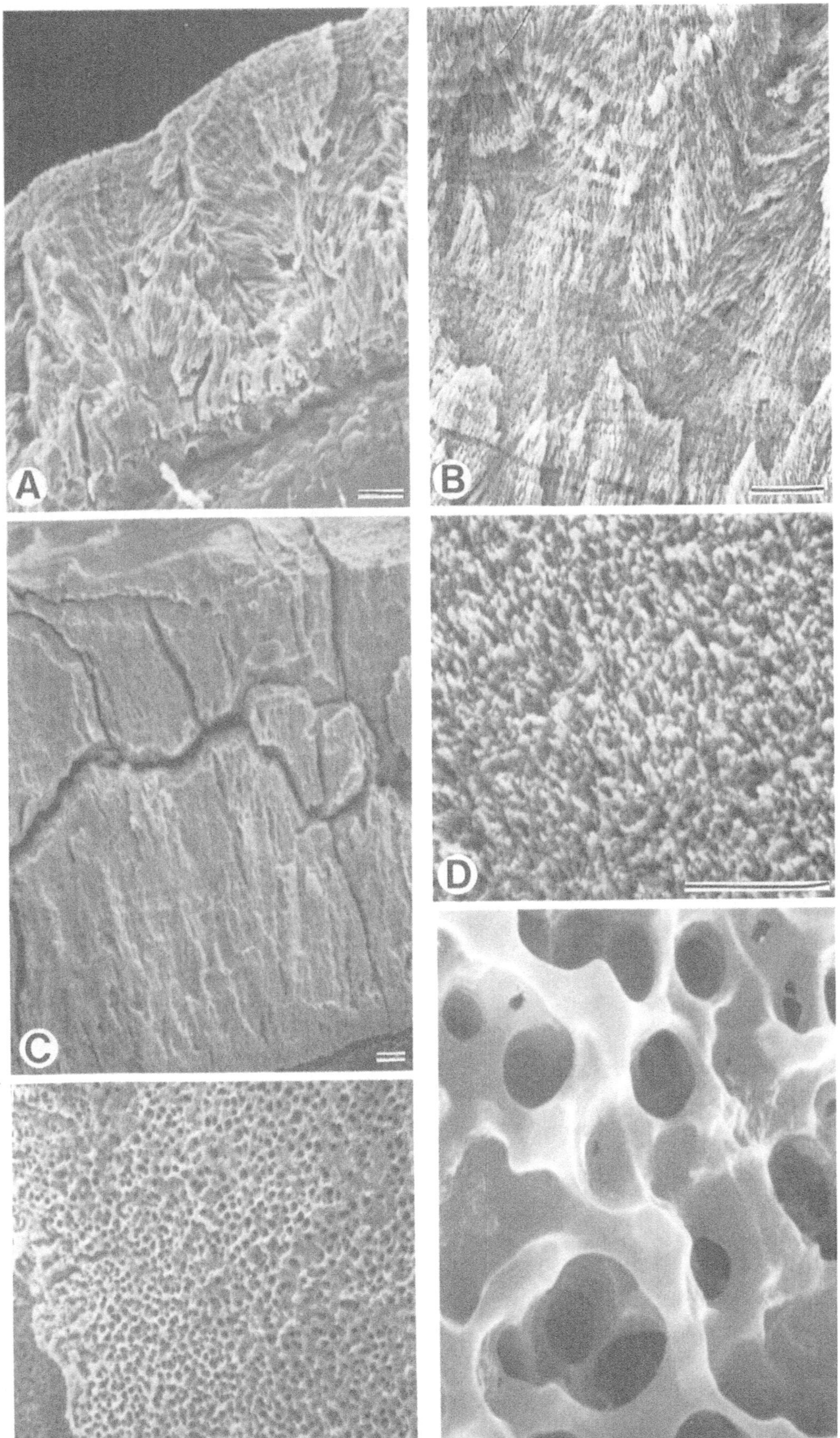
A
B
C
D
E
F

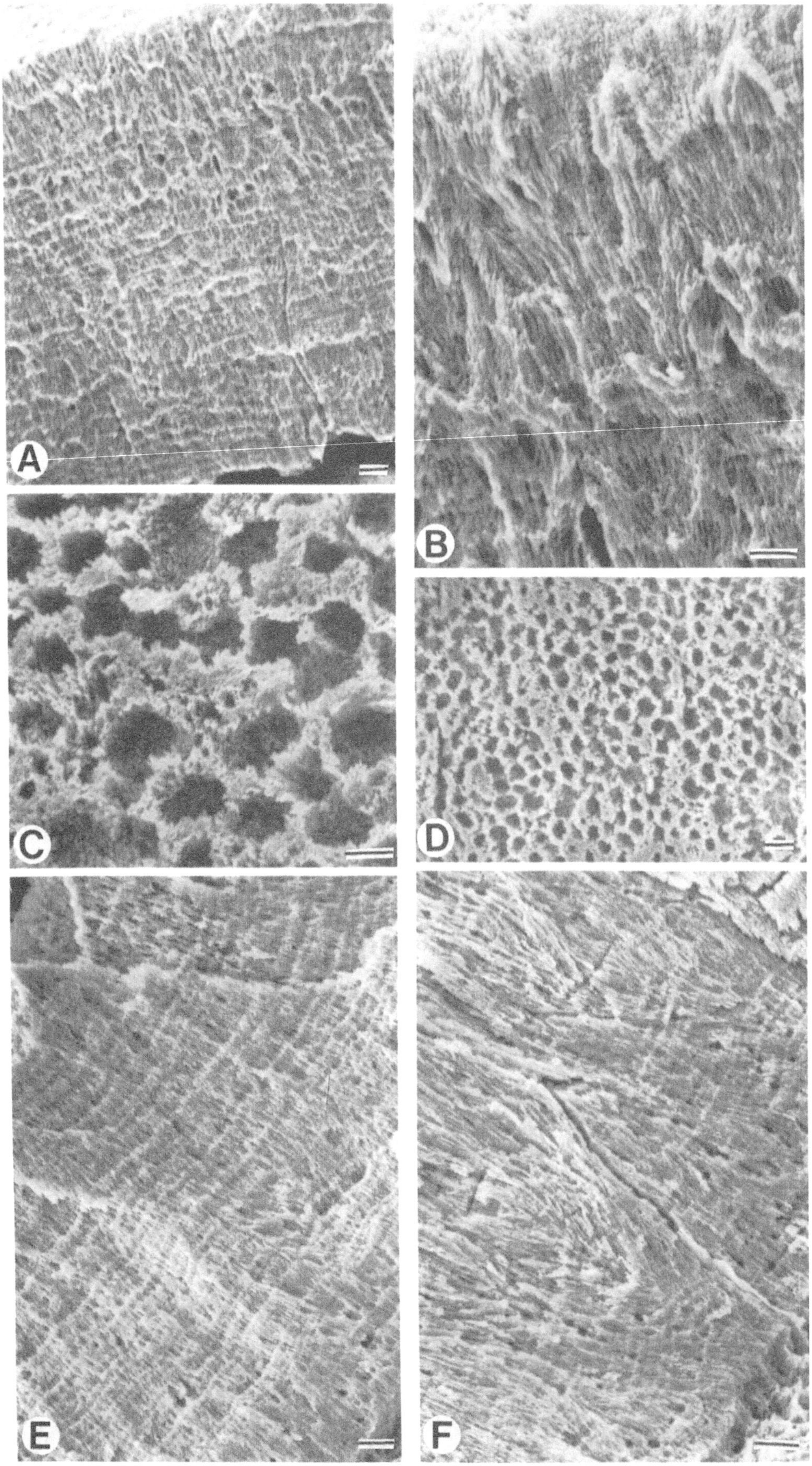
A
B
C
D
E
F

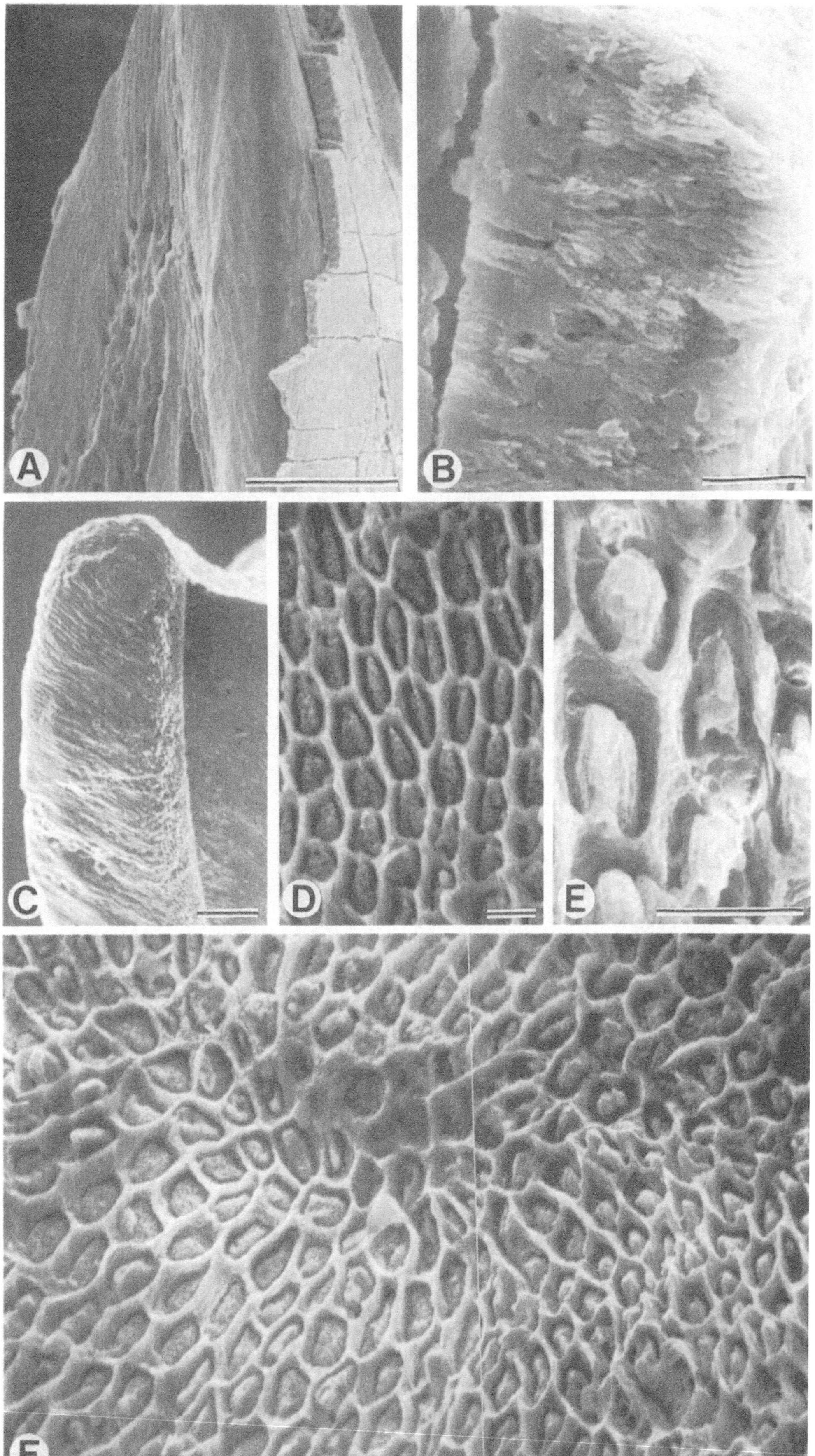
A
B
C
D
E
F

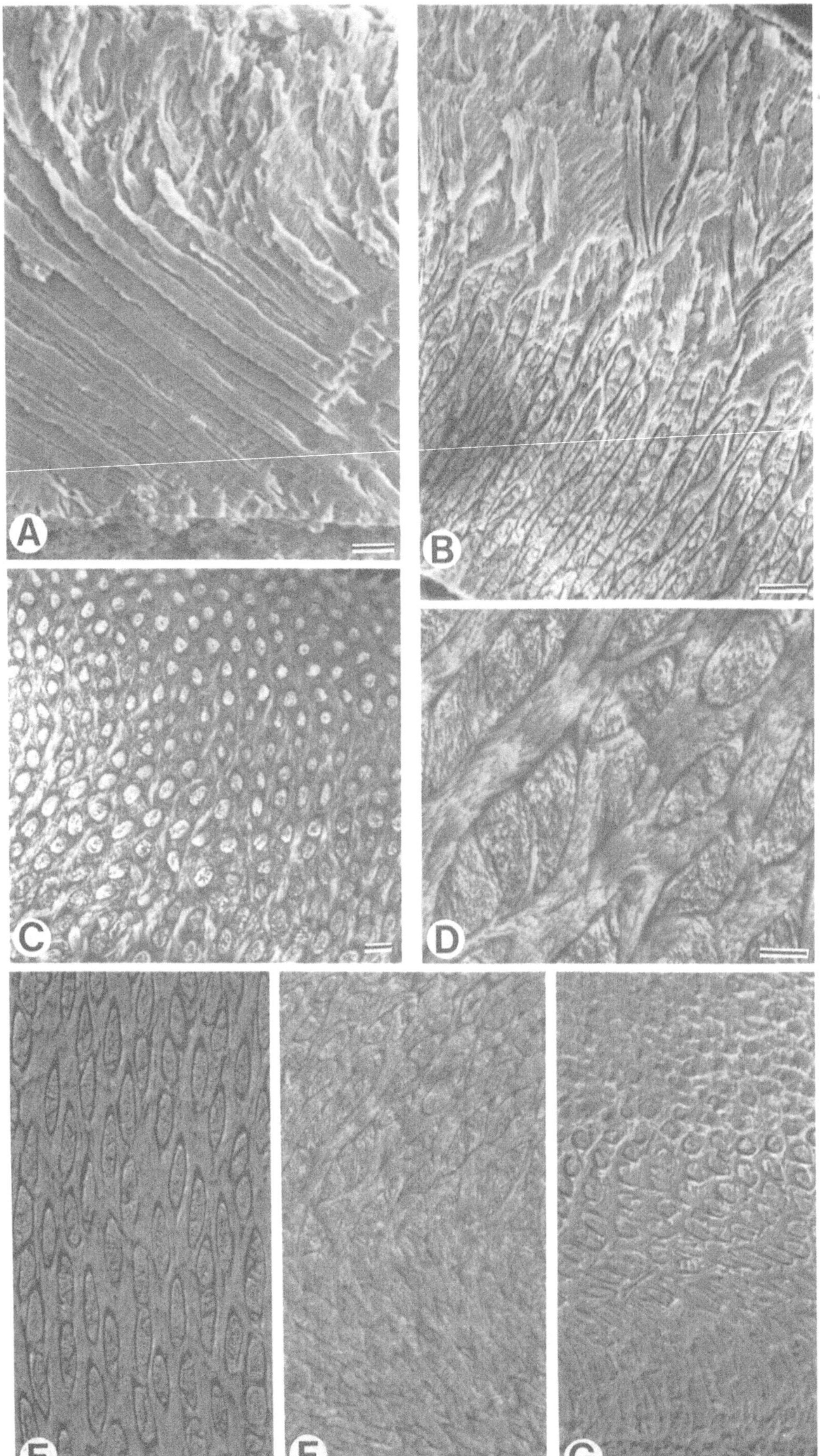
A
B
C
D
E
F
G

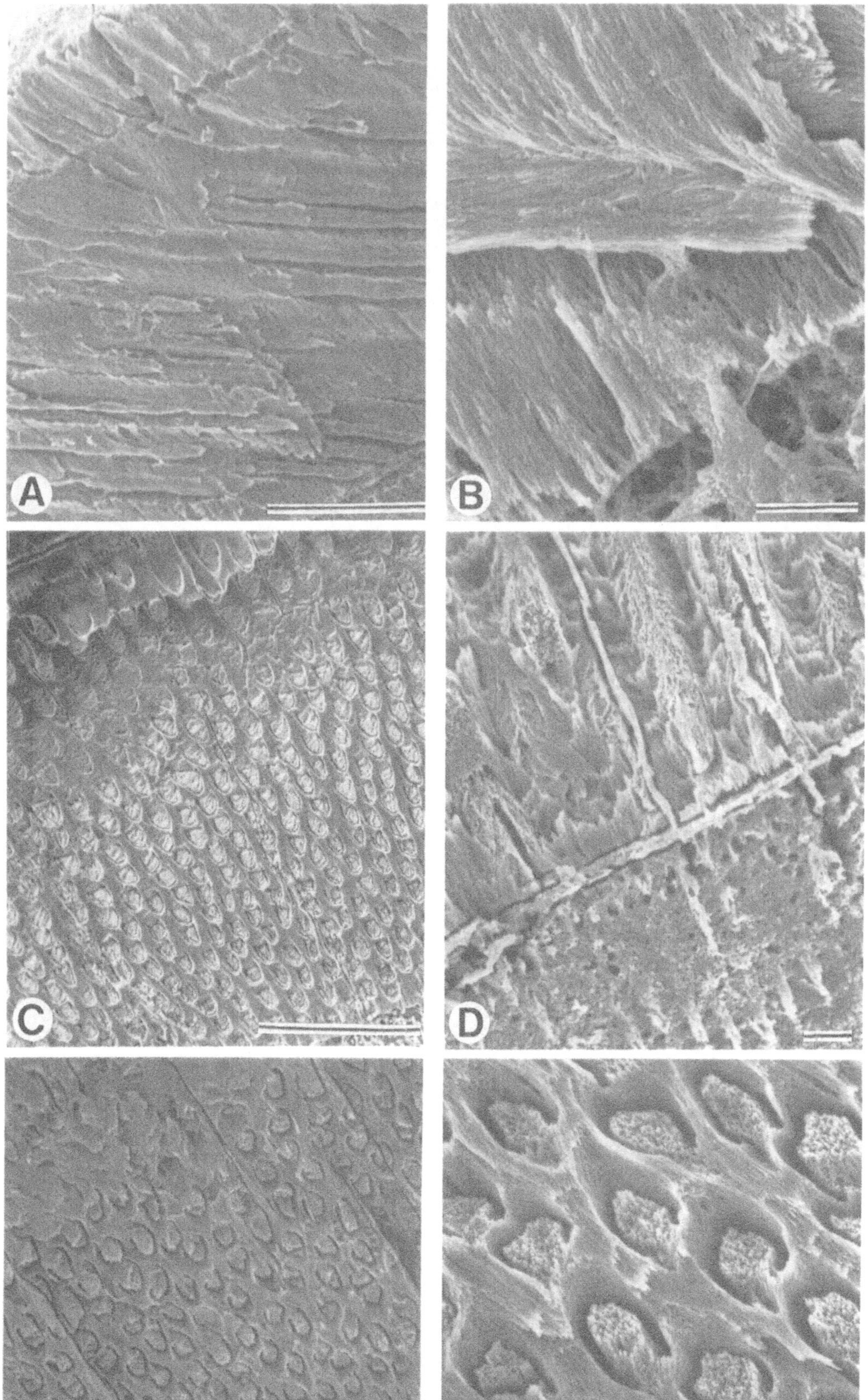
A
B
C
D
E
F

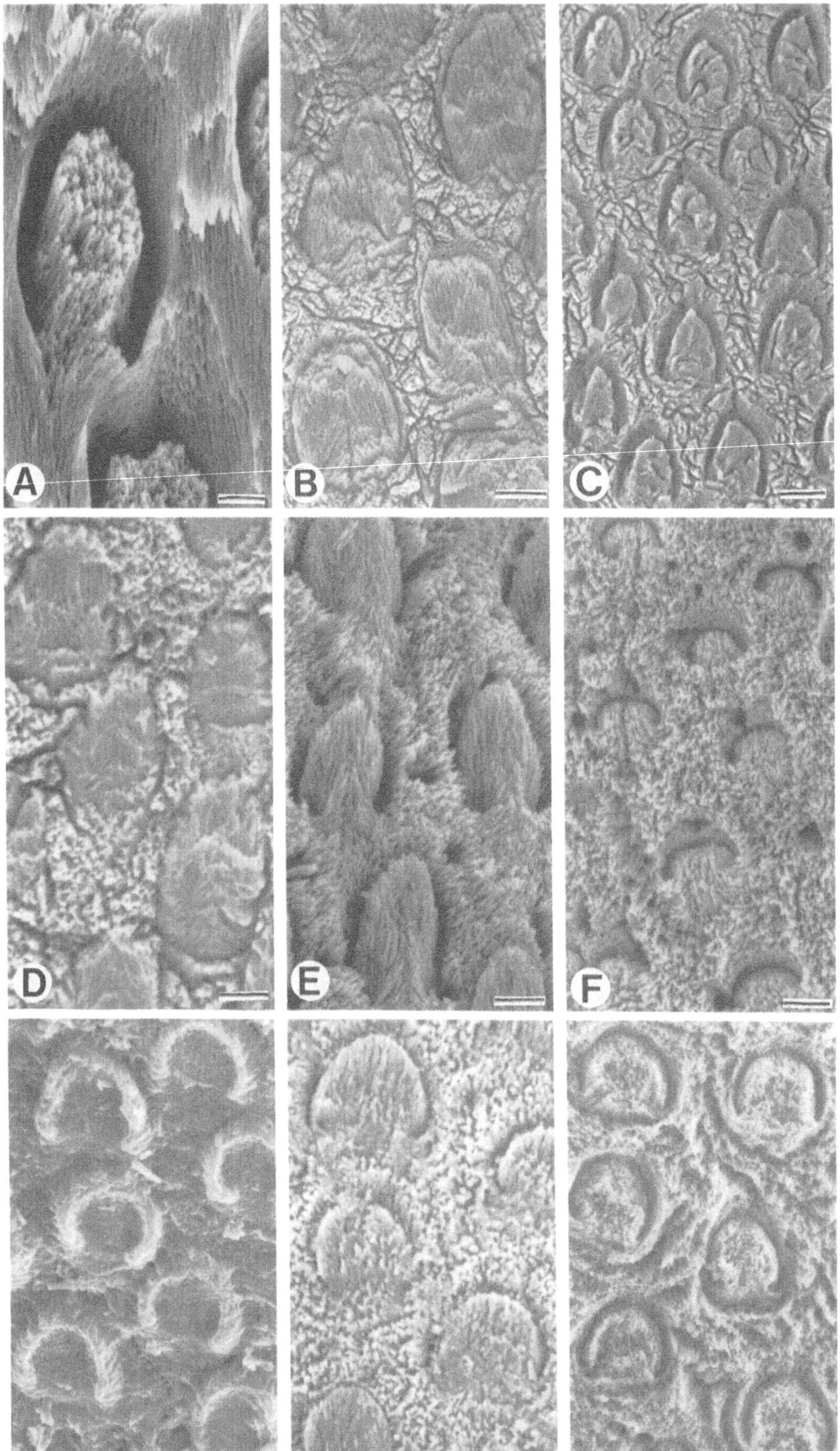
A
B
C
D
E
F
G
H
J

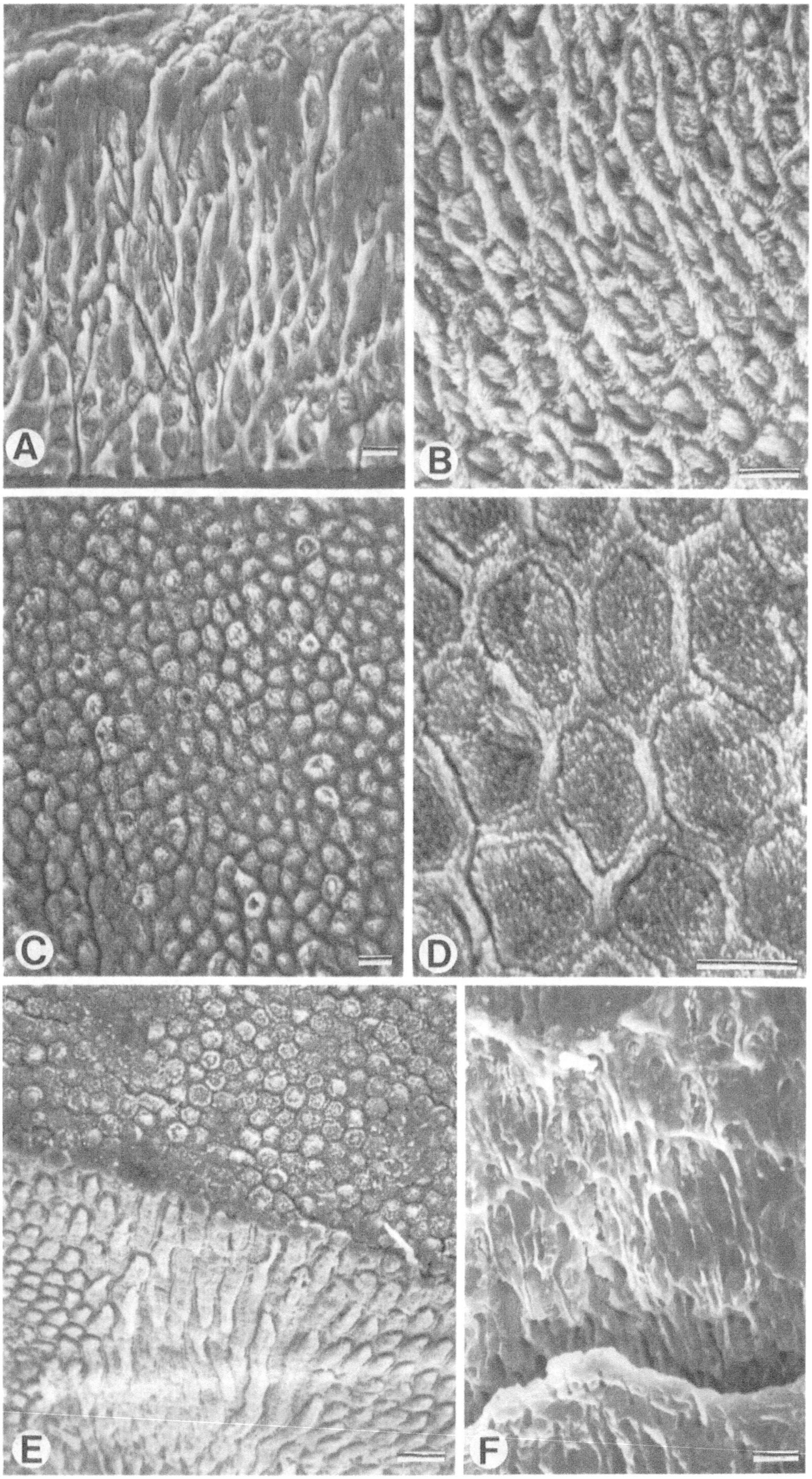
A
B
C
D
E
F

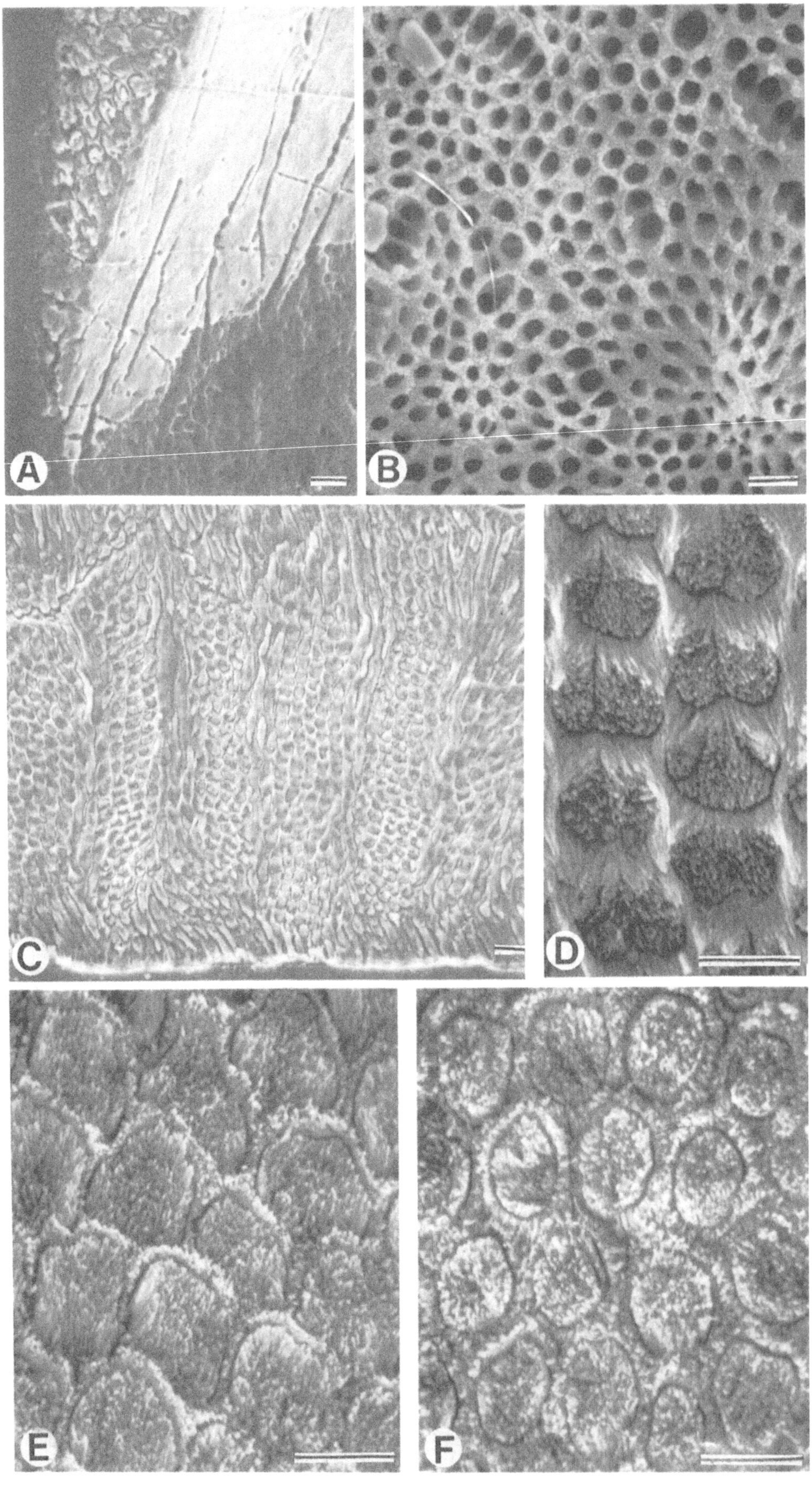
A
B
C
D
E
F

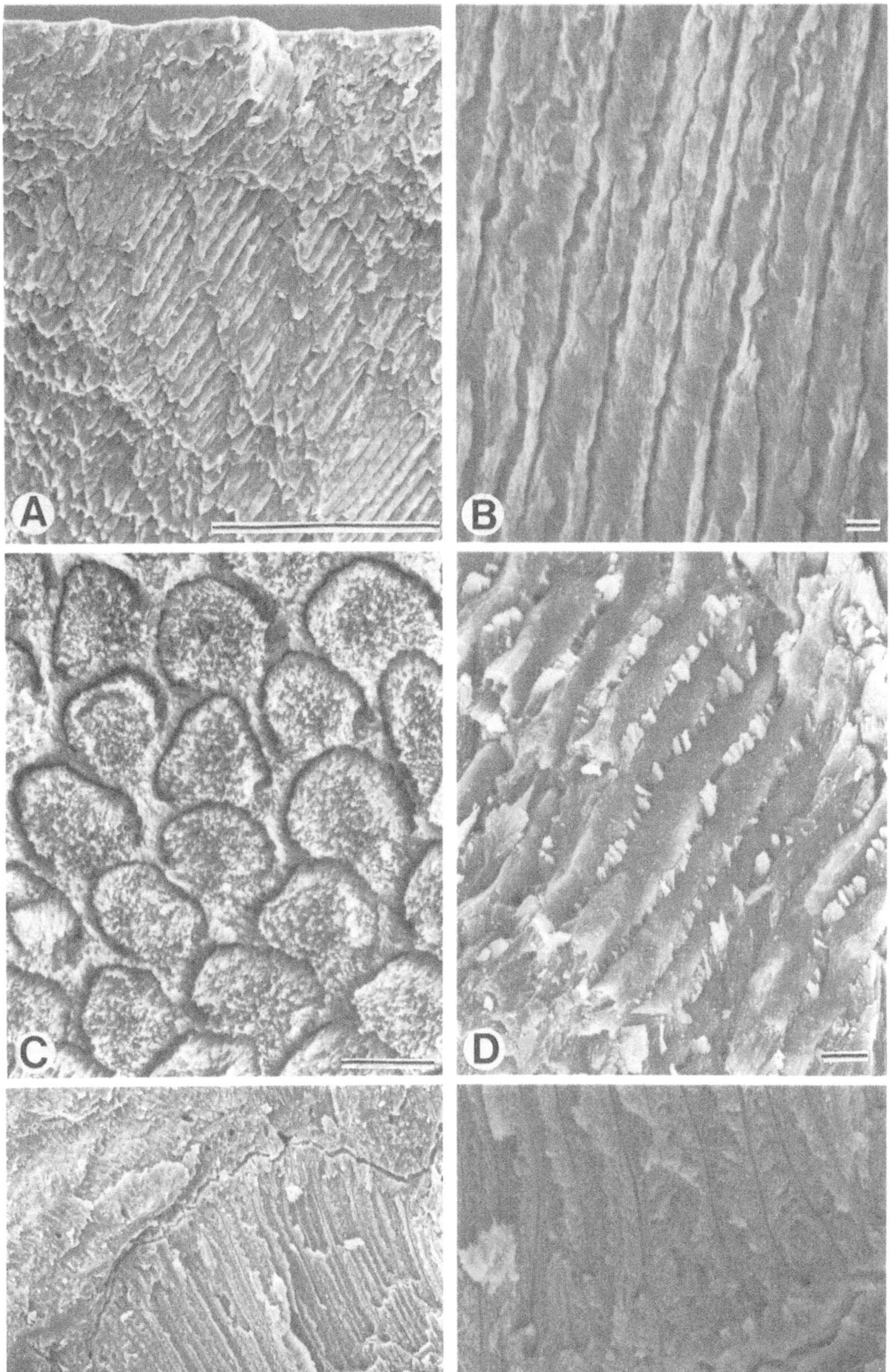
A
B
C
D
E
F

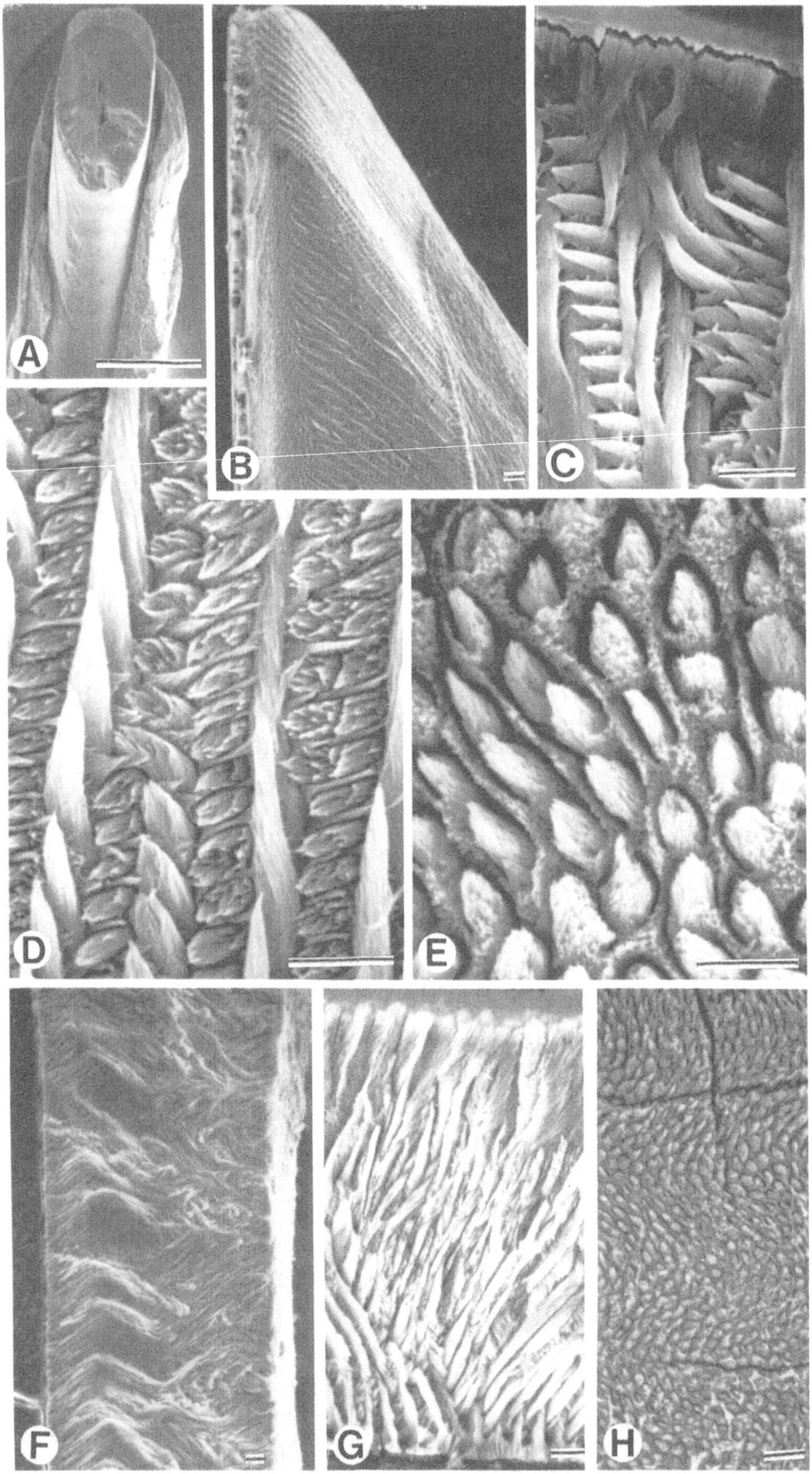
A
B
C
D
E
F
G
H

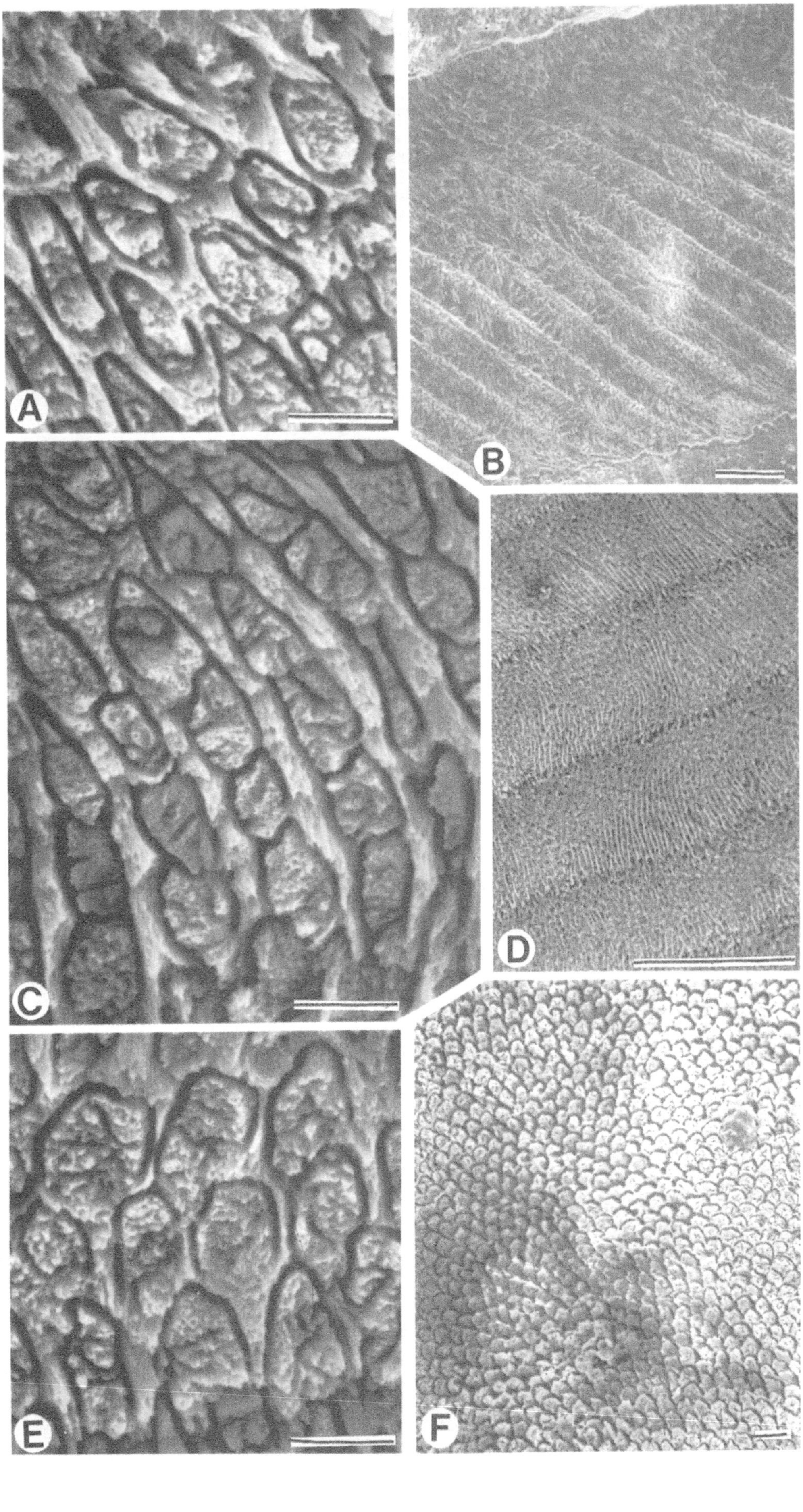
A
B
C
D
E
F

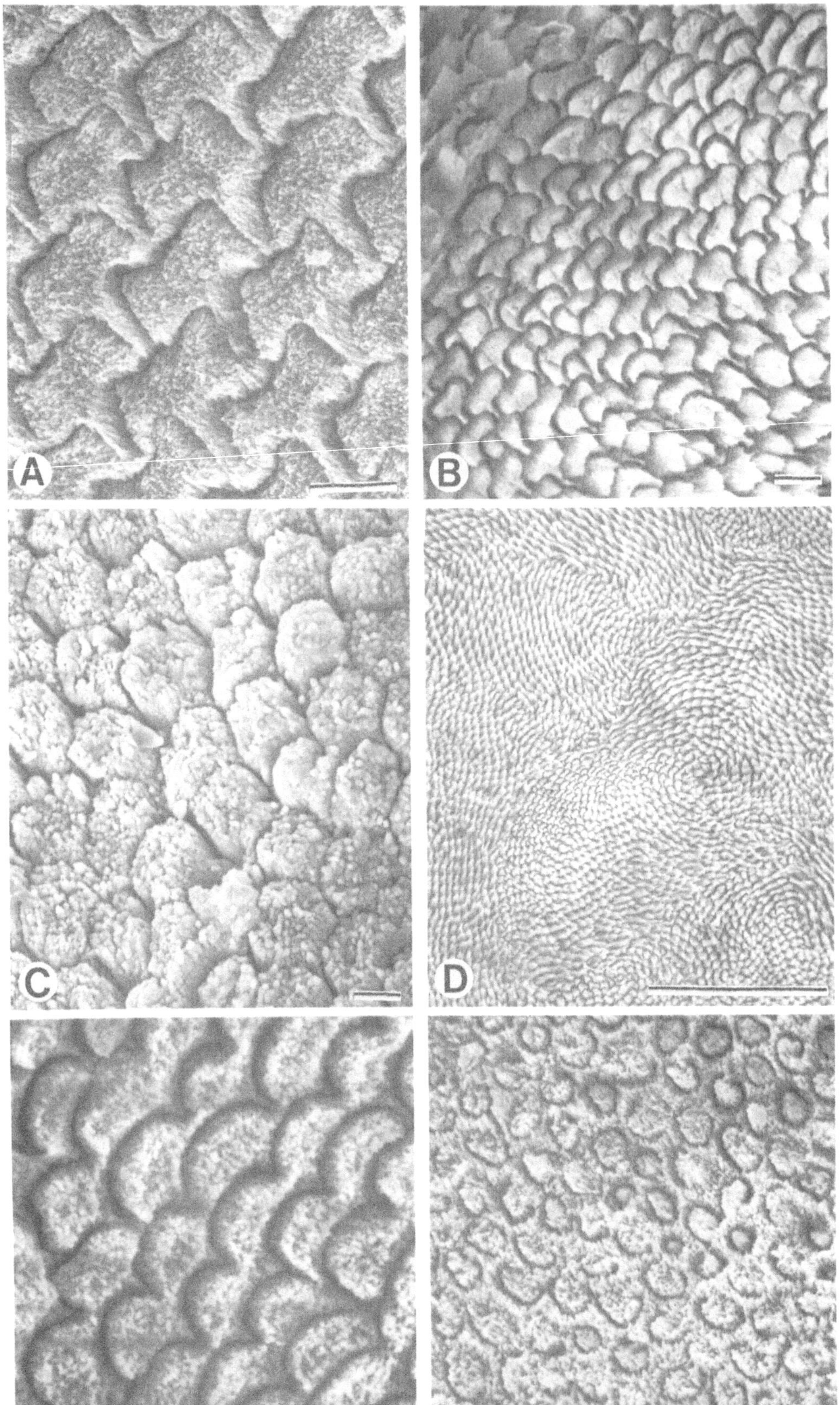
A
B
C
D
E
F

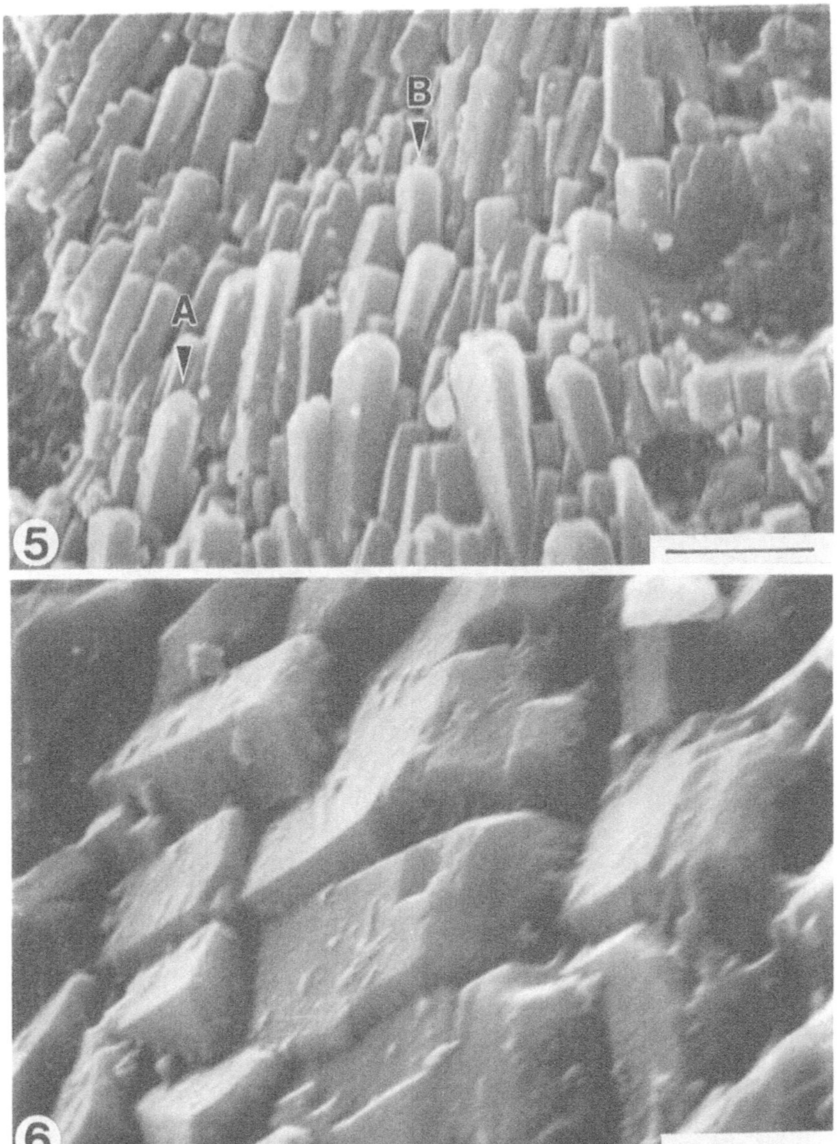
B
A
5
6
7

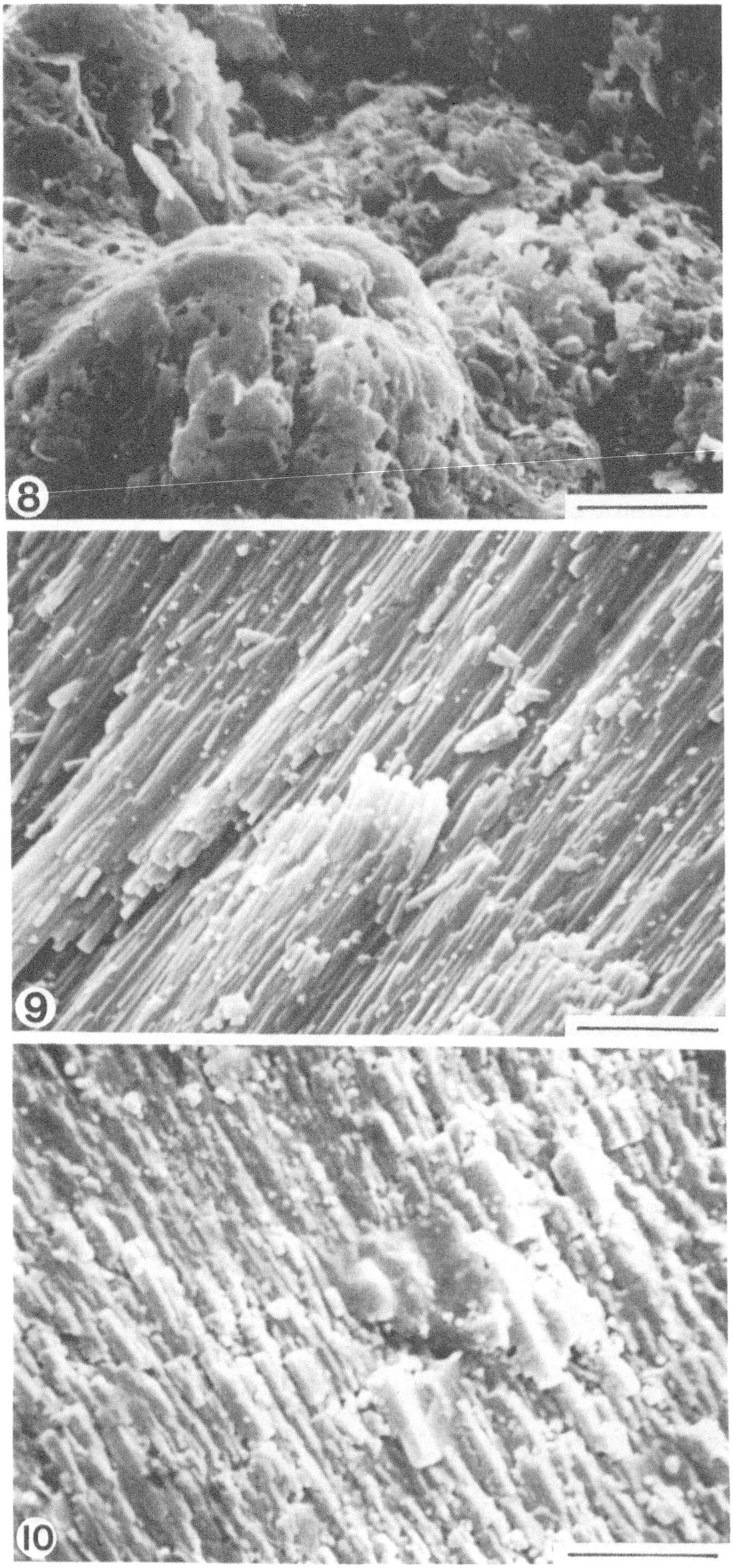
8
9
10

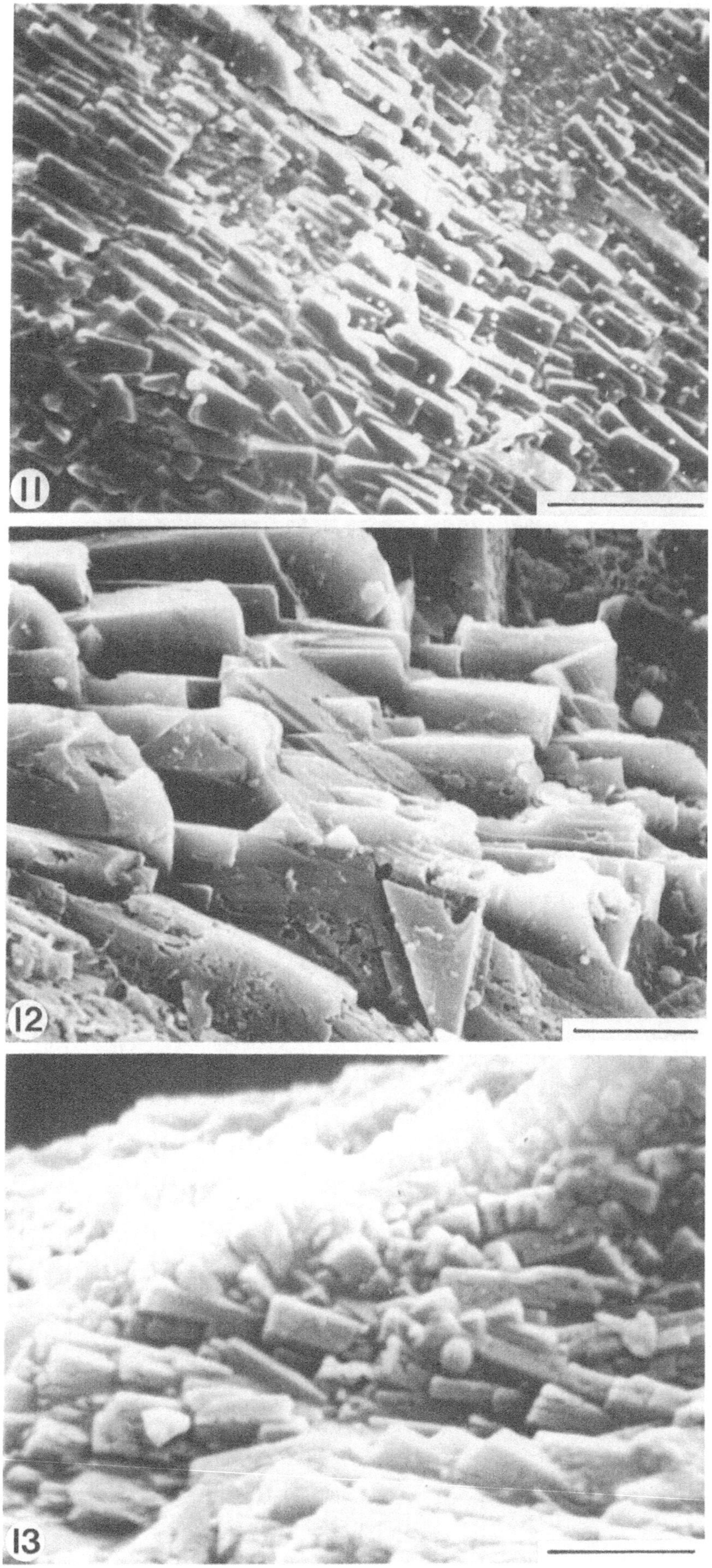
11
12
13

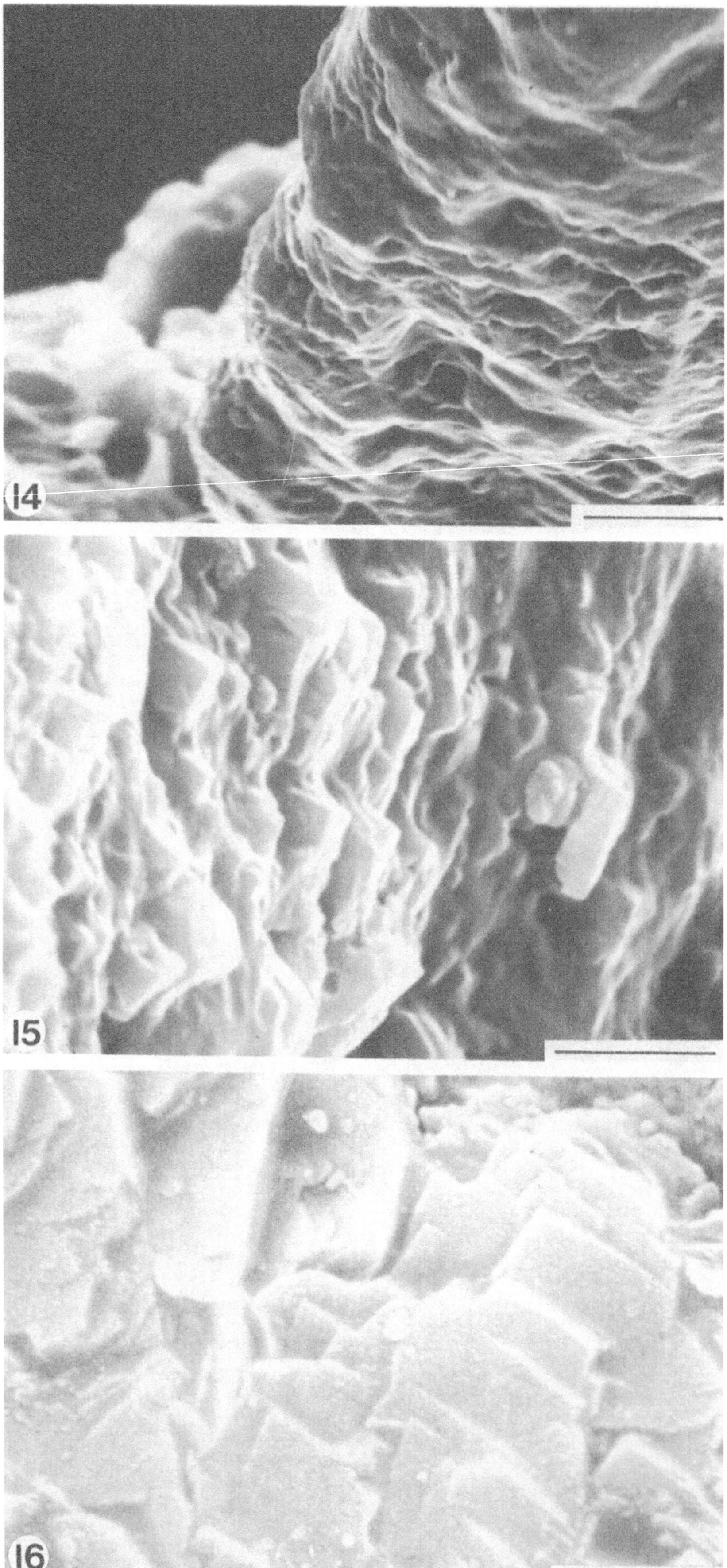
14
15
16

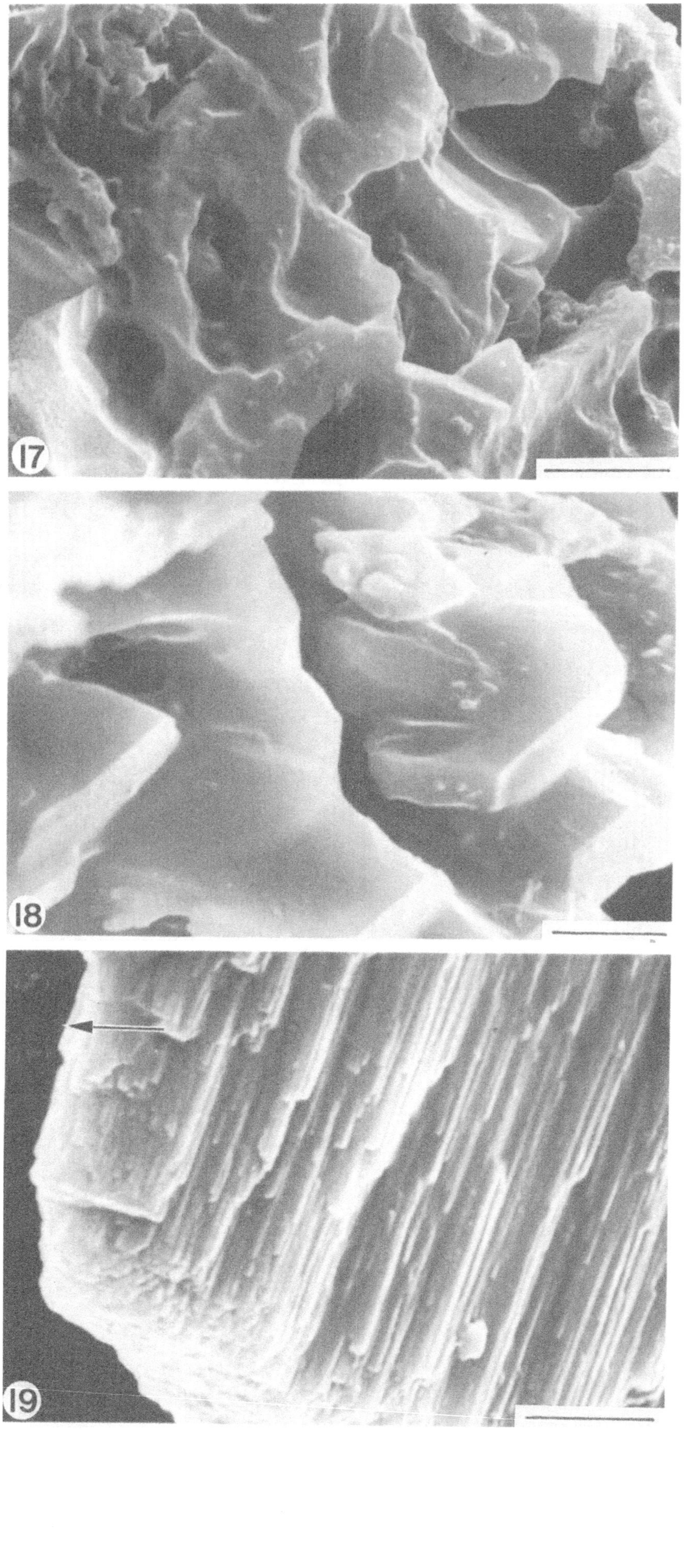
17
18
19

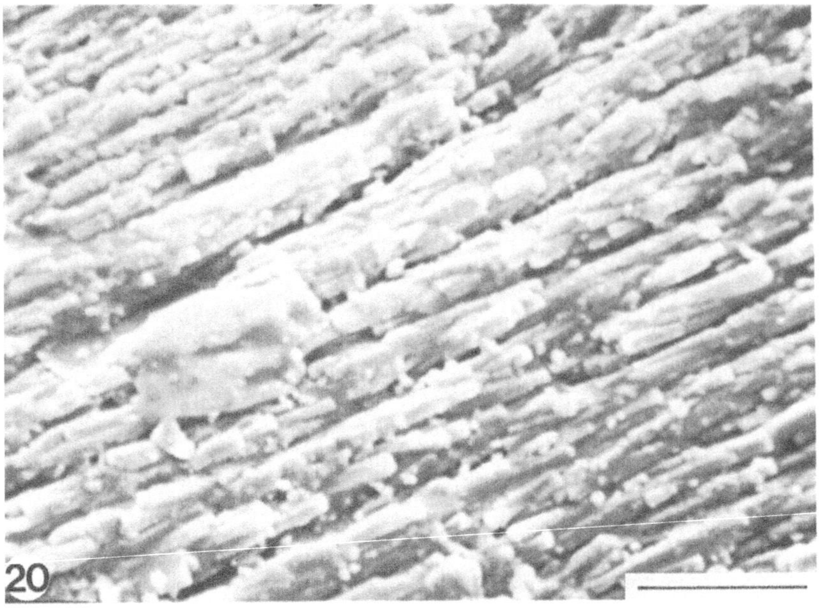
20

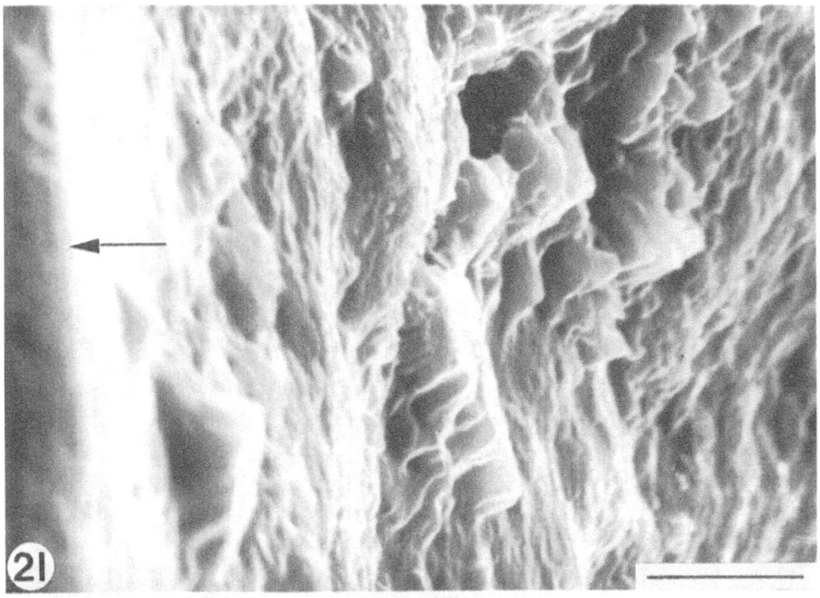
21

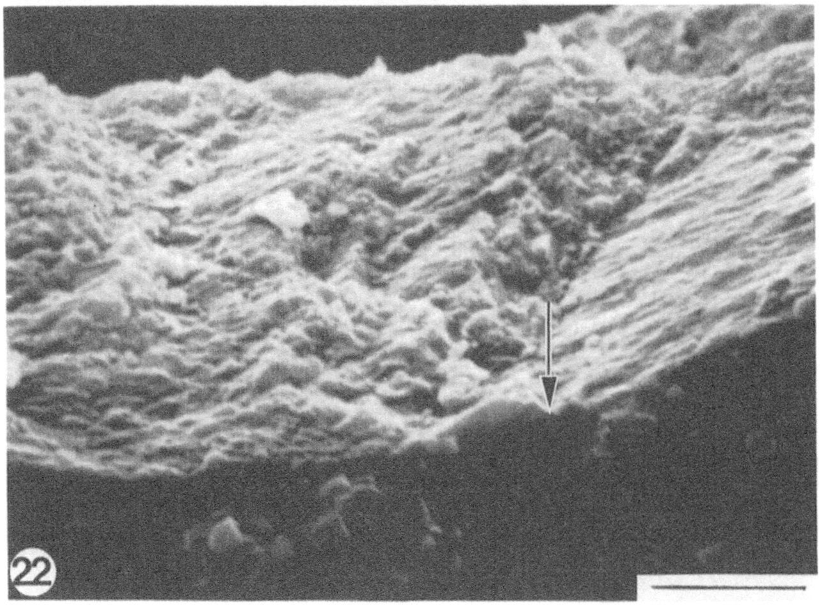
22

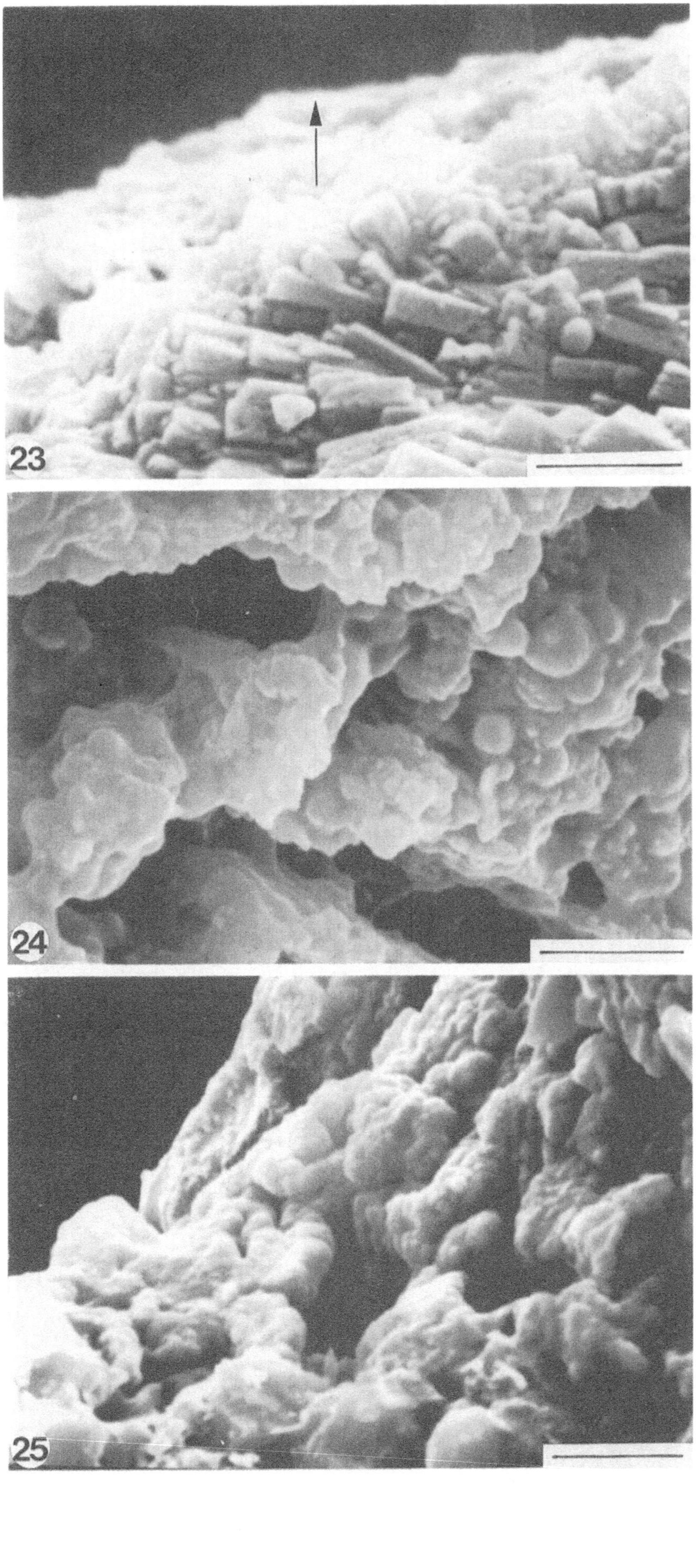
23
24
25

26

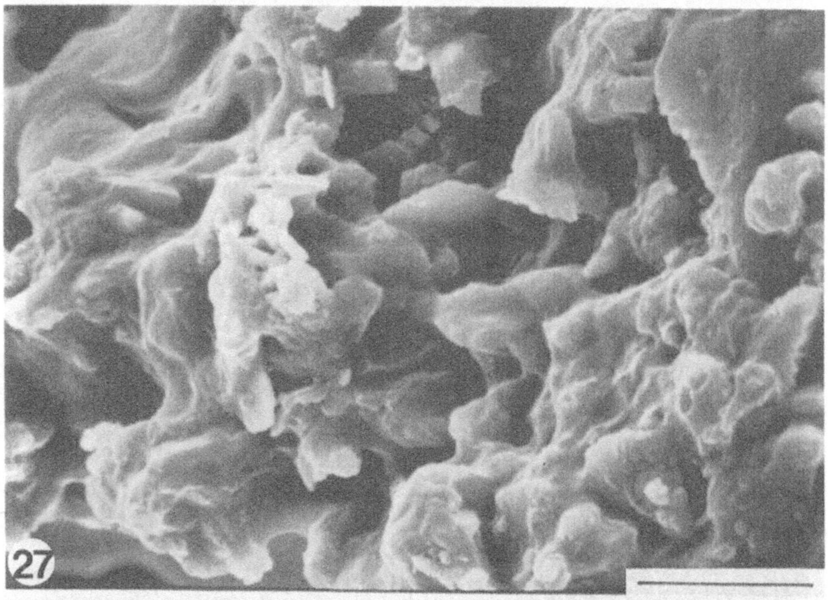
27

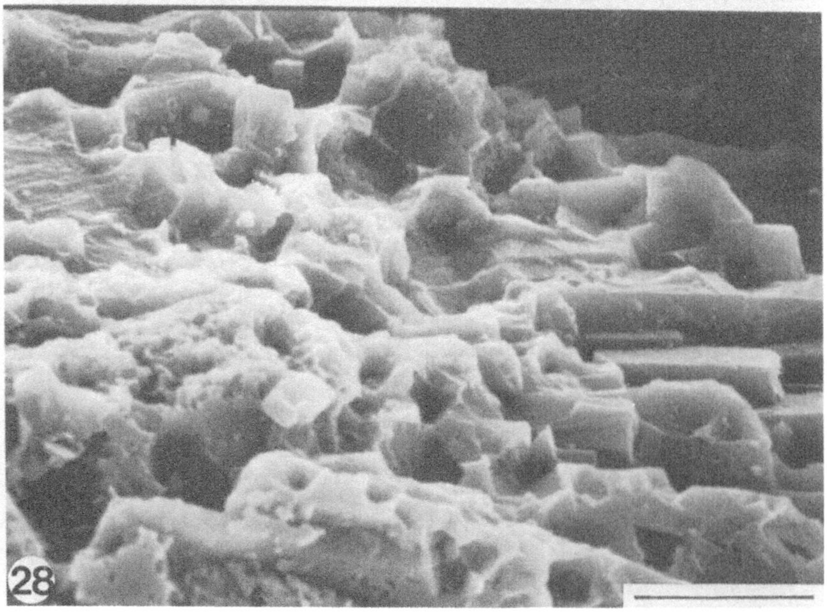
28

GPSR Compliance

The European Union's (EU) General Product Safety Regulation (GPSR) is a set of rules that requires consumer products to be safe and our obligations to ensure this.

If you have any concerns about our products, you can contact us on ProductSafety@springernature.com

In case Publisher is established outside the EU, the EU authorized representative is:

Springer Nature Customer Service Center GmbH
Europaplatz 3
69115 Heidelberg, Germany

Batch number: 10838288

Printed by Printforce, the Netherlands